Biotreatment of Industrial Effluents

Biotreatment of Industrial Effluents

Mukesh Doble
Department of Biotechnology
I.I.T. Madras, Chennai,
India

and

Anil Kumar
Department of Chemistry
Sri Sathya Sai Institute of Higher Learning
(Deemed University)
Puttaparthi, Ananthapur District, Andhrapradesh,
India

AMSTERDAM • BOSTON • HEIDELBERG • LONDON
NEW YORK • OXFORD • PARIS • SAN DIEGO
SAN FRANCISCO • SINGAPORE • SYDNEY • TOKYO

Elsevier Butterworth–Heinemann
30 Corporate Drive, Suite 400, Burlington, MA 01803, USA
Linacre House, Jordan Hill, Oxford OX2 8DP, UK

∞ Recognizing the importance of preserving what has been written, Elsevier prints its books on acid-free paper whenever possible.

Library of Congress Cataloging-in-Publication Data
Application submitted

British Library Cataloguing-in-Publication Data
A catalogue record for this book is available from the British Library.

ISBN: 0-7506-7838-0

For information on all Elsevier Butterworth–Heinemann publications visit our Web site at www.books.elsevier.com

Printed in the United States of America
05 06 07 08 09 10 10 9 8 7 6 5 4 3 2 1

To my parents

M.D.

To Bhagawan Sri Sathya Sai Baba

A.K.

Contents

Foreword

Industrialization has led to growth of manufacturing industries and the associated waste generated by them. Although green technologies that are devoid of waste would be the ideal solution, it is certain that industries will continue to generate effluent well into the foreseeable future. Environmental activism, stricter legislation, and improved awareness of environmental issues on the part of industries have collectively led to a serious effort to identify best solutions to the problem of waste management. Biochemical means of effluent treatment provide an attractive option that makes use of mild biological conditions for the treatment of the waste and does not produce new effluents. Moreover, identification of new microbial systems, including extremophiles, has opened up new possibilities for such treatment, and concerted efforts are being made in industries, academic institutions, and research labs in the areas of bioremediation and biodegradation of waste.

This book covers the treatment of effluents from manufacturing industries as diverse as chemical and electronic. It also looks at other complex wastes such as hospital waste. Comparisons are drawn between current chemical methods and biochemical methods of treatment, including their economics. Several of the biotreatment techniques are still in the infant stage and need sustained research and development before they will be accepted as viable technology options. The book also discusses succinctly the synergies between various effluent treatment techniques, a particularly useful contribution. I compliment the authors for the efforts they have made to bring out this timely publication.

Dr Pushpito Ghosh
Director, Central Salt and Marine Research Institute
Bhavnagar, Gujarat, India

Preface

As industrialization progressed rapidly in the Western world in the 20th century, chemical, petrochemical, iron, and steel industries mushroomed at a tremendous pace. The solid, liquid, and gaseous waste generated from these industries was disposed in public places with very little thought given to its treatment and detoxification. Local and federal governments were interested in the economic progress such industries were bringing to their community and were not aware of their short-term and long-term effects on the ecology or on the health of the general public. In the late 1950s and early 1960s, evidence mounted that waste that had been dumped affected the environment and public health. In response, industries started treating their effluent, using physical and chemical methods, before discharging it. Governments also laid down standards for the quality of waste leaving an industry (U.S Environmental Protection Agency guidelines are found in the http://www.epa.gov/ web site). As the electronics and communication industries grew rapidly during the 1970s and 1980s, highly toxic waste of a different kind was produced and had to be disposed of. Metal recovery and leather processing industries generated effluents with high concentrations of heavy metals and cyanides. Industries in the West tried to shift their manufacturing base to underdeveloped countries to avoid government regulations, but that was only a short-term solution, because newer laws made manufacturers responsible irrespective of the location of the manufacturing site. Although the physical and chemical methods detoxified the waste, they created waste of a different kind that also had to be disposed of in public places without causing damage. Hence industries started adapting to biochemical treatment techniques since they were mild and natural, and did not themselves generate waste.

Regulatory requirements have become stringent, and pressure from citizen groups to improve the safety of waste that is being disposed of has also increased, forcing manufacturing industries to spend more resources on newer effluent treatment procedures. In addition, sites that had been contaminated at the beginning of the industrialization era have to be cleaned as well due to pressure from local residents.

Biochemical treatment, although offering several potential advantages, cannot be used for treating all types of industrial effluents because many of these treatment techniques are still in the research stage. Most of the microorganisms are very specific to a particular pollutant, whereas a typical effluent is a mixture of several pollutants and toxic chemicals. Hence a combination of chemical/physical treatment followed by biochemical treatment methods appears to be the best alternative today. There is plenty of scope for scientists and engineers to undertake research both in the laboratory and during technology development. Only a very small percentage of microorganisms and bacteria have been tested, and there are still large amounts of untapped organisms both in the land and in the ocean that could help in degrading many of the recalcitrant wastes.

In the meantime, industries are carrying out research to recycle waste streams and solvents in order to minimize the effluents that leave their premises and achieve "zero discharge." Also, new process technologies are being developed with an eye to the concept of "atom efficiency"; namely, developing processes in which the raw materials are completely converted to the desired product without the formation of side or wasteful products. The figure on the following page sums up the changes that have been taking place both in manufacturing and in the field of effluent treatment during the past fifty years. Industries have realized that minimizing and eliminating waste at its source is more effective than generating it and treating it later. In the future, a combination of newer manufacturing and treatment strategies will be the most effective and the least expensive. These newer manufacturing strategies may include reducing the amount of solvent used, recycling and recovery of solvents, optimum operating conditions to avoid formation of side products and degraded products, and the use of chemicals which are known to biodegrade.

The book covers treatment of most of the effluents from manufacturing industries. Interestingly, industries that appear to be unrelated produce effluents that are similar in nature. Simple aerobic activated sludge process is used as a last step in most of the industrial units, since the unit is easy to construct, maintain, and operate. Although several new bioreactor designs are undergoing laboratory testing, full-scale technologies are slow to develop. As will be shown in this book, the knowledge gained by treating effluents from one industry has been extended to treat effluents from other industries. At times mixing of effluents may have a beneficial effect, but at the same time segregating and treating effluents may be easier. A colony of microorganisms brings in synergy, which is absent when single organisms are used for treatment.

As I was teaching a course on Environmental Biotechnology to the students in the class of 2005 at B. Tech. Industrial Center for Biotechnology, Anna University, Chennai, India, I realized that there was no single source that focused on biochemical treatment of all types of industrial effluents. I undertook the writing of this book in response to that need, and I would

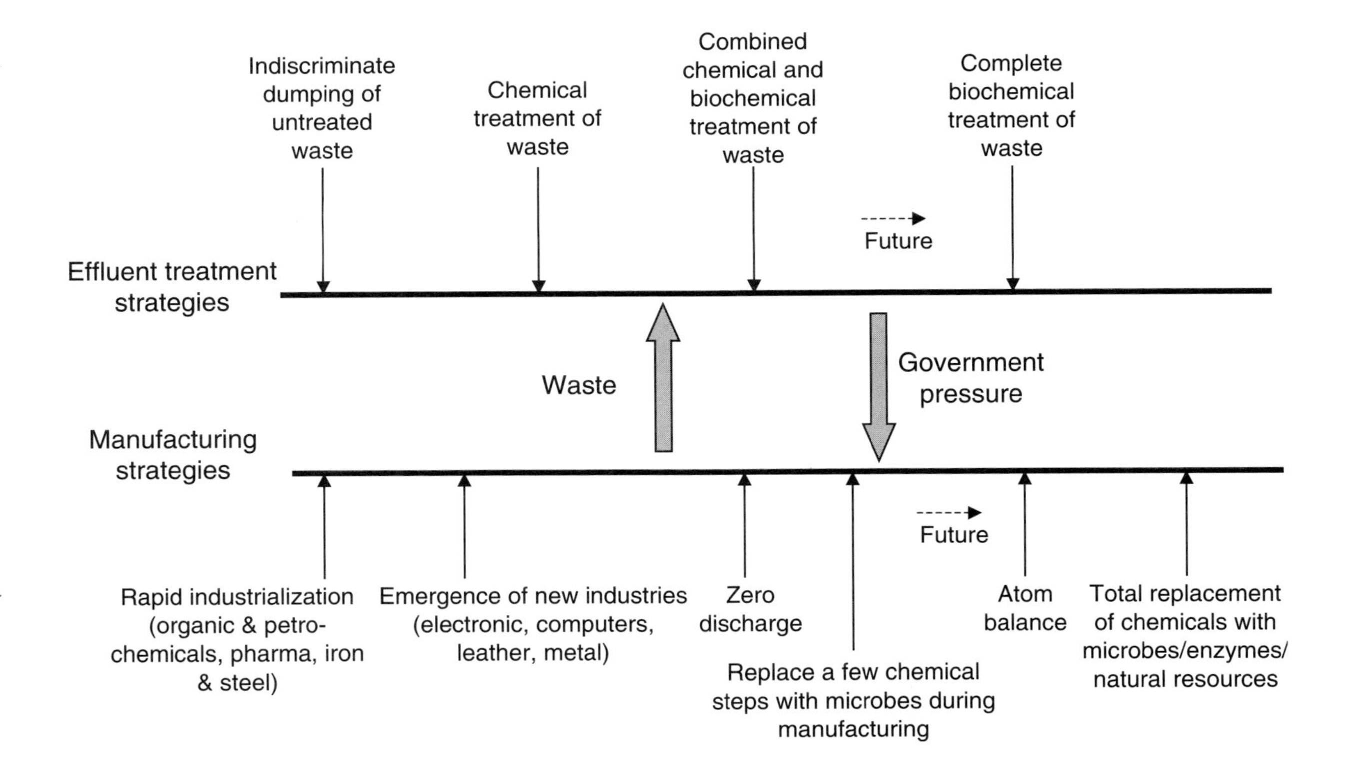

Indiscriminate dumping of untreated waste
Chemical treatment of waste
Combined chemical and biochemical treatment of waste
Complete biochemical treatment of waste
Future
Effluent treatment strategies
Waste
Government pressure
Manufacturing strategies
Future
Rapid industrialization (organic & petro-chemicals, pharma, iron & steel)
Emergence of new industries (electronic, computers, leather, metal)
Zero discharge
Replace a few chemical steps with microbes during manufacturing
Atom balance
Total replacement of chemicals with microbes/enzymes/ natural resources

like to thank my students for sowing the seeds in my mind to take on such a venture. This book can be used for the upper-level undergraduate as well as graduate level courses. It is also well suited as a first point of reference for practicing environmental engineers and researchers. Practitioners of environmental biotechnology come from a wide variety of disciplines, including agronomists, biochemists, microbiologists, botanists, chemical engineers, geneticists, enzymologists, molecular biologists, protein technologists, process chemists, and technologists. I hope this book has useful and relevant information for all of them.

Mukesh Doble August 2004

CHAPTER 1

Introduction

Movement of Pollutants from the Source

A pollutant is defined as "a substance that occurs in the environment, at least in part as a result of human activities, and has a deleterious effect on the environment." The term *pollutant* is a broad term that refers to a wide range of compounds, from a superabundance of nutrients giving rise to enrichment of ecosystems to toxic compounds that may be carcinogenic, mutagenic, or teratogenic. Pollutants can be divided into two major groups, namely, those that affect the physical environment and those that are directly toxic to organisms, including human beings. The movement of pollutants and toxic compounds through the environment is called pollution (Fig. 1-1) and is very similar to the movement of energy and nutrients within the ecosystem or, on a larger scale, through the biosphere.

Rapid industrialization in the twentieth century had led to the generation of vast amounts of gas, liquid, and solid waste that were introduced into the environment without much thought by the manufacturers of that waste. This has affected the ecosystem and has caused health problems for the inhabitants residing near the factories. As people became more aware of the toxic effects of this waste, as they saw the destruction of the ecosystem due to the indiscriminate discharge of the pollutants, and as federal and local laws imposed more stringent discharge norms, efforts were made to treat these wastes so as to make them innocuous before discharge into public systems. Initially, the treatment procedures were based on physical and chemical methods, which proved to be inadequate and costly. Biochemical methods, which have inherent advantages, are still in their early stages of development.

Effluent Discharge—Points to Keep in Mind

Several points related to the discharge of pollutants into public waterways or land have to be kept in mind, such as: local laws on discharge limits,

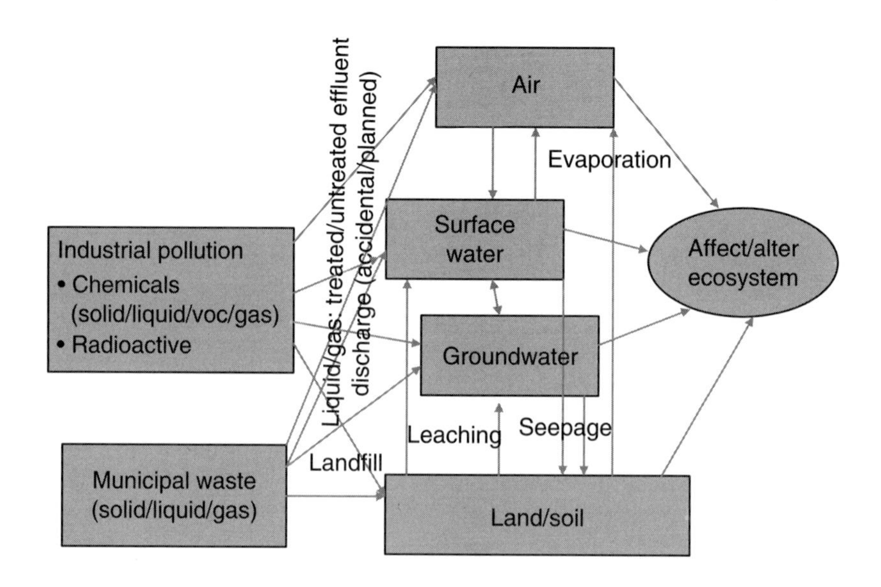

FIGURE 1-1. Movement of pollutants. VOC, volatile organic compound.

the effect of the pollutants on the ecosystem (short- and long-term data may not be available), toxicity of the secondary metabolites, discharge of secondary waste generated (such as sludge, inorganics, etc.), and the impact of modifying the existing microorganism population in the soil or water.

Different Treatment Procedures and Factors Affecting Technology Selection

The treatment of solid, liquid, and gaseous pollution can be carried out either in situ (i.e., at the contaminated site) or ex situ (i.e., removing the polluted material, transporting it to another site or plant, treating it, and then bringing it back to the site). Both approaches have several advantages and disadvantages. The former is cost effective but could be slow and nonuniform. The latter involves several steps and exposes the workers to the pollutants.

Physical, chemical, biological, and phytoremediation methods have been attempted to destroy pollutants. The selection of the treatment technology (Fig. 1-2) will depend on several factors, including cleanup time, maturity of the technology, capital and operating costs, residual product toxicity after treatment, local discharge norms, reliability of the process, ease of facility maintenance, company image, generation of volatile organic compounds (VOCs), and treatment of halogenated compounds or explosives.

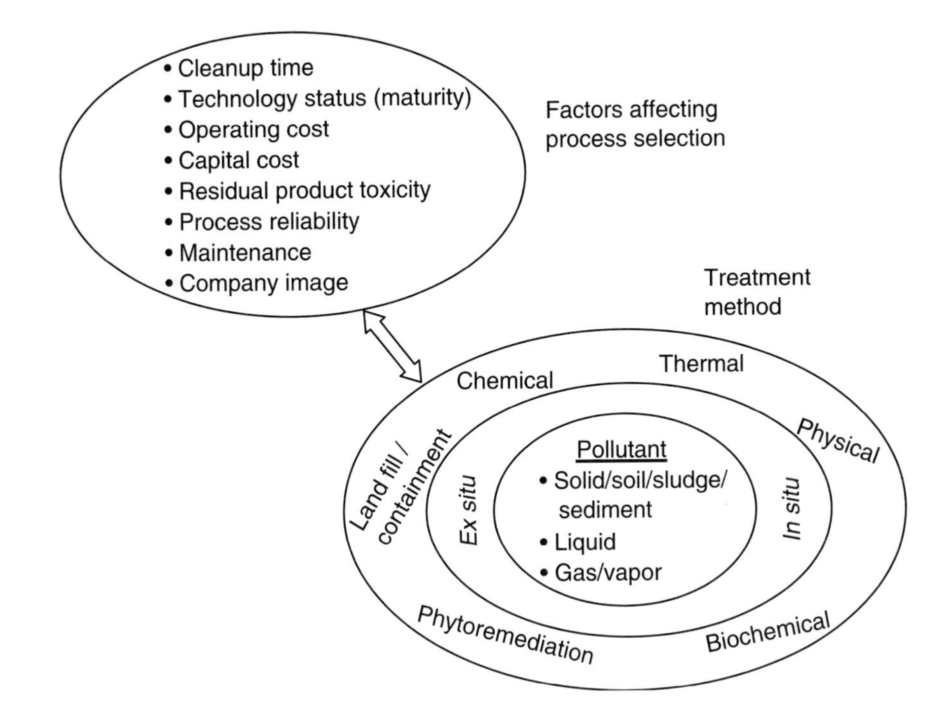

FIGURE 1-2. Selection of treatment technology.

In situ physical and chemical treatment techniques include:

- Chemical oxidation
- Electrokinetic separation
- Pneumatic or explosive fracturing of the soil
- Soil flushing using solvents, cosolvents, or surfactants
- Vapor extraction
- Thermal treatment using electric resistance, injecting hot air or steam, or electromagnetic heating
- Containment by creating physical barrier
- Landfill

Ex situ physical and chemical treatment techniques include:

- Extraction with acid, alkali, solvent, or surfactants
- Mechanical separation such as magnetic, sieving, filtration, etc.
- Stripping using air or steam
- Chemical or thermal oxidation, reduction, or dehalogenation
- Incineration

- Absorption or adsorption of liquid or gaseous contaminants
- Separation of liquid contaminants by distillation, ion exchange, crystallization, or membrane partition

An organic compound could be biodegraded by four different mechanisms, namely, (a) aerobic oxidation of an organic primary growth substrate (e.g., natural organic material, hydrocarbon fuels, chlorobenzenes, and the less oxidized chlorinated ethenes and ethanes); (b) anaerobic reduction, where the organic compound serves as the electron acceptor (e.g., highly oxidized chlorinated hydrocarbons, less chlorinated ethenes and ethanes such as trichloroethane); (c) coupled oxidation and reduction of an organic compound by a fermentation pathway; and (d) cometabolism of an organic compound, which occurs when the degradation is catalyzed by an enzyme cofactor produced by microorganisms for some other purpose. The microbial processes involved in biodegradation are linked to the extraction of chemical energy for microbial growth, which comes from coupling the oxidation and the reduction reactions.

Biological treatment technologies include:

- Enhanced bioremediation (enhancement achieved by addition of nitrate, oxygen, or metabolites)
- Bioventing
- Bioaugmentation (organisms that have been specifically grown in the lab on the contaminants)
- Biopiles
- Composting (windrow or static pile)
- Land farming
- Slurry phase bioreactors
- Biosorption

Phytoremediation techniques include:

- In situ phytoextraction/accumulation (removes toxins from the soil and concentrates them in the harvestable part of the plant)
- In situ phytodegradation (plants and associated microbes degrade the pollutants)
- In situ phytostabilization (the mobility of the pollutant is reduced through chemical modification or immobilization)
- In situ phytovolatilization (volatilization of the pollutant into the atmosphere)
- Enhanced rhizosphere biodegradation (plant roots absorb the pollutants)

Cost of phytoremediation has been estimated to be $25 to $100 per ton of soil treatment and $0.60 to $6 per 1000 gallons for treatment of aqueous waste streams. The cost of this remediation technique is estimated to be half that of any other treatment. According to 1997 U.S. Environmental

Protection Agency estimates, the cost of using phytoremediation in the form of an alternative cover (vegetative cap) ranges from $10,000 to $30,000 per acre, which is thought to be two- to fivefold less expensive than traditional capping (Macek et al., 2000).

Environmental Engineering—An Interdisciplinary Subject

Environmental engineering is an interdisciplinary subject encompassing microbiology, biotechnology, chemical engineering, chemistry, analytical chemistry, environmental chemistry, engineering design, and mechanical engineering. Knowledge of atmospheric sciences, oceanography, geology, and civil engineering is also essential.

Environmental engineering helps to maintain the quality of life through the betterment of the environment. This means the development of sufficient clean water supplies; the prevention of air, soil, river, lake, ocean, and groundwater pollution; the maintenance of good air quality; and the remediation by natural means of land and water contaminated with hazardous chemicals. The use of "natural means" is an important prerequisite, else we would be converting one type of pollution to another.

Various Chapters in the Book and How They Are Interrelated

The five most polluting industries in the United States (1987 data) are iron and steel, nonferrous metals, industrial chemicals, nonmetallic mineral products, and pulp and paper. The polluting industries were classified on the basis of the comprehensive index of emissions per unit of output. The index includes conventional air, water, and heavy metals pollutants (Mani and Wheeler, 1998). Table 1-1 compares the organic water pollution intensity

TABLE 1-1
Comparison of Organic Water Pollution Intensity

	Index
Food	1.00
Pulp and paper	0.87
Chemicals	0.29
Textiles	0.26
Wood products	0.13
Metal products	0.08
Metal	0.03
Nonmetallic minerals	0.02

TABLE 1-2
Classification of Industries Based on Gaseous Pollution and Volatile Organic Compounds

	CO	*NO$_X$*	*SO$_2$*	*VOC*
High	Iron	Industrial chemicals Other chemicals Petrol refinery Petrol	Metals	Industrial chemicals Other chemicals Petrol refinery Petrol
Medium	Industrial chemicals Other chemicals Petrol refinery Petrol	Iron Metal Paper and pulp	Industrial chemicals Other chemicals Petrol refinery Petrol Minerals Paper and pulp	Metals Paper and pulp

of various manufacturing sectors (Hettige et al., 1998). Table 1-2 groups the various manufacturing industries based on the intensity of gaseous pollution (Hettige et al., 1994; Rösner, 2003).

This book focuses on the biochemical treatment of gas, liquid, and solid effluents from a wide range of manufacturing industries such as dye, textile, paint, explosive, semiconductor, metal processing, pharmaceutical, organic chemical, petroleum, food and dairy, paper and pulp, pesticide, sugar and alcohol distillery, and polymer. Also discussed is treatment of solid waste, including hospital and municipal waste; ground water decontamination, including fluoride removal; denitrification and biodesulfurization of petroleum; and cyanide degradation. Several biodegradation techniques described in the book are still in the research and laboratory stage; in such cases the physical and the chemical treatment techniques are being followed by the relevant industries. Although the industries chosen in the book appear to be disjointed, they all are interconnected, as shown in Fig. 1-3. For example, effluents produced by organic chemical, pharmaceutical, pesticide, and dye stuff industries all have several common characteristics. Tannery, electrochemical, and semiconductor industries have metals in their effluents. Dye chemicals are present in textile and tannery effluents. Effluents from the food industry may have similar issues (Table 1-3 lists the effluent characteristics of various agriculture-based industries). The book also briefly compares the chemical and physical treatment procedures for these pollutants for the sake of completeness. Both aerobic and anaerobic techniques and various types of reactors that have been used for treatment are also discussed in detail. This book can be used as a ready reference for physical, chemical, and biochemical treatment of industrial pollutants as well as a source to understand the mechanism of biodegradation of a variety of contaminants.

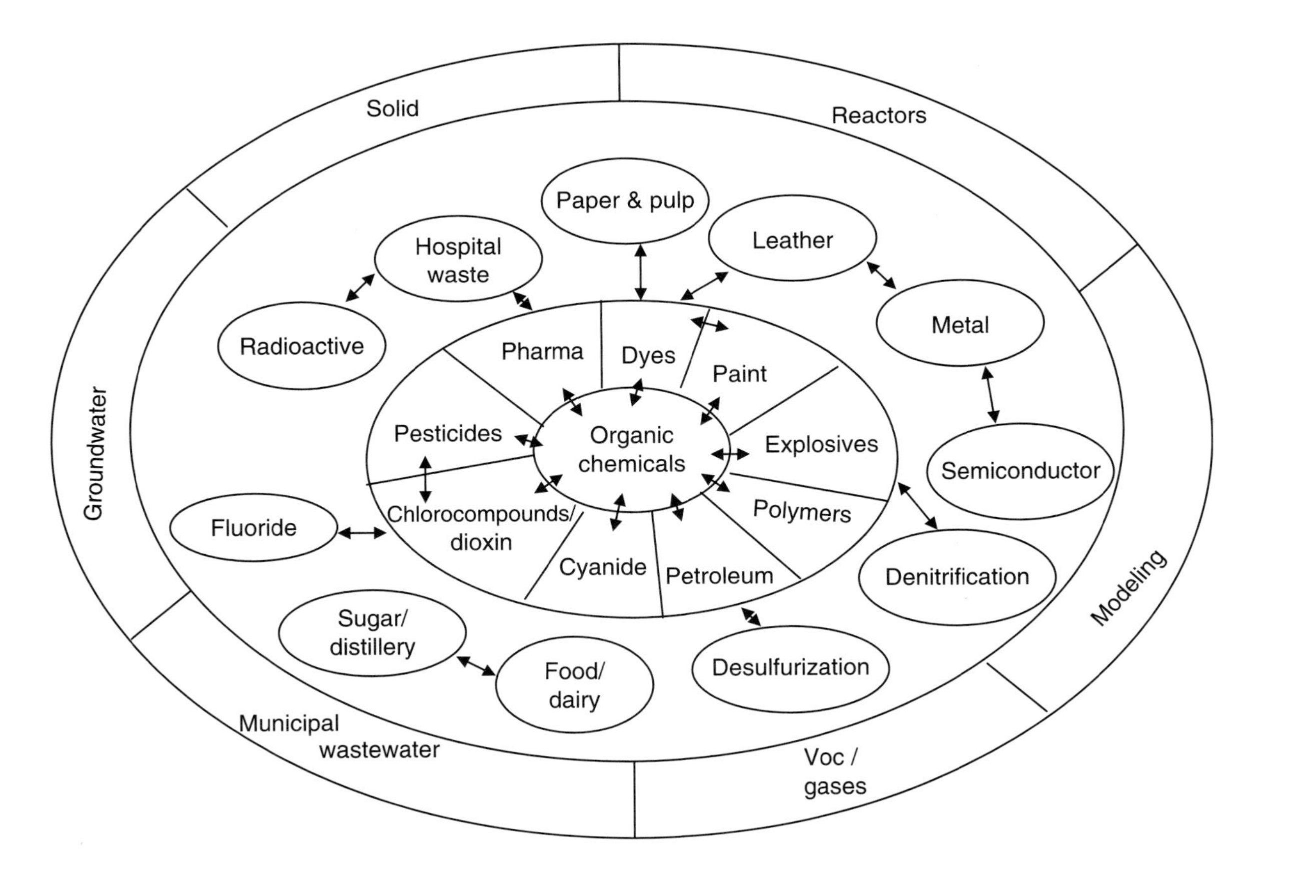

FIGURE 1-3. Various industrial pollutions considered in the book. VOC, volatile organic compound.

TABLE 1-3
Comparison of Agricultural Industry Effluents

Effluent source	*COD* (mg/L)	*BOD* (mg/L)
Wool scouring effluent	45,000	17,500
Distillery	60,000	30,000
Dairy	1,800	900
Tannery	13,000	1,270
Textile	1,360	660
Kraft mill	620	226

COD, chemical oxygen demand; BOD, biological oxygen demand.

Major Findings

A few broad conclusions can be listed based on the survey carried out by the authors:

- The majority of the treatment procedures followed by industries is still based on physical and chemical methods.
- Biodegradation techniques are not yet broad based and have several unsolved problems.
- Most of the wastewater from the industries studied is very complex and cannot be treated by a single microorganism, so microbial colonies appear to have good potential.
- Combined methods (physical, chemical, and biological or aerobic and anaerobic) show good potential.
- Biofilters are a cost-effective and ubiquitous reactor for treating VOCs and effluent gasses.
- Dynamic operation (as in sequential batch reactor) is able to achieve very high degradation rates compared with steady state operation.
- Membrane reactors have very good potential, but their cost factor makes them unpopular.
- White rot fungi (or any other extracellular organisms) are effective for general purpose degradation.
- Biosorption using dead surface-modified microorganisms is an effective technique for biodegradation and adsorption of metals from effluent streams.
- The majority of the biodegradation studies reported in the literature have been carried out with synthetic wastewater. These findings and conclusions have to be validated with real effluent, which may be more complex.

New Research Frontiers

A few of the areas that need focused study are:

- Degradation pathways
- Water-solvent interaction
- Use of biosurfactants for enhanced degradation
- Development of microorganisms tolerant to toxic effluents
- Development of microorganisms tolerant to principal and secondary pollutants
- Identification of bacterial consortiums
- Transport of aromatic compounds through solid and liquid phases
- Engineering of bacteria to do various ecological tasks
- Determination of enzyme crystal structure
- Development of analytical tools
- Development of tools to follow the active organisms
- Identification of new microorganisms from the environment or from the contaminated sites
- Identification of novel supports
- Long-term studies of the effects of pollutants on the ecology and the effects of engineered organisms on the ecosphere
- Wetland ecosystems

References

Hettige, H., P. Martin, M. Singh, and D. Wheeler. 1994. IPPS: The Industrial Pollution Projection System. Policy Research Working Paper, p. 1431.

Hettige, H., M. Mani, and D. Wheeler. 1998. Industrial Pollution in Economic Development: Kuznets Revisited. World Bank Development Research Group Working Paper. January. No. 1876.

Macek, T., M. Macková, and J. Ká. 2000. Exploitation of plants for the removal of organics in environmental remediation. *Biotech. Adv.* 18(1):23–34.

Mani, M. and D. Wheeler. 1998. In Search of Pollution Havens? Dirty Industry in the World Economy, 1960–95. *J. Environ. Devel.* 15(9):40–43.

Rösner, A. Pollution, Industrial Composition and Trade, Department of Economics Columbia University, USA, 2003. <http://www.columbia.edu/~ar378/rosner_paper.pdf>.

CHAPTER 2

Environmental Disasters

Environmental disasters occur because of natural or humanmade causes. The latter could be due to the release of pollutants into the environment, either accidentally or because of negligence or insufficient knowledge about the material. A flood that leads to human and material loss because of the building of a dam can also be considered a humanmade environmental disaster. Sometimes it may be difficult to connect a disaster to the cause, but humankind has slowly started realizing that the subsystems in our ecosystem are intricately interconnected and every action can lead to a disaster at a later point in time. Droughts, torrential floods, and other environmental disasters cost the world about $70 billion in 2002 (Reuters, 2002). In the United States, the number of incidents related to leakage or spills of chemicals and oil from pipelines, mobile units, storage tanks, railroads, and fixed manufacturing units increased from 25,700 per year in 1991 to 32,200 per year in 2003 (National Response Center, U.S. Coast Guard; http://www.nrc.uscg.mil/nrchp.html). The present generation, which has benefited from past progress, has also inherited past environmental mistakes. So it is the responsibility of the current generation to ensure that future generations inherit only the benefits of the progress, that they do not inherit any past environmental mistakes, and that the mistakes of the past generations be corrected.

The United Nations Environment Programme (UNEP) and the UN Office for the Coordination of Humanitarian Affairs monitor environmental occurrences (natural and humanmade) around the globe. Several of their reports on humanmade disasters are listed in Table 2-1. Pollution abatement expenditures in the European community vary from 0.5% of GDP in Greece and Spain to 1.6% of GDP in Germany. The expenditure in the United Kingdom is about 1.2% of its GDP (1990 data, Ecotec Research and Consulting Co., 1993), whereas the cost of pollution control in Japan is on the order of 2% of its GDP.

Various Disasters

This chapter briefly describes the various disasters that have occurred in the past few years and the problems caused by them. The study of various

TABLE 2-1
Humanmade Environmental Disasters [United Nations Environment Programme (UNEP) and the UN Office for the Coordination of Humanitarian Affairs]

Year	Disaster
2003	Pakistan—Oil spill in the Port of Karachi
	Morocco—Inland oil spill
	Kosovo, Serbia, and Montenegro—Phenol spill into a river system
2002	Republic of Djibouti—Toxic chemical spill
	Nigeria—Munitions dump explosion
2000	Venezuela—Mudslides and chemical spill, floods, chemical spill and the organic contamination in the coastal zone
	Romania, Hungary, Federal Republic of Yugoslavia—Baia Mare cyanide and heavy metal tailings spill
	Hungary and Romania—Baia Borsa mining waste spill
1999	Kenya—Aviation fuel spill
1997	Chile—Acute pollution of the Loa River
	Somalia—Hazardous waste, alleged dumping of hazardous substances
1996	Philippines—Mine tailings spill

incidents should help us to develop manufacturing processes that do not pollute, evolve safe means of disposing of toxic effluents, and avoid the hazards involved in storing and transporting toxic materials.

Acid Rain

Acid rain is rain with a pH of less than 5.7, which results from high levels of atmospheric nitric and sulfuric acids that get washed down to earth. Oxidation of sulfur and nitrogen in coal or other fossil fuels leads to the generation of acidic pollutants in the atmosphere. Acid rain has caused considerable damage to forests in many developed countries. Use of low-sulfur coal and gasoline can prevent acid rain.

Carbon Dioxide

CO_2 is not a pollutant in the conventional sense, since it is essential for plant growth. Combustion of fossil fuels, including coal-fired thermal power stations and forest fires, has increased the background levels of CO_2 from 315 ppm in 1960 to 405 ppm in 2000, which leads to an atmospheric greenhouse effect, which in turn increases the average temperature.

Ozone Layer Depletion and Human Health

In 1993 the atmospheric ozone layer surrounding the earth thinned to the lowest levels ever recorded. With the loss of stratospheric ozone, the atmosphere became more transparent to radiation, resulting in an increase in the amount of ultraviolet (UV) solar radiation reaching the earth. It has been found that an increase in UV radiation can lead to an increase in human diseases, including skin cancers, eye damage, and reduction in the effectiveness of the body's immune system. White skin is more prone to burning than black or brown skin. The Arctic ozone holes within the next 10 to 20 years could affect inhabited areas of northern Europe, Canada, and Russia (*New Scientist*, Oct. 10, 2000).

The ozone layer is affected because of emissions from earth and deposition in the stratosphere of compounds such as bromofluorocarbons (halons), chlorofluorocarbons (CFCs), hydrochlorofluorocarbons (HCFCs), CO_2, nitrogen oxides (NO_x), chlorinated carbons, methyl bromide (CH_3Br), methane (CH_4), and nitrous oxide (N_2O). CFCs are among the most important ozone-depleting substances; they are used in aerosol propellants, coolant agents in refrigerators, cleaning agents, and plastic foam-blowing agents. An international agreement was reached (the Montreal Protocol and its amendments, signed by 148 countries) that banned the production of most CFCs by the year 2000, and the Copenhagen amendment to the Montreal Protocol called for the cessation of HCFC (an alternate to CFCs) production by 2030.

Global Warming—Petrol versus Diesel

Based on some theoretical studies, it is believed that CO_2 produced by petrol engines could be less harmful to the planet than the soot and dust produced by diesel engines. A climate model showed that the soot produced by diesel engines will warm the climate more over the next century than the extra CO_2 emitted by petrol-powered vehicles. In addition, the soot particles would alter the humidity of air by allowing water droplets to condense around them, causing pollutants to accumulate in the air and change weather patterns. Hence 1 g of black carbon is 360,000 to 840,000 times as powerful a global warming agent as 1 g of CO_2.

Pollution Reducing Sunshine

Airborne pollutants have led to a steady decline in sunshine in vast polluted regions of eastern China. The amount of sunshine has fallen by between 2 and 3% a decade, and the maximum summer temperatures have also fallen by around 0.6°C a decade. In Zambia and the Brazilian Amazon, pollution blots out around a fifth of the sun's radiation at certain times of year (Freeman, 1990; *New Scientist*, 2002e).

Polychlorinated Biphenyls in the Environment

About 80 million pounds of polychlorinated biphenyls (PCB) are produced annually, and they find applications in capacitors, transformer oils, and heat transfer fluids. Half the amount is used as plasticizers, hydraulic fluid, and adhesives, as well as in carbon paper. About 10 million pounds escape annually and become environmental contaminants. These are very stable compounds, do not degrade, and accumulate in animal tissues. PCBs have been found in polar bears in the Arctic and penguins in Antarctica, creating havoc. Killer whales in the Gulf of Alaska are among the most heavily PCB-laden marine mammals in the world, and their numbers are in rapid decline.

Methyl Tertiary Butyl Ether (MTBE)

MTBE is an "oxygenate" that makes gasoline burn cleaner and more efficiently, but it is also identified as a probable carcinogen that spreads rapidly when gasoline escapes from leaky underground storage tanks, contaminating sources of groundwater and drinking water from New York to California in the United States. At least 16 states already have passed measures to ban or significantly limit the use of MTBE in gasoline.

Exxon Valdez Spill

The grounding of the oil tanker *Exxon Valdez* on Bligh Reef on March 24, 1989, released almost 11 million U.S. gallons of North Slope crude oil into the waters of Prince William Sound, Alaska. A major storm a few days later spread the oil into the shorelines of the numerous islands in the western part of the Sound and out into the Gulf of Alaska. Bioremediation was carried out by the application of an oleophilic liquid fertilizer, a micro emulsion of a saturated solution of urea in oleic acid containing tri(laureth-4)-phosphate and butoxy-ethanol to stimulate the activity of the oil-degrading bacteria. Two weeks after application of the fertilizer, the cobbles on the treated section of the shoreline were substantially clean. But most seabird populations hit by the oil spill have not shown signs of recovery even a decade after the disaster (*New Scientist*, May 2001).

Pipeline spills reported to the U.S. Department of Transportation average 12 million gallons of petroleum products a year. The U.S. General Accounting Office says an average of 16,000 small oil spills seep into waterways each year, half of them during loading or unloading operations, and the real number could be three to four times that.

Cyanide Spill at Baia Mare, March 2000

On January 30, 2000, following a breach in the tailing dam of the Aurul SA Baia Mare Company, a major spill of cyanide-rich tailings waste from the extraction of precious metals was released into the river system near Baia Mare in northwest Romania. The contaminant traveled via tributaries into the Somes, Tisza, and finally the Danube rivers before reaching the Black Sea (UNEP/OCHA Environment Unit, 2000).

Corals Affected by Human Waste

Human wastewater containing undegraded drugs and antibiotics is having a bad effect on the aquatic environment, especially on the corals off the coast of Florida, which form the world's third largest barrier reef (*New Scientist*, 2002b). It has been found that half of the live coral off the Florida coast has disappeared in the past 5 years. The fish that feed on these corals have developed deformities and died in much higher numbers than usual.

Movement of Pollutants into Coastal Aquifers

Wells located near coasts could be more polluted than the ones located inland because the pollutants dumped into the sea diffuse faster through the soil barrier (they are less soluble in salty water because of the "salting out effect") (*New Scientist*, 2003b). This phenomenon can be observed in the movement of pollutants from sea to the coastal aquifers—natural reservoirs of freshwater held in porous rock and also toward coastal agricultural land.

Chernobyl Accident

The Chernobyl accident in the Ukraine in 1986 was the result of a flawed nuclear reactor design. The reactor was operated with inadequately trained personnel and without proper regard for safety, leading to a steam explosion and fire that released ~5% of the radioactive reactor core into the atmosphere and downwind of the plant. Some 31 people were killed, and there have since been around 10 deaths from thyroid cancer attributed to the accident.

Bhopal Disaster

On December 2, 1981, more than 40 tonnes of methyl isocyanine (MIC) and other lethal gases, including hydrogen cyanide, leaked from a pesticide factory at the northern end of the Bhopal, the capital of Madhya Pradesh, India. More than 8,000 people were killed, and more than 500,000 people suffered multisystemic injuries. Toxic gas exposure was found to have had a detrimental effect on the immune system (Lepkowski, 1985).

Bashkiria Train–Gas Pipeline Disaster

The Bashkir train–gas pipeline disaster occurred in June 1989. At least 400 people were killed when a pipeline transporting a methane-propane mixture exploded as two trains were passing, causing 400 immediate deaths and more than 800 casualties, mostly with burns (Kulyapin et al., 1990).

Seveso Dioxin Accident

Dioxins and furans are halogenated aromatic hydrocarbons that are commonly produced by combustion of fossil fuels and incineration of municipal waste, as a byproduct of pulp and paper bleaching, and in the production of other chemicals. 2,3,7,8-Tetrachlorodibenzo-*p*-dioxin is the most toxic member of this family. It is an endocrine disrupter as well as a potent animal carcinogen and teratogen that persists in both the environment and biological tissues. On July 10, 1976, a valve broke at the Industrie Chimiche Meda Societa Azionaria chemical plant in Meda, Italy, releasing about 3,000 kg of dioxin-containing chemicals into the atmosphere. Approximately 4% of local farm animals died, and roughly 80,000 animals were killed to prevent the contamination from moving up the food chain. It is believed that this exposure affected the sex ratio in future progeny.

Czech Plant Leaked Hundreds of Kilos of Deadly Gas

Several hundred kilograms of highly poisonous chlorine gas leaked into the air in an accident at a flooded chemical plant in the Czech Republic on August 23, 2002. The accident happened when workers at Spolana, a unit of the chemicals group Unipetrol, pumped fluid chlorine gas out of a storage unit that had been damaged in the flood. There were no casualties in the accident.

Superfund

The Superfund Program in the United States was initiated in 1980, when the Environmental Protection Agency started identifying contaminated sites where hazardous waste was dumped on the land or in the rivers, or buried as landfill by industries, polluting the environment and causing harm to animals and humans (U. S. Environmental Protection Agency, Washington, DC). There are over 1,400 Superfund sites across the United States that were contaminated by chemicals such as acetone, benzene, 2-butanone, carbon tetrachloride, chlordane, chloroform, 1,1-dichloroethane, 1,2-dichloroethane, methylene chloride, naphthalene, pentachlorophenol, PCBs, polycyclic aromatic hydrocarbons (PAHs), tetrachloroethylene, toluene, trichloroethylene, vinyl chloride and xylene; metals such as arsenic, barium, cadmium, chromium, cyanide, lead, mercury, nickel, and zinc; and pesticides such as DDT, DDE, and DDD.

Conclusions

Several approaches could be adapted to mitigate environmental disasters such as:

- Use biodiesel as fuel, and use nonpolluting means of transportation.
- Manage waste properly, and install waste treatment plants in every industry.

- Develop processes that generate little or no waste.
- Avoid storage of toxic and hazardous chemicals.
- Determine the toxic nature of all chemicals and materials that are being used by humankind.
- Develop better analytical techniques for monitoring pollution.
- Use environmental resources judiciously.
- Manage hazardous wastes properly.
- Bring about public awareness.

Society should be willing to bear the extra cost involved in efforts to reduce or eliminate pollution at its source. Otherwise we end up paying for cleaning the polluted ecosystem and/or for our medical bills. In many cases the long-term impact of pollution on the ecosystem and human health is not fully understood, and it is too late by the time extra knowledge is gained. Estimating the cost of pollution is not easy and straightforward. There are several direct and hidden costs: the decrease in the market value of the resource caused by pollution, the cost of pollution prevention or environmental remediation, the cost of the impact of environmental pollution on ecosystems and human health, and the cost of environmental protection. Another important factor is society's "willingness to pay" for the reduction of environmental pollution.

References

Freeman, H. 1990. *Geograph. Res. Letters*. 29:2042.
Lepkowski W. 1985. Bhopal. *Chem. Eng. News*, 23 (Dec. 2).
New Scientist. 2002b. Future of corals is going down. Aug. 10.
———. 2002e. Pollution is plunging us into darkness. Dec. 14.
———. 2003b. Seawater pumps pollutants into the coastal aquifers. June 17.

Bibliography

Batterbee, R. W., R. J. Flower, A. C. Stevenson, and B. Rippey. 1985. Lake acidification. *Nature* 314:87–88.
Clark, R. B. *Marine Pollution*. 1992. 3rd ed. Oxford: Clarendon Press.
Ecotech Research & Consulting, Ltd. 1993. A Review of UK Environmental Expenditure: A final report to the Department of Environment. London: HMSO.
Freeman, H. (ed.) 1990. *Hazardous Waste Treatment and Minimization*. New York: McGraw-Hill.
Hawksworth, D. and F. Rose, 1970. Qualitative scale for estimating SO_2 air pollution. *Nature* 227:145–148.
Holdgate, M. W. 1979. *A Perspective of Environmental Pollution*. Cambridge: Cambridge University Press.
Kiely, G. 1998. *Environmental Engineering*. Chemical and Petrochemical Engineering Series. International Edition. New York: McGraw-Hill.
Kulyapin, A. V., V. G. Sakhahtdinov, V. M. Temerbulatov, W. K. Becker, J. P. Waymack: 1990. The Bashhkiria train/gas pipeline disaster, June 1989 Bashkirian Republic. *Burns* 16: 339–342.

New Scientist, 2000. Ozone layer thins over Europe. Oct. 10.
———. 2001. Exxon oil spill still affecting the birds. May 3.
———. 2002a. Trees cannot solve the problem of global warming. April 13.
———. 2002c. Diesel's dirty green surprise. Nov. 2.
———. 2002d. When is an oil spill an environmental disaster? Nov. 30.
———. 2002f. Pollution triggers genetic defects. Dec. 14.
———. 2003a. Heat will soar as haze fades. Jun. 7.
Society for General Microbiology. 1997. Symposium 48, *Microbial Control of Pollution*. Eds. J. C. Fry, G. M. Gadd, R. A. Herbert, C. W. Jones, and I. A. Watson-Crick. Cambridge: Cambridge University Press.
UNEP/OCHA Environment Unit. 2000. <http://ochaonline.un.org/GetBin.asp?DocID=1695>.
U. S. Environmental Protection Agency. <http://www.epa.gove/Compliance/resources/newsletters/cleanup/cleanupb.pdf>.

CHAPTER 3

Aerobic and Anaerobic Bioreactors

Introduction

The aerobic biodegradation process can be represented by:

$$C_xH_y + O_2 + (\text{microorganisms/nutrients}) \rightarrow H_2O + CO_2 + \text{biomass}$$

The anaerobic bioprocess can be represented by:

$$C_xH_y + (\text{microorganisms/nutrients}) \rightarrow CO_2 + CH_4 + \text{biomass}$$

Aerobic Degradation

Bacteria that thrive in oxygen-rich environments break down and digest waste. The mixed aerobic microbial consortium uses the organic carbon present in the effluent as its carbon and energy source. The complex organics finally get converted to microbial biomass (sludge) and carbon dioxide (CO_2). Food here is limiting, which results in the microorganisms consuming their own protoplasm to obtain energy for cell maintenance reactions (endogenous respiration). Therefore, the biomass concentration continuously decreases until the energy content reaches a minimum so as to be considered biologically stable and suitable for disposal in the environment. Sludges with a specific oxygen uptake rate of less than or equal to 1 mg O_2/h/g can be considered stabilized.

Digestion Pathway

During this oxidation process, contaminants and pollutants are broken down into CO_2, water, nitrates, sulfates, and biomass (microorganisms). In the conventional aerobic system, the substrate is used as a source of carbon and energy (Fig. 3-1). It serves as an electron donor, resulting in bacterial growth. The extent of degradation is correlated with the rate of oxygen consumption, as well as the previous acclimation of the organism in the same substrate.

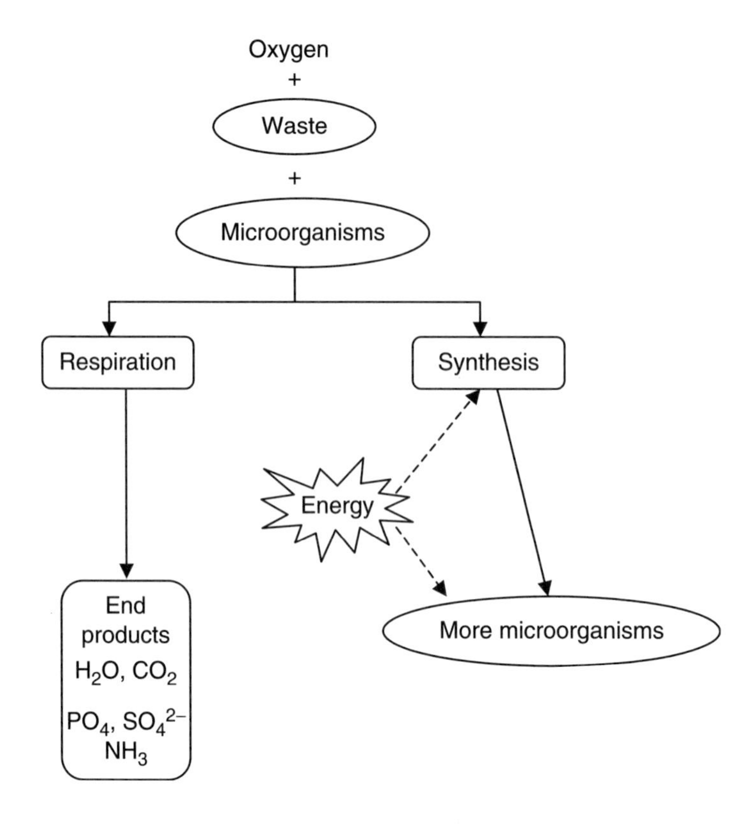

FIGURE 3-1. Aerobic degradation pathway.

Two enzymes primarily involved in the process are di- and mono-oxygenases. The latter enzyme can act on both aromatic and aliphatic compounds, while the former can act only on aromatic compounds. Another class of enzymes involved in aerobic degradation is the peroxidases, which have been receiving attention recently for their ability to degrade lignin.

Anaerobic Degradation

In the anaerobic process, the complex organics are first broken down into a mixture of volatile fatty acids (VFAs), mostly acetic, propionic, and butyric acids. This is achieved by "acidogens," a consortium of hydrolytic and acidogenic bacteria (Gottschalk, 1979). The VFAs are in turn converted to CO_2 and methane by acetogenic (acetogens) and methanogenic (methanogens) bacteria, respectively (Zehnder et al., 1982).

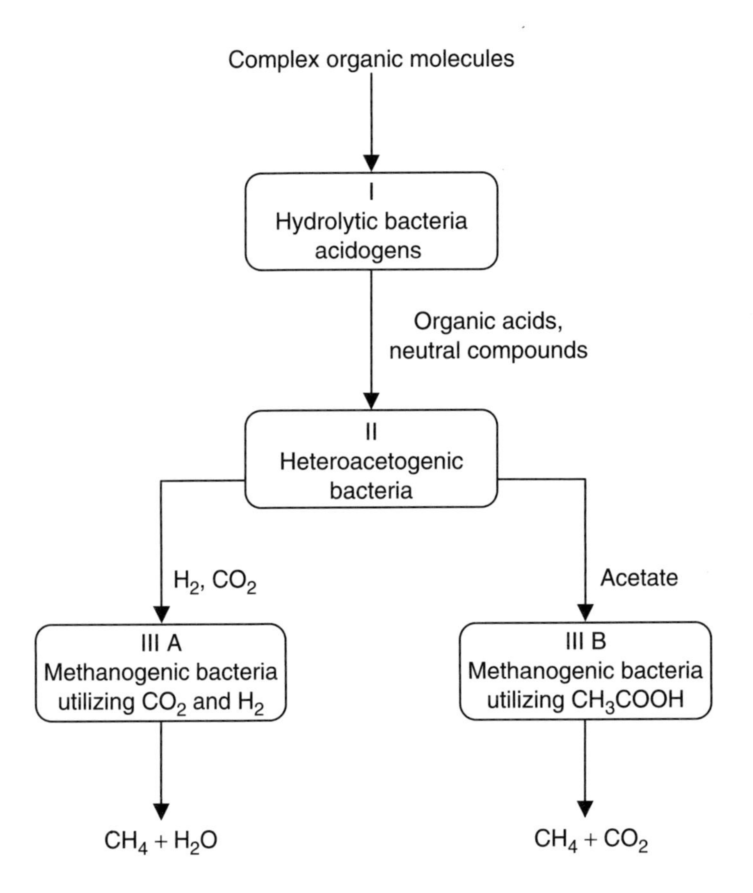

FIGURE 3-2. Anaerobic degradation pathway.

Anaerobic Digestion Pathway

Anaerobic digestion is a biological process in which organic matter is converted by several independent, consecutive, and parallel reactions. In the absence of oxygen, close-knit communities of bacteria cooperate to form a stable, self-regulating fermentation that transforms organic matter into a mixture of methane and CO_2 (Fig. 3-2). The amount of methane gas produced varies with the amount of organic waste fed to the digester and the operating temperature. Anaerobic digestion occurs in six main stages (Jeyaseelan, 1997):

- Hydrolysis of complex organic biopolymers (proteins, carbohydrates, and lipids) into monomers (amino acids, sugars, long chain fatty acids) by hydrolytic bacteria (group I) (acidogens)
- Fermentation of amino acids and sugars by hydrolytic bacteria (group I)

- Anaerobic oxidation of volatile fatty acids and alcohols by heteroacetogenic bacteria (group II)
- Anaerobic oxidation of intermediary products such as volatile fatty acids by heteroacetogenic bacteria (group II)
- Conversion of hydrogen to methane by methanogenic bacteria utilizing hydrogen (group IIIA)
- Conversion of acetate to methane by methanogenic bacteria utilizing acetate (group IIIB)

The hydrolysis of undissolved carbohydrates and proteins follows separate paths. The heteroacetogenic bacteria grow in close association with the methanogenic bacteria during the final stages of the process. The reason for this is that the conversion of the fermentation products by the heteroacetogens is thermodynamically possible only if the hydrogen concentration is kept sufficiently low. This requires a close symbiotic relationship among the classes of bacteria.

Comparison between Aerobic and Anaerobic Degradation Pathways

While both aerobic as well as anaerobic degradation routes can equally remove complex organics from the effluents, the anaerobic route has an obvious advantage because it produces methane, a combustible biogas with a reasonably good calorific value of 24 MJ/m^3. Aerobic treatment produces 2.4 kg CO_2/kg COD, while an anaerobic process produces only 1 kg CO_2/kg COD. Sludge disposal is an important consideration since it represents about 60% of the total treatment cost. The cost to dispose of the sludge produced by an anaerobic plant is only 10% that of a corresponding aerobic plant. Nutrient requirements are 20% lower for anaerobic plants than for aerobic plants, and the aeration process also involves possible volatilization of some of the organic contaminants, turning water pollution into air pollution. A comparison of electron acceptors, type of reaction, and metabolic byproducts of aerobic and anaerobic processes is shown in Table 3-1. Oxygen is the only electron acceptor in the aerobic process that produces water and carbon dioxide as the products of reaction. In the anaerobic process, several electron acceptors are possible, giving rise to several different metabolic byproducts; hence this is preferred for aromatic compounds.

Several major differences exist between the aerobic and anaerobic degradation pathways of aromatic compounds. They are listed in Table 3-2 (Jothimani et al., 2003).

Aerobic Reactors

The effectiveness of the design and operation of a biological treatment system depends on several parameters. They include: amount of nutrients

TABLE 3-1
Electron Acceptors and Byproducts in Aerobic and Anaerobic Processes

Electron acceptor	*Type of reaction*	*Metabolic byproduct*
Oxygen	Aerobic	Carbon dioxide, water
Nitrate (NO_3^-)	Anaerobic respiration	Nitrogen gas, carbon dioxide
Manganese (Mn^{4+})	Anaerobic	Manganese (Mn^{2+})
Ferric iron (Fe^{3+})	Anaerobic	Ferrous iron (Fe^{2+})
Sulphate (SO_4^{2-})	Anaerobic respiration	Hydrogen sulfide
Carbon dioxide	Anaerobic respiration	Methane

TABLE 3-2
Comparison of Aerobic and Anaerobic Degradation of Aromatic Compounds

	Anaerobic	*Aerobic*
Channeling	$+H_2O$, 2H, −2H, +CO2, $+CO_4$	O_2
Central intermediates	Benzoyl CoA, resorcino phloroglucinol	Catechol, proto catechuate gentisate
Ring attack	2 or 4 H + H_2O	O_2
Central intermediates	Easy to reduce or hydrate	Easy to oxidize
Cleavage of the ring	Hydrolysis of 3-oxo compound	Oxygenolysis

available for the organism to grow, dissolved oxygen concentration (for aerobic treatment), food-to-microorganism ratio (this ratio applies to only activated sludge systems and is a measure of the amount of biomass available to metabolize the influent organic loading to the aeration unit), pH, temperature, cell residence time, hydraulic loading rate (the length of time the organic constituents are in contact with the microorganisms), settling time (time for separating sludge from liquid), and degree of mixing.

The design parameters for aerobic reactors are tabulated in Table 3-3. The gas-to-liquid mass transfer and bubble size governs the rate of the aerobic biodegradation process. Hence, the reactor designs strive to achieve high gas transport rates and generate small air bubbles (see Chapter 4 for design). In the breeding of aerobic organisms, adequate amounts of dissolved oxygen must be ensured in the medium. Since the solubility of oxygen in the medium is very low, it must be continuously supplied and gas-to-liquid mass transfer should be maintained high. A minimal critical concentration of dissolved oxygen must be maintained in the substrate to keep the microorganisms active, and the values are in the range of 0.003 to 0.05 mmol/L, which is 0.1 to 10% of the oxygen solubility values in water. But the presence of salts such

TABLE 3-3
Design Parameters for Aerobic Digestion

Parameter	*Value*
1. Retention time	
Activated sludge only	15–20 days
Activated sludge with primary treatment	20–25 days
2. Solids loading	1.6-3.2 kg VSS/m^3·day
3. Air required	
Activated sludge only	20–35 L/min·m^3
Activated sludge with primary treatment	55–65 L/min·m^3
4. Power required	0.02–0.03 kW/m^3

VSS, volatile suspended solids.

as NaCl decreases oxygen solubility by a factor of 2. Microbes use oxygen for cell maintenance, respiratory oxidation for further growth (biosynthesis), and oxidation of substrates into related metabolic end products. The oxygen uptake rate for bacteria and yeast are the highest (0.2 to 2 × 10^{-3} kg/m^3/s), followed by fungi (0.1 to 1) and the rest of the biocultures (0.01 to 0.001).

The simplest and the cheapest design uses open lagoons and oxidation ponds, which are nothing but open pits where the effluent is stored and oxygen is either bubbled through the liquid medium using blowers or the liquid is churned using slowly rotating disks (Fig. 3-3). The sludge that is generated settles to the bottom of the tank and is drained from time to time. Oxygen transfer rates are low, and the residence time is generally on the order of 2 to 7 days. Each disk is covered with a biological film that degrades dissolved organic constituents present in the wastewater. As the disk slowly rotates, it carries a film of the wastewater into the air, where oxygen is available for aerobic biological decomposition. The excess biomass produced disengages from the disk and falls into the trough. Several contactors are often operated in series.

Another common aerobic reactor design is the mechanically stirred tank reactor. In this design, air is introduced from the bottom through a sparging arrangement. Airlift reactors are reactors without any mechanical stirring arrangements for mixing. The turbulence caused by the gas flow ensures adequate mixing of the liquid. The inner draft tube is provided in the central section of the reactor (Fig. 3-4). The introduction of the fluid (air or liquid) causes upward motion and results in circulatory flow in the entire reactor. Because of the low liquid velocities, energy consumption is also low. These reactors can be used for both free and immobilized cells. The oxygen mass transfer coefficient in this reactor is high in comparison to stirred tank reactors. Deep shaft reactors are 50 to 150 m long and are made of concrete (Fig. 3-5). They are buried underground and are used for

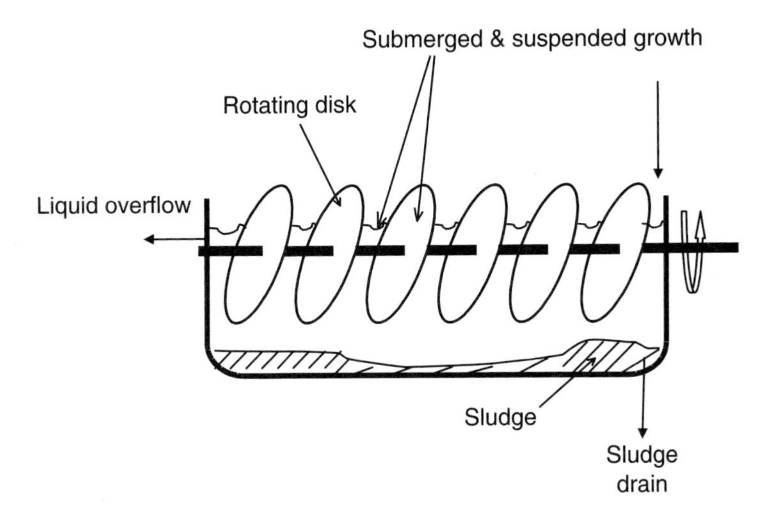

FIGURE 3-3. Rotating disk contactor.

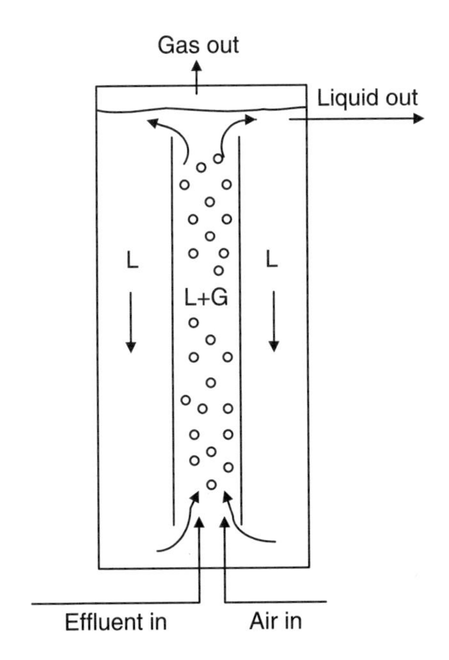

FIGURE 3-4. Airlift reactor for aerobic operation.

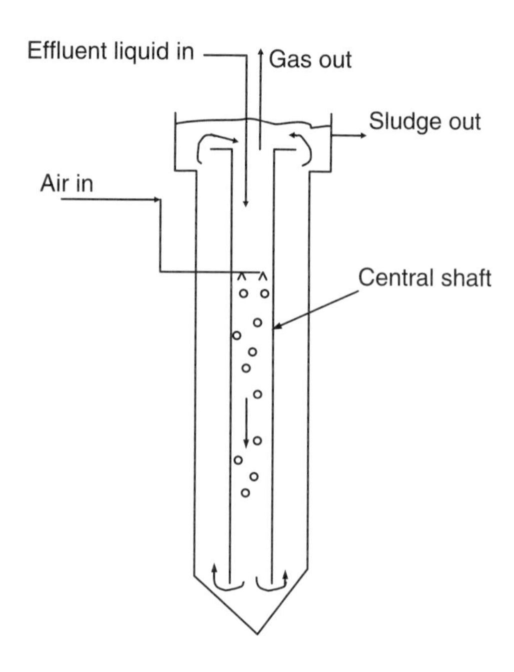

FIGURE 3-5. Deep shaft reactor for aerobic operation.

sewage treatment. The air in this reactor is not introduced at the bottom but in the middle. These reactors are also suitable for shear sensitive, foaming, and flocculating organisms.

Other widely used aerobic reactors in wastewater treatment are packed bed or fixed bed bioreactors with attached biofilms (Fig. 3-6). These reactors are widely used with immobilized cells. Wastewater is pumped into the top of the reactor and made to flow downward through the packed bed or sometimes vice versa. Air is pumped from the bottom. Microorganisms in the aerated packed bed grow and degrade organic matter contained in the wastewater. The disadvantages of packed beds are the change in the bed porosity and bed compaction with time, resulting in high pressure drop across the bed, and channeling due to turbulence in the bed. Several modifications such as tapered beds to reduce the pressure drop across the length of the reactor, inclined beds, horizontal beds, and rotary horizontal reactors have been tried with limited success.

Bubble columns are slender, tall columns with a gas distributor at the bottom (Fig. 3-7). The construction of bubble columns is very simple, and higher mass transfer coefficient can be achieved than with loop reactors. These reactors can be as large as 5,000 m^3. Since they have broad residence time distribution and good dispersion properties, they can be used for aerobic wastewater treatment. The liquid column provides high pressure at the

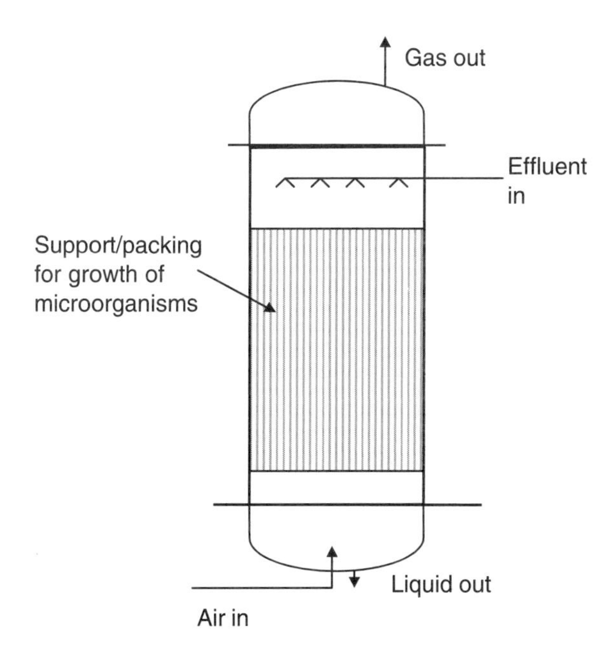

FIGURE 3-6. Packed bed reactor for aerobic operation.

reactor bottom, giving rise to increased oxygen solubility. Hence, gas holdup in such reactors is generally very high.

An inverse fluidized bed is used in aerobic wastewater treatment (Fig. 3-8), where the solid phase is an inert particle coated with a biofilm, the gas phase is oxygen/air, and the liquid phase is the wastewater that needs treating. The bed of solids has a density lower than that of the liquid phase, but a fluidized state is created by the downward flow of the liquid. The gas flows countercurrent to the liquid. This mode of operation improves the mass transfer rate, reduces the attrition rate of solids, and helps the bed to refluidize easily after shutdown. Low-concentration synthetic and municipal wastewaters are treated at residence times ranging from 0.6 to 3 h in an anaerobic inverse fluidized bed. Sufficient care should be taken during start-up, when the biofilm is forming on the inert support.

The slurry-phase bioreactor is a stirred tank in which soil is suspended in water (greater than 40% solids) and mixed with air, microbial cells, and nutrients. In the solid-phase bioreactor (also known as biopile), water is just sprinkled over the soil to adjust the soil moisture content; otherwise it is similar to a slurry-phase reactor. Air and nutrients are fed through perforated pipes. These reactors are a very cost-effective ex situ treatment if the bioremediation time is not limiting. Apart from slow degradation time, another

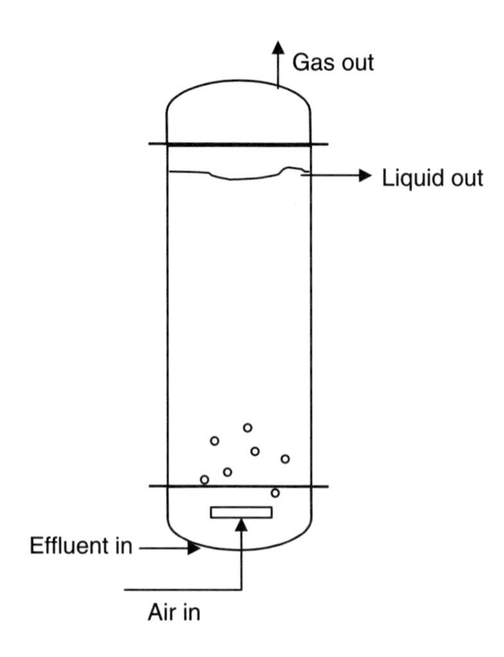

FIGURE 3-7. Aerobic bubble column reactor.

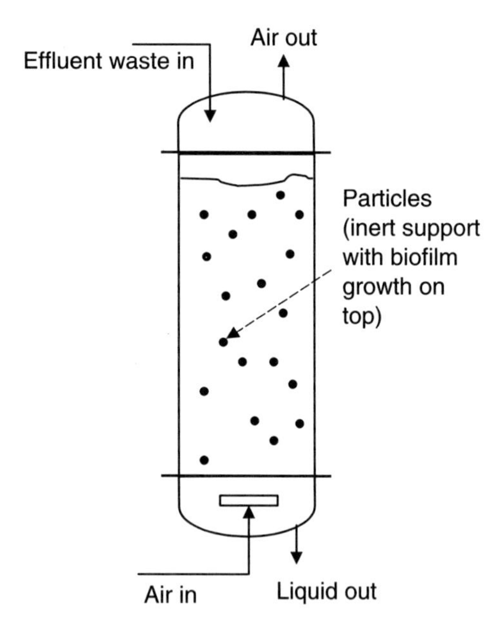

FIGURE 3-8. Inverse fluidized bioreactor (for aerobic operation).

disadvantage of the latter is nonhomogeneous degradation. At the beginning of the treatment when the effluent concentration is high, conversion capabilities of the microbes are exploited to the fullest, while at low contaminant concentrations, the mass transfer rate toward the degrading microbes becomes the rate limiting step, thus significantly reducing the overall rate of contaminant degradation. The air and the nutrients can be fed into the biopile either externally or internally.

Anaerobic Reactors

Since the overall rate of the anaerobic process is controlled by the methanogenic step, the rate of biomethanation can be accelerated only by enhancement of the rate conversion of VFAs to methane. In addition, proper attention must be paid to the safety aspects during the reactor design, because large quantities of potentially explosive gas are generated. Table 3-4 lists the broad design parameters for the anaerobic reactor (Praveen and Ramachandran, 1993; Canovas-Diaz and Howell, 1988; Hickey and Owens, 1981; Marchaim, 1992; Kosaric and Blaszczyk, 1991; Rajeswari et al., 2000).

Anaerobic rotating biological contactors are similar to their aerobic counterparts except that they are maintained at anaerobic conditions. Similarly, anaerobic ponds are comparable to their aerobic counterparts; if the land cost is less than U.S. $5 per square meter then they can be economical.

The fixed film reactors are so named because the biomass is made to grow on a fixed support material such as PVC, wood, carbon, rock, etc. These reactors have a simple construction, good for higher loading rates, and can withstand higher toxic and organic loadings (Fig. 3-9). The liquid flows upward, and the biogas generated is collected at the top and piped to storage tanks. The main disadvantages of these reactors are clogging of the reactor due to the increase in biofilm thickness and/or an increase in the concentration of high suspended solids in the waste. Additionally, a large portion of the reactor volume is occupied by the media.

TABLE 3-4
Design Parameters for Anaerobic Digestion

Parameter	*Normal/standard rate*	*High rate*
1. Solids retention time, days	30–90	10–20
2. Volatile solids loading, $kg/m^3/day$	0.5–1.6	1.6–6.4
3. Digested solids concentration, %	4–6	4–6
4. Volatile solids reduction, %	35–50	45–55
5. Gas production, m^3/kg VSS added	0.5–0.55	0.6–0.65
6. Methane content, %	65	65

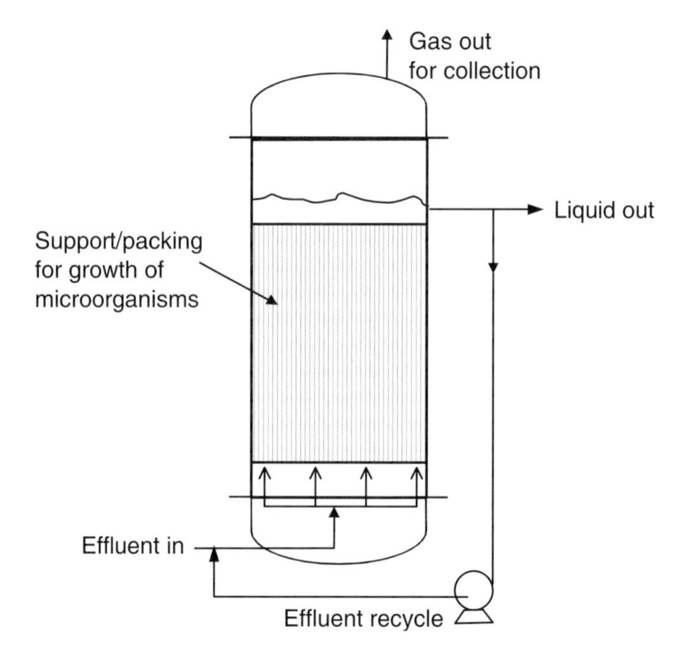

FIGURE 3-9. Anaerobic fixed biofilm reactor.

The anaerobic bacteria, especially methanogens, have a tendency to form self-immobilized granular structures with good settling properties, which are employed in an upflow anaerobic sludge blanket (UASB) reactor. The effluent flows up the reactor through a sludge blanket that is formed by the anaerobic bacterial granules (Fig. 3-10). The effluent substrate diffuses into the sludge granules and is degraded by an anaerobic pathway. Because of the higher biomass concentrations, a UASB reactor can achieve conversions several times higher than conventional anaerobic processes and also tolerate pH, temperature, and influent feed fluctuations. Additionally, since no support medium is required for attachment of the biomass, the capital cost is decreased. There is no mechanical mixing, recirculation of sludge, or, recirculation of effluent within the reactor; hence, the energy requirement is less. This design is becoming popular for treating effluents from various industries including distilleries, tanneries, and food processing. The disadvantages of this reactor are its long start-up time and the need for skilled operators, since washout of the sludge is possible if the reactor is not controlled properly. A variation of the UASB reactor is the expanded sludge bed in which higher superficial liquid velocities are employed to achieve fluidization at the bed top, leading to good mixing and retention of granular sludge material only.

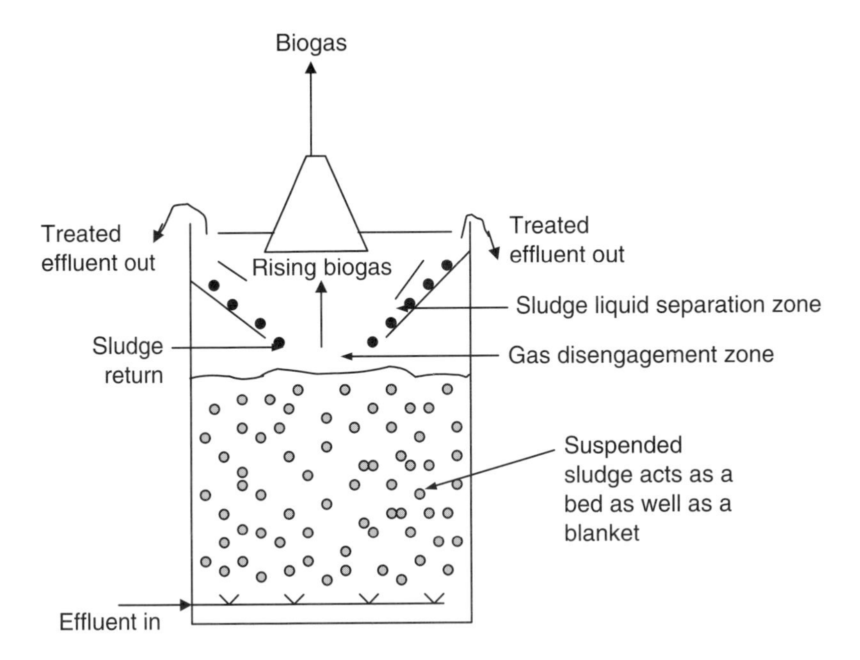

FIGURE 3-10. Upflow anaerobic sludge blanket reactor.

In anaerobic fluidized bed (AFB) reactors, a mixed culture of bacteria in the form of a film is made to grow on the surface of some inert carrier particle (Fig. 3-11). These particles are then fluidized using the energy of the influent stream. The linear velocity of the effluent is kept above the minimum fluidization velocity so that the film-covered particles are always in motion. The effluent substrate diffuses into the "biofilm" and gets converted into VFAs and ultimately to methane, which diffuses out through the biofilm into the bulk liquid. The mixing and mass transfer attained in these reactors is excellent, so the resulting conversions are comparable or even superior to those obtained with UASB reactors. These reactors have typical loading rates of 25 kg COD/m^3 day. However, as the biofilm grows, the film-covered particles increase in size, and hence their composite density decreases. This causes the particle to move up in the bed, ultimately resulting in its leaving the reactor, thereby leading to a reduction in the carrier particle concentration inside the reactor. This problem is overcome by removing the biofilm from the carrier particle that has exited the reactor and then recycling it back into the reactor. Another drawback of AFB reactors is the high energy requirement resulting from the large recycle rates employed in these systems (Forster and Wase, 1990).

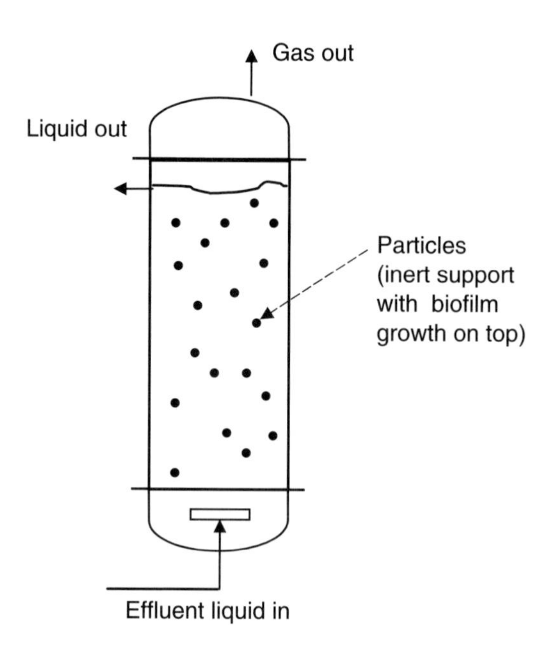

FIGURE 3-11. Anaerobic fluidized bed reactor.

The upflow anaerobic reactor allows the waste to flow up from the bottom, and the gas collection chamber is located at the top. In the anaerobic activated-sludge process, the bioreactor and clarifier are placed in series. Submerged media anaerobic reactors (SMAR) are similar to the upflow bioreactor, but they also have a packed bed of rocks that supports bacterial growth and that is submerged in the effluent. The fluidized-bed SMAR uses smaller particles as support material, which can be fluidized during operation. In both systems the gas and liquid effluent are separated at the top of the reactor. Hence, efficient liquid disengaging systems have to be designed to prevent effluent carryover.

A hybrid reactor is a combination of a UASB and a fixed biofilm reactor. It has a support section at the top of the fluidized section (Fig. 3-12). This helps the microbes to attach themselves firmly, preventing washout and also increasing the microbial population in the reactor. Also, the microbes are more resistant to shock load changes. Sludge loss is minimal, and mass transfer rates are higher.

Membrane Reactors

Several membrane reactor designs are possible. In the first design, the soluble enzymes are suspended in solution inside the reactor and the product

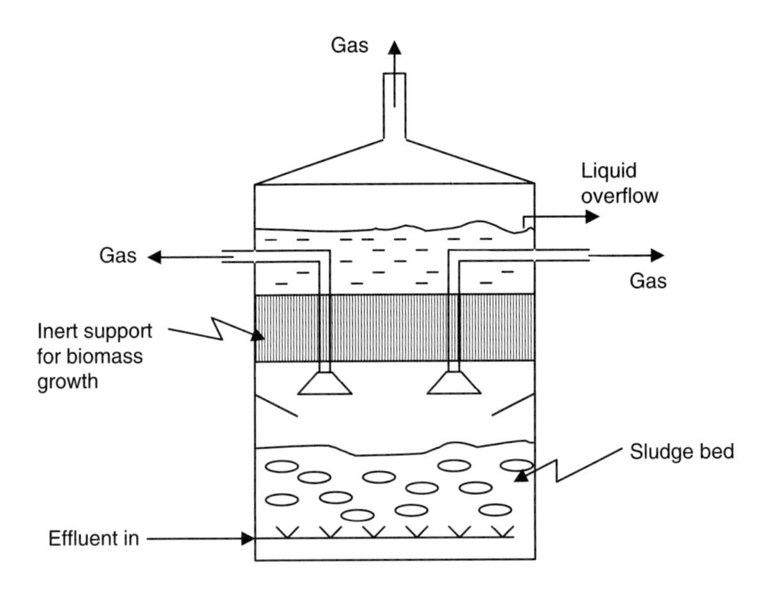

FIGURE 3-12. Hybrid reactor.

mixture, including the enzymes, is withdrawn and passed through the membrane filter, which retains the enzyme and lets the product pass through. The semipermeable membrane creates a physical barrier between the enzyme and the reactants and/or products. The low molecular weight (or smaller sized) products are separated from the reaction mixture by the action of a driving force across the membrane, which could be a chemical potential, a pressure differential, or an electric field. The retained enzyme is recycled back into the reactor (Fig. 3-13A). This design is also called a direct contact membrane reactor.

In the second design, the membrane filter is submerged inside the reactor as shown in Fig. 3-13B so that the permeate flowing out of the reactor will be free from the enzyme; the latter is retained inside the reactor. The disadvantage of this design is that in the case of fouling, the membrane material has to be removed and cleaned.

In the third design, the enzyme is immobilized on the membrane material (immobilized enzyme membrane bioreactor) so that reaction and separation happen simultaneously. Generally these reactors have tubular designs as shown in Fig. 3-13C.

In the fourth type, the enzymes are immobilized or entrapped in a support membrane–like fiber or gel and hence are retained inside the reactor (extractive membrane bioreactor) as shown in Fig. 3-13D. The rate-limiting step is the diffusion of the substrate through the membrane material to reach

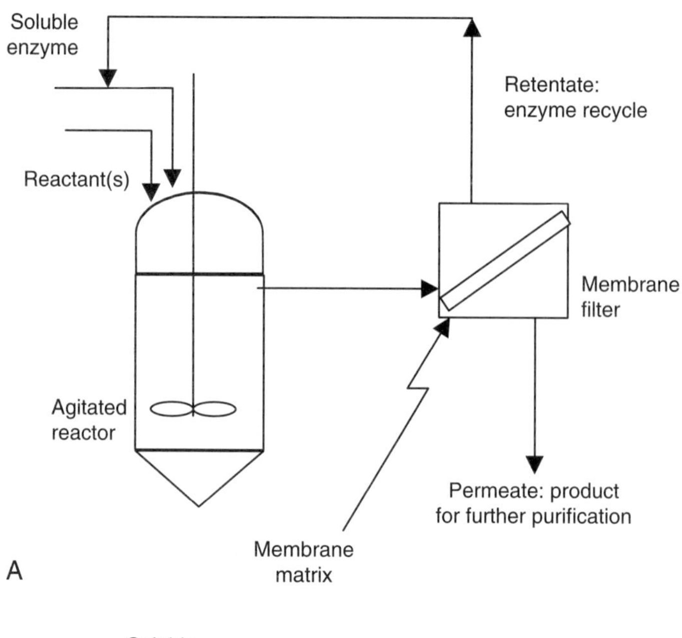

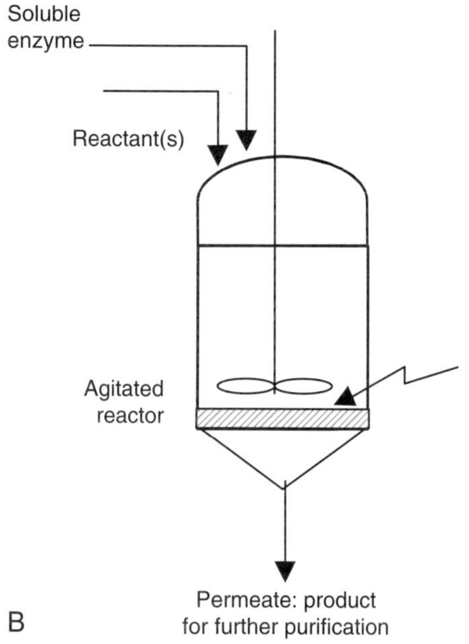

FIGURE 3-13. Membrane reactor designs: (A) with external membrane filter, (B) with internal membrane filter—submerged membrane bioreactor, (C) with enzyme immobilized within the membrane matrix, and (D) enzyme trapped in gel, fiber, or microcapsules (immobilized enzymes).

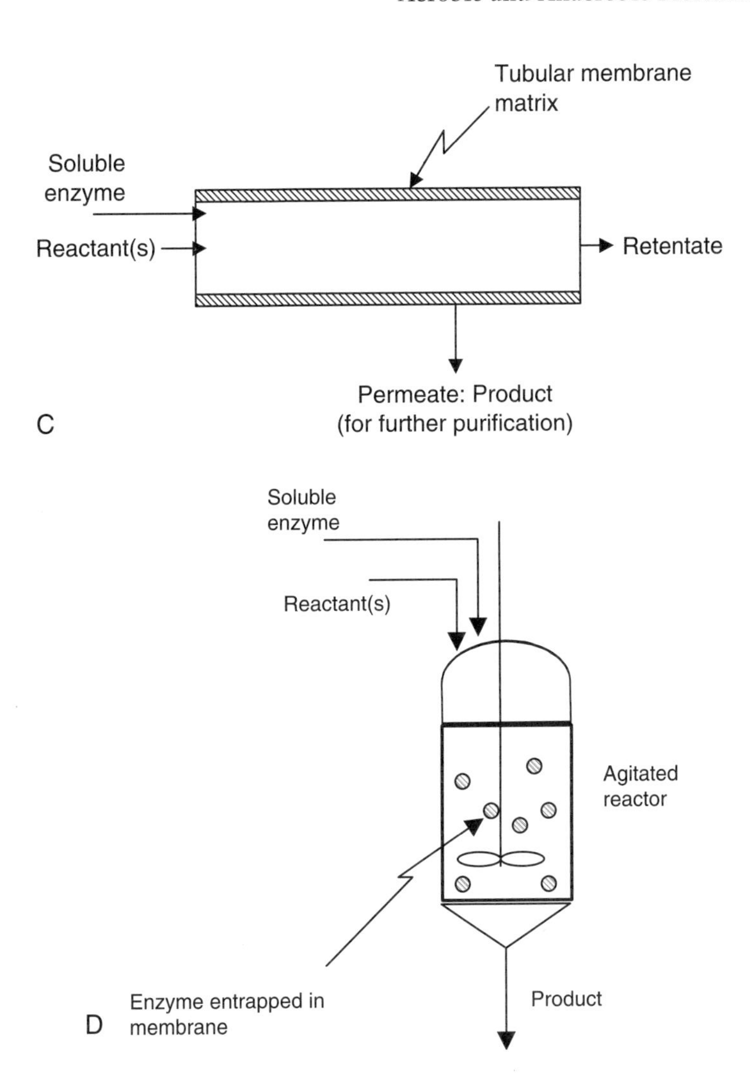

FIGURE 3-13. Continued.

the enzyme. Enzyme-membrane reactors can be operated in batch or continuous mode. Examples of enzyme-membrane systems studied are listed in Table 3-5 (López et al., 2002).

Membrane bioreactors have several advantages over the activated sludge process, including low sludge production and a lower land area requirement. The sludge production of a submerged membrane reactor is between 0.0 and 0.3 kg/kg BOD, whereas it is 0.6 for a conventional activated sludge process and 0.3 to 0.5 for a trickling filter. The main disadvantages of these reactors are fouling of the membranes and high operating cost.

TABLE 3-5
A Few Enzyme-Membrane Bioreactor Systems Mentioned in the Literature

Design	*Enzyme*	*Source*	*Effluent*
1	Soybean peroxidase	Ground soybean seed hulls	Phenolic wastewater
1	Manganese peroxidase	*Bjerkandera sp.*	Dye decolorization
3	Laccase	*Trametes versicolor*	Phenylurea pesticide in wastewater
3	Polyphenol oxidase	*Agaricus bisporum*	Coal gas conversion plant
4	Glucose oxidase	*Aspergillus niger*	Synthetic effluent containing glucose
4	Glycerol dehydrogenase	*Enterobacter aerogenes*	Ethanol oxidation

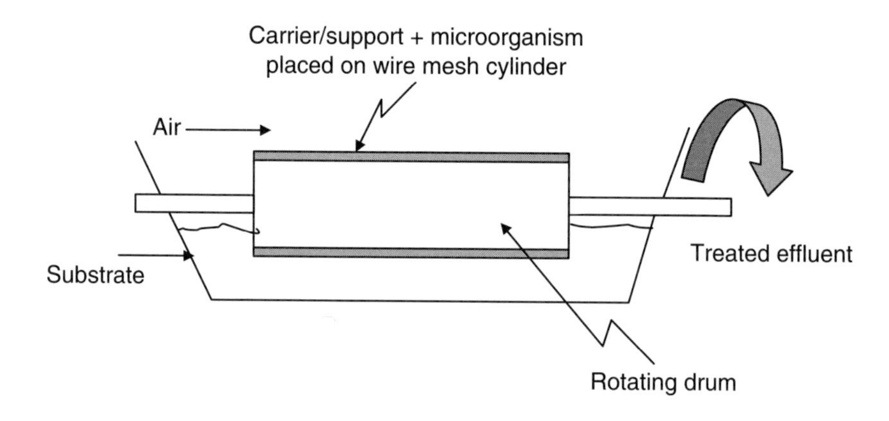

FIGURE 3-14. Rotating drum bioreactor.

The submerged system utilizes approximately half the energy of the side stream direct contact system (design 1) (Gander et al., 2000).

Rotating drum bioreactors (Fig. 3-14) have a rotating drum immersed in the substrate, which revolves at 2 to 3 r/min. The support material is held onto the drum with the help of wire mesh. The microorganism grows on the support, and it comes in contact with air at the upper part of the vessel. The reactor operates as a solid-state process.

At times excavated soil is taken to a reactor containing water, and the biotreatment is performed after addition of the required nutrients and microorganisms. This ex situ bioslurry treatment has been carried out in mixing tank, airlift, fluidized bed, rotating drum, and lagoon type of reactors. Hazardous wastes treated effectively with bioslurry technology include petroleum hydrocarbons, solvents, polycyclic aromatic hydrocarbons (PAH), pesticides, and pentachlorophenol and associated chlorinated aromatics used in wood preservation.

Gas-phase bioreactors that include biofilters, bioscrubbers, and biotrickling filters are discussed in Chapter 30, Gaseous Pollutants and Volatile Organics.

Mode of Operation

The reactors are operated in batch, continuous, or semicontinuous mode. The latter includes semibatch, where one or more of the substrates are added initially in one lot and one or more of the remaining substrates or nutrients are added during the course of the reaction time, either at a fixed rate (extended fed batch) or in lots (fed batch). The concept of the sequencing batch reactor (SBR) has gained considerable interest, where the sequence of operations like fill, react, and part discharge are carried out in the same reactor.

Laboratory- and pilot-scale slurry treatment has been carried out using a soil slurry–sequencing batch reactor (SS–SBR), continuous-flow stirred tank reactor (CSTR), and tanks in series. In the field, SS-SBR and CSTR are the most common. Both modes of operation have advantages and disadvantages. Because a CSTR dilutes the feed, the reaction rate (if it is concentration dependent) decreases, but this may be desirable if the contaminants are toxic to the organisms or if they exhibit substrate inhibition. CSTR requires continuous use of one vessel, which means higher operating and maintenance costs. Some disadvantages of SS–SBR are longer batch times because of fill and discharge times, and the formation of excessive foam. However, SS–SBR provides better operational flexibility, and the volume of slurry replaced per treatment cycle can be adjusted to provide optimal concentrations of contaminants and acclimated microorganisms. In addition, each treated batch can be tested before it is discharged, and the process can be fine-tuned to achieve optimum operation. At the beginning of each cycle, a larger amount of substrate is available for biomass growth (*feast* conditions), and at the end, the low contaminant concentration reduces bioavailability and establishes famine conditions. This cycling of feast and famine conditions modifies the metabolic potential of the microorganisms, and hence improves their performance in contaminant removal.

Conclusions

A plethora of reactors are available for treating wastewater either aerobically or anaerobically. The anaerobic reactors are advantageous ecologically, energetically, and economically when compared with aerobic reactors.

But complete mineralization and handling of toxins can be handled more effectively with aerobic processes.

References

Canovas-Diaz, M., and J.A. Howell. 1988. Stratified mixed culture biofilm model for anaerobic digestion. *Biotech. Bioeng.* 32:348.

Forster, C., and D. Wase. 1990. *Environmental Biotechnology*. Chichester: Ellis Horwood.

Gander, M., B. JeVerson, and S. Judd. 2000. Aerobic MBRs for domestic wastewater treatment: a review with cost considerations, *Sep. Purif. Technol.* 18:119–130.

Gottschalk, G. 1979. *Bacterial Metabolism*. New York: Springer-Verlag.

Hickey, R.F., and R.W. Owens. 1981. Biotechnology and Bioengineering Symposium No. 11, 399–405.

Jeyaseelan, S. 1997. A simple mathematical model for anaerobic digestion process. *Wat. Sci. Tech.* 35:185–191.

Jothimani, P., G. Kalaichelvan, A. Bhaskaran, D. A. Selvaseelan, and K. Ramasamy. 2003. Anaerobic biodegradation of aromatic compounds. *Indian J. Biotech.* 41:1046–1067.

Kosaric, N., and R. Blaszczyk. 1991. Aerobic granular sludge and biofilm reactors. *Adv. Biochem. Eng.* 41:28–31.

López, C., I. Mielgo, M. T. Moreira, G. Feijoo, and J. M. Lema. 2002. Enzymatic membrane reactors for biodegradation of recalcitrant compounds. Application to dye decolourisation, *J. Biotech.* 99:249–257.

Marchaim, U., 1992. *Biogas Processes for Sustainable Development*. Rome, Italy: Food and Agriculture Organization of the United Nations, Viale delle Terme di Caracalla, M-09, ISBN 92-5-103126-6.

Praveen, V.V., and K.B. Ramachandran. *Proceedings of the Ninth National Convention of Institution of Engineers*. C. Ayyanna (Ed.), Vishakhapatnam, India, 1993. Kolkata-700020: Institution of Engineers.

Rajeswari, K.V., M. Balakrishnan, A. Kansal, K. Lata, and V.V.N. Kishore. 2000.State of the art of anaerobic digestion technology for industrial waste water treatment, *Renew. & Sustainable Energy Rev.* 4: 135–156.

Zehnder, A. J. B., K. Ingvorsen, and T. Marti. 1982. In: *Anaerobic Digestion-1981*. Ed. D. E. Hughes. Amsterdam, Netherlands: Elsevier Biomedical.

CHAPTER 4

Mathematical Models

This chapter describes mathematical models for the design of basic batch and continuous reactors; the design of aerobic activated-sludge process; the mass transfer and diffusion correlations needed for design; diffusion through landfill; and diffusion of airborne pollutants.

In order to develop a reactor model, the following information is required:

- Kinetics of reaction (including inhibition kinetics)
- Mass transfer from gas to liquid
- Mass transfer from liquid (substrate or nutrient) to surface of the microorganism
- Heat transfer
- Mixing and agitation
- Type of reactor
- Physical properties of the medium
- Physical properties of the microorganism
- Mode of operation of the reactor

Modeling the pollutant transport process requires knowledge of the following subjects:

- Diffusion coefficients of the pollutant in soil, liquid, or atmosphere
- Medium conditions (temperature, pH, etc.)
- Flow and turbulence
- Concentration of the contaminant
- Physical properties of the medium and the pollutant

Basic Reactor Models

The basic design parameters for various types of reactors are as follows;

Batch Reactor

$$t = \int_{S_o}^{S(t)} \frac{dS}{-r_S} \tag{4-1}$$

Continuously Stirred Tank Reactor

$$\theta = \frac{V}{q_o} = \frac{(S_o - S)}{r_S} \tag{4-2}$$

Monod Chemostat

The Monod Chemostat is an extension of the continuously stirred tank reactor (CSTR) model, which considers both substrate utilization and the cell growth.

The cell balance is

$$X(\mu - D) + DX_o = 0 \tag{4-3}$$

where $D = V/q_o$. If the number of cells entering the reactor is approximately 0, then

$$D = \mu \tag{4-4}$$

The substrate balance is given by

$$D(S_o - S) = \mu\, X/Y \tag{4-5}$$

where μ is the specific growth rate and is given by the Monod equation,

$$\mu = \frac{\mu_{max} S}{K_s + S} \tag{4-6}$$

This leads to equations for exit substrate and cell concentrations, respectively,

$$S = DK_s/\mu_{max} - D \tag{4-7}$$

$$X = Y\left(S_o - \frac{DK_s}{\mu_{max} - D}\right) \tag{4-8}$$

CSTR with Recycle

$$D(1 - f) = \mu \tag{4-9}$$

Plug Flow Reactor

$$\theta = \frac{V}{q_o} = \int_{S_o}^{S(t)} \frac{dS}{-r_s} \tag{4-10}$$

Fed Batch Reactor

In this mode of operation there is no outflow, but after the initial reactor charge, nutrient(s) addition is intermittent, causing the substrate concentration and reactor volume to vary with time.

$$V\frac{dS}{dt} + S\frac{dV}{dt} = q_o S_o - r_S V \tag{4-11}$$

Sequential Batch Reactor

A sequential batch reactor operates in the fed batch mode; both concentration and reactor volume vary with time.

Extended Fed Batch

In this method the feed to the reactor is constant, leading to a constant substrate concentration inside the reactor (i.e., $dS/dt = 0$).

Design of Biotrickling Filter

The performance of a biotower (a tall biotrickling filter with well-structured packing that uses a modular plastic media, leading to high porosity) is given by the following correlation:

$$\frac{\text{Effluent to influent substrate biological}}{\text{oxygen demand (BOD) (mg/L)}} = e^{-kH/Q^n}$$

where Q = hydraulic loading rate, $m^3/m^2{\cdot}min$
k = treatability constant, a function of wastewater and medium characteristics (per minute), = 0.01 to 0.1 (0.06 for modular plastic media)
H = bed height, m
n = 0.5 for municipal waste and modular plastic media (an empirical constant)

For a trickle bed air biofilter, a performance equation similar to the previous one can be written as

$$\frac{\text{Effluent to influent volatile}}{\text{organic compound concentration}} = \alpha \exp\left(-\beta D A_s H K_s m / u_g \delta S_i\right)$$

D = mass diffusivity in biofilm, m^2/min
A_s = biofilm surface area per unit volume of packing material, m^2/m^3
H = height of the biofilter, m
m = thermodynamic distribution coefficient
u_g = superficial gas velocity, m/min
δ = biofilm thickness, m
S_i = influent concentration, mol/m^3

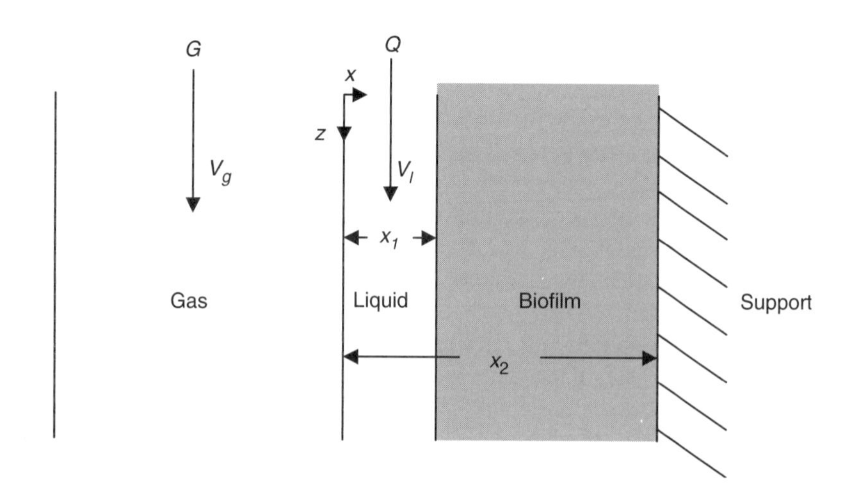

FIGURE 4-1. Three-phase system.

K_s = Michaelis Menten constant, mol/m^3
α and β = correlation constants

A detailed model for the bed can be written from the basic mass balance equations. Steady state mass transfer combined with reactions in the gas, liquid, and biomass phases are written as follows (see Fig. 4-1). There is no reaction in the gas phase. Substrate flux resulting from convection in the *x* direction is negligible when compared with diffusion. Inside the liquid and the biofilm, liquid transfer in the *z* direction is neglected.

$$V_g \frac{dS}{dz} - D_g \frac{d^2S}{dx^2} = 0 \quad \text{for } -x_3 \leq x \leq 0 \tag{4-12}$$

$$V_l \frac{dS}{dz} - D_l \frac{d^2S}{dx^2} - X_l \mu_{max} \frac{S}{K_s + S} = 0 \quad \text{for } 0 \leq x \leq x_1 \tag{4-13}$$

$$-D_b \frac{d^2S}{dz^2} - D_b \frac{d^2S}{dx^2} - X_b \mu_{max} \frac{S}{K_s + S} = 0 \quad \text{for } x_1 \leq x \leq x_2 \tag{4-14}$$

with boundary conditions,

$$S(x = 0, z) = S_{o,l} \qquad \text{for } 0 < z < L$$

$$S(x, z = 0) = 0 \qquad \text{for } 0 < x < x_2$$

$$S(x, z = 0) = S_{o,g} \qquad \text{for } x_3 < x < 0$$

$$dS_{x=x_2}/dx = 0 \qquad \text{for } 0 < z < L$$

l and *g* represent liquid and gas, respectively. D_g, D_l, and D_b are the diffusion coefficients in the gas, liquid, and biofilm phases, respectively.

Reaction Kinetics

The simple kinetic equation for a single substrate reaction is

$$r_s = kS^n \tag{4-15}$$

where n is the order of the reaction and k is the rate constant, which is a function of temperature and could be of the form

$$k = k_o e^{-E/RT} \tag{4-16}$$

Michaelis Menten Equation

The rate equation for enzyme catalyzed reactions is given by the Michaelis Menten equation.

$$r_s = \frac{V_{\max} S}{K_m + S} \tag{4-17}$$

The rate equation changes in the presence of inhibitors. The most important of these are as follows.

$$r_s = \frac{V_{\max} S}{K_m (1 + I/K_i) + S} \tag{4-18}$$

Uncompetitive inhibitor

$$r_s = \frac{V_{\max} S}{K_m + S(1 + I/K_i)} \tag{4-19}$$

Noncompetitive inhibitor

$$r_s = \frac{V_{\max} S}{(K_m + S)(1 + I/K_i)} \tag{4-20}$$

Substrate inhibition (uncompetitive)

$$r_s = \frac{V_{\max} S}{K_m + S(1 + S/K_i)} \tag{4-21}$$

Product inhibition (uncompetitive)

$$r_s = \frac{V_{\max} S}{(K_m + S)(1 + (S_o - S)/K_i)} \tag{4-22}$$

Monod Equation

There are several forms of modified Monod equation; the basic one is:

$$\mu = \frac{\mu_{max} S}{K_s S_o + S} \tag{4-23}$$

Rapid growth

$$\mu = \frac{\mu_{max} S}{K_s S_o + K_i + S} \tag{4-24}$$

Teisser model

$$\mu = \mu_{max} \left(1 - e^{-S/K_s}\right) \tag{4-25}$$

Moser model

$$\mu = \frac{\mu_{max}}{1 + K_s S^{-\lambda}} \tag{4-26}$$

Contois model The Contois model takes biomass concentration into account.

$$\mu = \frac{\mu_{max} S}{K_m X + S} \tag{4-27}$$

Substrate and product inhibition models are also possible. They are similar to the enzyme rate equations. If two substrates are rate limiting, then the Monod equation becomes:

$$\mu = \frac{\mu_{max} S_1}{K_{S_1} + S_1} \frac{S_2}{K_{S_2} + S_2} \tag{4-28}$$

Growth rate equation with maintenance of the cells

$$\mu = \frac{\mu_{max} S}{K_s + S} - k_e \tag{4-29}$$

Oxygen Transfer Rates

The oxygen transfer rate per unit liquid volume is $k_L a \left(C_L^* - C_L\right)$. The term within parentheses is the driving force, k_L is the mass transfer coefficient, and a is the interfacial area per unit liquid volume.

Mass Transfer and Diffusion Coefficients

There are several equations available for estimating the mass transfer and diffusion coefficients.

Mass transfer coefficient from fluid to solid or gas to liquid for particles less than 0.60 mm

$$k_L = 2D_{AB}/d_p + 0.31N_{\mathrm{Sc}}^{-2/3}\left(\Delta\rho\,\mu_c g/\rho_c^2\right)^{1/3} \tag{4-30}$$

where k_L is in m/s, diameter in m, density in kg/m^3, viscosity in kg/m·s, and g in m/s^2.

Gas to liquid mass transfer for bubble swarms when d < 2.5 mm

$$N_{\mathrm{Sh}} = k_L d_p/D_{O_2} = 0.31N_{\mathrm{Sc}}^{1/3}N_{\mathrm{Gr}}^{1/3} \tag{4-31}$$

Gas to liquid mass transfer for bubble swarms when d > 2.5 mm

$$N_{\mathrm{Sh}} = k_L d_p/D_{O_2} = 0.31N_{\mathrm{Sc}}^{1/2}N_{\mathrm{Gr}}^{1/3} \tag{4-32}$$

Mass transfer into a free liquid surface or into a falling film

$$k_L = \sqrt{D_{O_2}/\pi\tau} \tag{4-33}$$

where τ is the surface renewal time, which is the stream depth per average velocity. Another equation is,

$$k_L = 1.46\sqrt{D_{O_2} \times \text{velocity/stream depth}} \tag{4-34}$$

Mass transfer during aeration in lakes and ponds (O'Connor-Dubbins)

$$k_L a = \sqrt{D_{O_2} \times \text{velocity}/\pi \text{ stream depth}^3} \tag{4-35}$$

Mass transfer correlation for an agitated aerated vessel under turbulent conditions (Calderbank) (Bailey and Ollis, 1977)

$$N_{\mathrm{Sh}} = 0.13N_{\mathrm{Sc}}^{1/3}N_{\mathrm{Re}}^{3/4} \tag{4-36}$$

(In this equation, we use the impeller diameter as the characteristic length for N_{Re})

Mass transfer correlation for agitated vessel under turbulent conditions [Kulov et al. (1983)]

Taking the power input by the agitator into consideration

$$N_{\mathrm{Sh}} = k_L D_T/D_{O_2} = 0.267N_{\mathrm{Sc}}^{1/4}N_{\mathrm{Re}}^{3/4}N_P^{1/4}\left(D_T^4/VD_i\right)^{1/4} \tag{4-37}$$

where the power number N_P is

$$N_P = P/\left(\rho_L N^3 D_i^5\right) \tag{4-38}$$

(In this equation, the characteristic length of N_{Sh} is the tank diameter and the characteristic length of N_{Re} is the impeller diameter.)

Interfacial area per unit liquid volume (a)

$$a = \varepsilon_g 6/d \tag{4-39}$$

Equations are found in the literature for calculating ε_g, bubble volume to reactor volume (which generally varies between 8 and 30%).

Diffusivities of small solutes in aqueous solutions with molecular weights less than 1000 (Geankoplis, 2003)

$$D_{AB} = \frac{9.96 \times 10^{-16} T}{\mu_B V_A^{1/3}} \tag{4-40}$$

Semiempirical equation of Polson for calculating diffusivities with molecular weights above 1000,

$$D_{AB} = \frac{9.40 \times 10^{-15} T}{\mu_B M_A^{1/3}} \tag{4-41}$$

Wilke-Chang equation for calculating diffusivities for dilute solutes in liquids

$$D_{AB} = \frac{1.173 \times 10^{-16} (\varphi M_A)^{1/2} T}{\mu_B V_A^{0.6}} \tag{4-42}$$

Mass transfer correlations for packed beds

$$J_D = 1.09 N_{Re}^{-2/3}/\varepsilon \qquad N_{Re} = 0.0016 \text{ to } 55, \quad N_{Sc} = 165 \text{ to } 70{,}600 \tag{4-43}$$

$$J_D = 0.25 N_{Re}^{-0.31}/\varepsilon \qquad N_{Re} = 55 \text{ to } 1500, \quad N_{Sc} = 165 \text{ to } 10{,}690 \tag{4-44}$$

$$J_D = 0.4548 N_{Re}^{-0.4069}/\varepsilon \qquad N_{Re} = 10 \text{ to} 10{,}000 \tag{4-45}$$

where $J_D = k_c N_{Sc}^{-2/3}/v'$

v' = empty tube velocity of the gas

Guedes de Carvalho et al. (1991) give an empirical correlation for mass transfer to or from the spheres in a fluidized bed reactor.

$$N'_{Sh} = [4 + 0.576 N'^{\,0.78}_{Pe} + 1.28 N'_{Pe} + 0.141\,(d_p/D_T)\, N'^{\,2}_{Pe}]^{0.5} \tag{4-46}$$

where $N'_{Sh} = \sqrt{2} k_L D_T/\varepsilon d_p$

$N'_{Pe} = \sqrt{2} U_{mf} D_T/\varepsilon d_p$

Activated Sludge Process

Most of the biomass occurs as flocs or aggregates, and the concentration of biomass is referred to as mixed liquid suspended solids (MLSS, ~1500 to 3500 mg/L). Sludge loading (kg BOD/m^3/day, ~0.4 to 1.2) is the ratio of biodegradable organic material to active biomass (Forster, 1990).

$$\text{Sludge loading} = \frac{\text{flow rate} \times \text{BOD}}{V \times \text{biomass}} \tag{4-47}$$

Organic loading rate (kg MLSS·day/OLR, kg BOD) =

$$\frac{\text{BOD (g/L)} \times \text{flowrate(m}^3\text{/day)}}{\text{MLSS (g/L)} \times V} = \frac{\text{BOD}}{\text{MLSS} \times \theta} \tag{4-48}$$

For high rate, conventional, and extended aeration systems, the OLR varies between 0.5 and 5, 0.25 and 0.45, and 0.05 and 0.2, respectively.

Sludge age (s)

$$= \frac{\text{MLSS (g/L)} \times \text{ aeration tank volume (m}^3\text{)}}{\text{sludge wasted (kg/day)} + [\text{flow rate (m}^3\text{/day)} \times \text{ effective solids (g/L)}]} \tag{4-49}$$

Minimum sludge age compatible with nitrification (t)*

$$1/[0.18 - 0.15\,(7.2 - \text{ pH})]\, e^{0.12\,(T-15)} \tag{4-50}$$

Daily carbonaceous oxygen requirement (kg/day)

$$0.75 \text{ BOD} + (0.048 \text{ MLSS})\, V \tag{4-51}$$

Air requirement for extended aeration

$$\text{BOD} + 4.34N_H - 2.85N_T + (0.024 \text{ MLSS})\, Vr \times 1.07^{T-20} \tag{4-52}$$

Air requirement for extended aeration and burning only carbonaceous material at 20°C

$$\text{BOD} + (0\ .048 \text{ MLSS})\, V \tag{4-53}$$

The mass balance equation for biomass production in a completely stirred, tank-activated sludge reactor is (see Fig. 4-2):

Biomass in the influent + biomass produced

= biomass in the effluent + sludge wasted

$$q_o X_o + V\frac{dX}{dt} = (q_o - q_W)X_e + q_W X_W \tag{4-54}$$

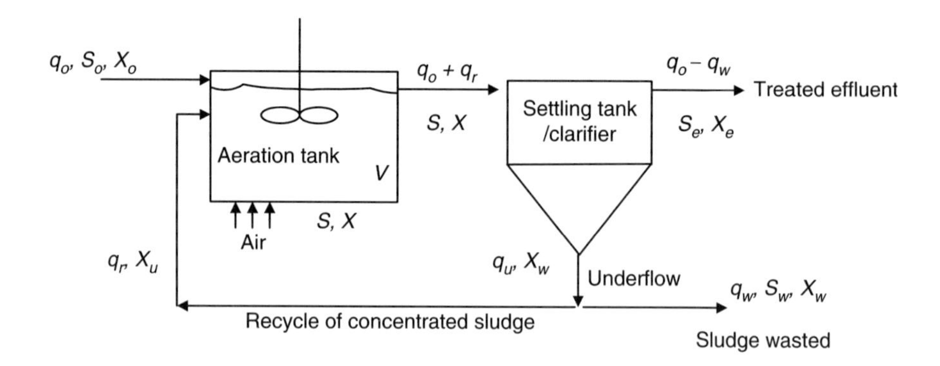

FIGURE 4-2. Activated sludge process.

Biomass production rate

$$\frac{dX}{dt} = \left[\frac{\mu_m S}{K_s + S}\right] X - k_e X \tag{4-55}$$

When the biomass amount in the inflow and outflow liquid streams is negligible, then X_o and $X_e = 0$. Equation (4-54) becomes

$$\frac{\mu_m S}{K_s + S} = \frac{q_W X_W}{VX} + k_e \tag{4-56}$$

Mass balance equation for substrate utilization

$$q_o S_o - V(dS/dt) = (q_o - q_W)S_e + q_W S_W \tag{4-57}$$

Substrate consumption rate

$$\frac{dS}{dt} = \left[\frac{\mu_m S}{K_s + S}\right] \frac{X}{Y} \tag{4-58}$$

Since degradation is taking place only in the aeration tank,

$$S = S_e = S_W \tag{4-59}$$

Substituting Eqs. (4-58) and (4-59) into Eq. (4-57)

$$\frac{\mu_m S}{K_s + S} = \frac{q_o Y(S_o - S)}{VX} \tag{4-60}$$

Cell residence time (sludge age)

$$\theta_c = VX/q_W X_W \tag{4-61}$$

or

$$\frac{1}{\theta_c} = \frac{\mu_m S}{K_s + S} - k_e \tag{4-62}$$

Hydraulic residence time

$$\theta = V/q_o \tag{4-63}$$

Combining Eqs. (4-56) and (4-60), i.e., cell and substrate utilization balances,

$$\frac{q_w X_w}{VX} + k_e = \frac{q_o Y(S_o - S)}{VX} \tag{4-64}$$

Substituting Eqs. (4-61) and (4-63) into (4-64) gives an equation for the amount of biomass in the exit stream of the activated sludge plant.

$$X = \frac{Y(S_o - S)\,\theta_c}{(1 + k_e\theta_c)\,\theta} \tag{4-65}$$

Volumetric loading rate is the ratio of the mass of BOD in the influent to the volume of reactor.

$$V_L = q_o S_o/V \tag{4-66}$$

Food to mass ratio is the ratio of mass of BOD removed to the biomass in the reactor.

$$F/M = q_o(S_o - S)/VX \tag{4-67}$$

If the aeration vessel is a plug flow type (complete mixing in the radial direction and no mixing in the direction of flow), then

$$\frac{1}{\theta_c} = \frac{\mu_m(S_o - S)}{(S_o - S) + (1 - \alpha)\,K_s \ln S_i/S} - k_e \tag{4-68}$$

where α is the recycle ratio and S_i is the substrate concentration after mixing the feed with the recycled sludge.

$$S_i = \frac{(S_o + \alpha S)}{(1 + \alpha)} \tag{4-69}$$

Ponds and Lagoons

In the case of facultative systems, complete mixing is assumed for the liquid portion of the reactor. The solids that fall to the bottom are not resuspended; hence the balance considers only the soluble BOD. This soluble food is assumed to be distributed uniformly, and if the rate is assumed to be first order, then

$$\frac{S}{S_o} = \frac{1}{1 + k\theta} \tag{4-70}$$

where S and S_o = the soluble food entering and leaving the system, respectively
k = the first order rate constant
θ = the hydraulic residence time

If n reactors are arranged in series then,

$$\frac{S_n}{S_o} = \frac{1}{(1 + k\theta/n)^n} \tag{4-71}$$

where S_n is the concentration of the soluble food leaving reactor n.

Transport in Soils

When liquid organic pollutants are released into the soil, they can become physically bound within the soil phase, as well as at the pore spaces that separate the soil particles from one another. A single particle has an intraparticle porosity that characterizes the internal structure of the particle as well as an interparticle porosity that characterizes the packing of the particles.

An empirical relation that can be used to estimate effective diffusivity (D_{eff}) of liquid through soil bed is (Middleman, 1998)

$$\frac{D_{eff}}{R_K} = \frac{\varepsilon^2 D_f}{\varepsilon + (1 - \varepsilon)K_p\rho_s} \tag{4-72}$$

where R_K = retardation of diffusion due to the absorption of the solute on the surface of the particle
ρ_s = mass density of the solid phase
K_p = equilibrium constant relating the contaminant concentration in the fluid and solid phases
D_f = diffusion coefficient of the contaminant in the fluid phase
ε = bed porosity

The amount of contaminant remaining in a spherical particle at any time t is obtained by integrating the diffusion equation as follows:

$$\frac{M(t)}{M_o} = \frac{6}{\pi^2}\sum_{n=1}^{\infty}\frac{1}{n^2}\exp\left[-\frac{(D_{eff}/R_K)n^2\pi^2 t}{R^2}\right] \tag{4-73}$$

where M_o is the initial amount of contaminant and R is the radius of the particle. Except for early times, the second and subsequent terms in the infinite series can be neglected, which gives rise to a simple equation

$$M(t) = 6M_o \exp\left[-\frac{(D_{\text{eff}}/R_K)\pi^2 t}{R^2}\right] \tag{4-74}$$

This equation can also be used to estimate the time required for the concentration of the contaminant to reach half its initial value.

Municipal landfills require construction of a barrier that separates the contaminant storage region from the ground water supply. The barrier could be a polymeric film with low permeability to toxins placed on top of a thick layer of clay. Clay has a high resistance to hydraulic flow and hence prevents vertical flow of water due to gravity. Nevertheless, diffusion of the toxins does take place across the barrier.

According to Fick's law,

Flux through a polymeric film = (permeability of the film × concentration driving force)/thickness of the film

Transport through the clay layer is assumed to be diffusion through a semi-infinite planar region with an initial concentration $C_o = 0$ and a constant concentration C_s at the landfill-clay boundary. Then

$$\frac{C(x,t)}{C_s} = \operatorname{erfc}\left(\frac{B}{2\sqrt{D_{AB}t}}\right) \tag{4-75}$$

where erfc is the complementary error function. For example, if the thickness B = 10 m and $D_{AB} = 3.7 \times 10^{-8}$ cm^2/s, the time required for the lower surface of the liner to reach a value of 1% of the value of the upper surface will be 65,000 years. If the liner thickness is 1 m, then the time required becomes 650 years.

Suppose the landfill lies above an aquifer so that the water carries the pollutants away because of its flow. The contaminant gets fully mixed with the aquifer flow. Then the mass balance for the toxin will be

Landfill flux × area of interface with the aquifer
= contaminant concentration in the aquifer
× flow rate of the aquifer (Middleman, 1998)

$$\text{Landfill flux at the interface} = -D_{AB}\left.\frac{dC}{dy}\right|_{y=B}$$

$$\frac{C}{C_s} = \left(1 - \frac{y}{B}\right) - \frac{2}{\pi}\sum_{n=1}^{\infty}\frac{\sin n\pi(y/B)}{n}\exp(-n^2\pi^2 X_B) \tag{4-76}$$

where $X_B = D_{AB}t/B^2$

The boundary conditions for the diffusion are

$$C = 0 \text{ at } t \leq 0 \quad 0 \leq y \leq B$$
$$C = C_S \text{ at } y = 0 \quad t \geq 0$$
$$C = 0 \text{ at } y = B \quad \text{for } t > 0$$

Diffusion and Transport of Gases in Air

Accidental release of toxic gases or vapors can occur in rural or urban areas, during day or night and from ground level or from an elevation. Vertical mixing of pollutants is more vigorous in urban areas than in rural areas. The concentration of the toxic gas at any distance x released at ground level is given by Eq. (4-77) (Kumar, 1998, 1999).

$$C_X = \frac{Q}{\pi (aX^b)(cU^d)} \tag{4-77}$$

The values of a, b, c, and d are given in Table 4-1.

Nomenclature

a	gas-liquid interfacial area (gas area per unit liquid volume)
BOD	biological oxygen demand
C_L	liquid phase concentration
C_L^*	liquid phase concentration that is in equilibrium with the gas phase
C_X	concentration of pollutant at distance x
d	bubble diameter
d_p	particle diameter
D_i	impeller diameter
D_T	tank diameter
E	activation energy
f	recycle ratio
J_D	Coulburn j factor

TABLE 4-1
Values of Constants (for rural ground release)

	a	b	c	d
Daytime, $x < 500$ m	0.01082	1.78	2.46	−0.56
Daytime, $x > 500$ m	0.04487	1.56	2.80	−0.64
Nighttime, $x < 2000$ m	0.0049	1.66	1.28	−0.6
Nighttime, $x > 2000$ m	0.01901	1.46	1.32	−0.68

M_A	molecular weight of solute
M_B	molecular weight of solvent
N_H	Ammonical nitrogen removed (kg/day)
N_{Pe}	Peclet number
N_{Re}	Reynolds number
N_{Sc}	Schmidt number
N_T	Total nitrogen removed (kg/day)
q_o	feed flow rate
Q	toxic release rate, g/s
R	endogeneous respiration rate (3.9 mg O_2/g MLSS /h)
R	Gas constant
S	substrate concentration in the exit stream
S_o	substrate concentration in the exit stream
T	temperature
U	wind speed, m/s
U_{mf}	minimum fluidization velocity
V	reactor or aeration tank volume (m^3)
V_A	molar volume of solute
X	cell concentration
Y	yield coefficient
ε	bed porosity
ε_g	bubble volume to reactor liquid volume
μ	Monod specific growth rate
μ_B	viscosity of solvent
μ_{max}, K_s	Monod constants
ϕ	Association factor (1-2.6)
θ	residence time
θ_c	sludge age

References

Bailey, J. E., and D. F. Ollis, 1977. *Biochemical Engineering Fundamentals.* Tokyo: McGraw-Hill Kogakusha Tokyo.

de Carvalho. G. 1991. *Trans. Inst. Chem. Eng.* 69: 63–67.

Forster, C., and D. Wase. 1990. *Environmental Biotechnology.* Chichester, U.K.: Ellis Horwood. 1990.

Geankoplis, C. J. 2002. *Transport Processes and Unit Operations.* New Delhi: Prentice Hall of India.

Kulov, P. 1983. Chem. Eng. Comm., 21: 259–262.

Kumar A., 1998. Estimating hazard distances from accidental releases. *Chem. Eng.* August 121–124.

Kumar A., 1999. Estimate dispersion for accidental release in rural areas. *Chem. Eng.* July 91–94.

Middleman, S., 1998. *An Introduction to Mass Transfer and Heat Transfer, Principles of Analysis and Design.* New York: John Wiley.

Peavey, H. S., D. R. Rowe, and G. Tchobanoglous. 1985. *Environmental Engineering.* International Ed., New York: McGraw-Hill.

CHAPTER 5

Treatment of Waste from Organic Chemical Industries

Introduction

The organic chemical industry, which manufactures carbon-containing chemicals, produces an enormous number of materials that are essential to the economy and to modern life. This industry obtains raw materials from the petroleum industry and converts them to intermediate materials or basic finished chemicals. Based on the type and source of chemicals, this industry is classified into three categories (U.S. EPA, 2002), viz:

- Gum and wood chemicals (tall oil, rosin, turpentine, pine tar, acetic acid, and methanol)
- Cyclic organic crudes and intermediates (benzene, toluene, xylene, naphthalene, dyes, and pigments)
- Organic chemicals not elsewhere classified (ethyl alcohol, propylene, ethylene, and butylene)

From the viewpoint of the market, this industry is also categorized into:

- Bulk or commodity chemicals
- Fine or specialty chemicals

A wide range of chemicals is produced from common feedstock such as petrochemicals, coal, natural gas, and wood. Fossil fuels provide small (molecular size) chemicals such as benzene, ethylene, propylene, xylene, toluene, butadiene, methane, and butylene, which find end use in a large variety of industries ranging from agricultural chemicals to cosmetics (Table 5-1). Thus the organic chemicals industry forms the fulcrum for the needs of modern life (U.S. EPA, 2002).

TABLE 5-1
Major Organic Chemical Products

Category	*Example chemicals*	*Example end uses*
Aliphatic and other acyclic organic chemicals	Ethylene, butylenes and formaldehyde	Polyethylene plastic, plywood
Solvents	Butyl alcohol, Ethyl acetate, Ethylene glycol ether, perchloroethylene	Degreasers, dry cleaning fluids
Polyhydric alcohols	Ethylene glycol, sorbitol, synthetic glycerin	Antifreeze, soaps
Synthetic perfume and flavoring materials	Saccharin, citronellol, synthetic vanillin	Food flavoring, cleaning, product scents
Rubber processing chemicals	Thiuram, hexamethylene tetramine	Tires, adhesives
Plasticizers	Phosphoric acid, phthalic anhydride, stearic acid	Raincoats, inflatable toys
Synthetic tanning agents	Naphthalene sulfonic acid condensates	Leather coats and shoes
Chemical warfare gases	Tear gas, phosgene	Military and law enforcement
Cyclic crudes and intermediates	Benzene, toluene, mixed xylenes, naphthalene	Eyeglasses, foams
Cyclic dyes and organic pigments	Nitrodyes, organic paint pigments	Fabric and plastic coloring
Natural gas and wood chemicals	Methanol, acetic acid, rosin	Latex, adhesives

All the same, some unavoidable problems to our environment accompany this industry—toxic wastes. Organic chemical industries are among the largest producers of toxic wastes. According to the Toxic Release Inventory (TRI), USA data, 467 chemical facilities (industries) in the United States released (to the air, water, or land) and transferred (shipped offsite or discharged to sewers) a total of 594 million pounds of toxic chemicals during calendar year 2000 (U.S. EPA, 2002). Of the approximately 650 chemicals released into the environment, those released in the largest amounts were:

- Methanol
- Ammonia
- Nitric acid
- Nitrate compounds
- Acetonitrile

- Propargyl alcohol
- Chlorinated solvents

Some of the chemicals released into the environment during the year 2000 in the United States are given in Table 5-2.

Oil spills are one of the major problems of present society. Humans have long exploited the volume-dilution power of the sea to dispose of unwanted wastes. Although concern about waste accumulation in marine environments is increasing, especially for coastal waters, marine remediation efforts are nearly nonexistent. The notable exception to this rule is crude oil and refined petroleum product spills. Tanker spills account for only 13% of the estimated 3.2 million metric tons of annual marine petroleum hydrocarbon inputs (National Research Council, 1985). Yet tanker spills have remained the focus of research efforts related to remediation of marine oil contamination.

The potential for truly massive spills from modern supertankers and the readily visible direct impact on affected areas have captured the public's attention and sensitized regulatory and industry groups to the local destructive potential of such accidents. Petroleum is a complex mixture of thousands of individual compounds, and the degradation pathways of spilled oil are numerous and complex. Biodegradation, especially by microbes, is believed to be one of the primary mechanisms of ultimate removal of petroleum hydrocarbons from marine and shore environments. Acceleration of this natural process is the objective of bioremediation efforts.

Bioremediation has yet to become an established spill-response technology, but some attempts to implement it have been encouraging. The inability of established nonbiological techniques to cope with recent large spills has led to increased interest in bioremediation. Special problems associated with marine oil spills include the uncontained nature of the waste, the potential size of the contaminated area, and difficulty of access for remediative and monitoring activities. As with other forms of in situ bioremediation, natural biodegradation of marine oil spills may be enhanced by inducing changes in either the microbial population or the availability of microbial nutrients.

Most researchers have concluded that nutrient availability is the chief limitation of natural biodegradation, and most research has been directed toward enhancing nutrient availability. Marine oil-spill cleanups represent some of the largest in situ remediation projects ever attempted. The March 1989 spill of 11 million gallons of crude oil from the supertanker *Exxon Valdez* into Prince William Sound, Alaska, provided a testing ground for many nutrient enrichment technologies. The U.S. EPA and Exxon spent about $8 million on a joint program to test and apply such measures (Thayer, 1991). The results obtained indicate that for the conditions encountered, the bioremediative action of indigenous bacteria can safely be accelerated two- to fourfold over control beaches by a single addition of nutrients. A second application 3 to 5 weeks later boosted this figure to as high as five- to tenfold.

TABLE 5-2
Toxic Releases from Organic Chemicals Industries (United States) for the Year 2000

Chemical name	*Average release, pounds/year/facility*
Ethylene	149,941
1,2-Dichloro-1,1,2-trifluoroethane	108,518
2,4-Dimethyl phenol	50,449
Acetamide	439,090
Acetonitrile	289,850
Acetophenone	68,917
Acrylamide	356,087
Acrylic acid	169,875
Acrylonitrile	136,379
Ammonia	160,150
Biphenyl	233,233
Bromine	126,746
Bromomethane	54,716
Carbonyl sulfide	466,000
Chlorobenzene	55,228
Chlorodifluoro methane	70,099
Cyanide compounds	162,943
Cyclohexanol	233,104
Dichloro fluoromethane	59,855
Diethyl sulfate	1,461,723
Ethylene glycol	249,902
Formaldehyde	51,459
Formic acid	90,152
Hydrogen cyanide	61,354
Malanonitrile	255,157
Manganese	74,735
m-Cresol	61,458
Methanol	317,328
Naphthalene	105,382
Nitrate compounds	538,297
Nitric acid	304,713
Nitro benzene	230,417
N-Methyl-2-pyrrolidine	254,443
o-Cresol	72,216
Propargyl alcohol	206,965
Propylene	78,770
Pthalic anhydride	134,433
Pyridine	49,630
Sodium nitrite	77,853
t-Butyl alcohol	293,411
Toluene	155,039
Vinyl acetate	80,082

Biotreatment

By and large, biodegradation is the most suitable and economic way of mineralizing organic pollutants. In the case of ammonia, nitrate compounds, and cyanide compounds, biodegradation is the ideal choice because any of the chemical methods would produce a large volume of salts (sludge).

The industrial effluents in which these organic chemicals occur are frequently acidic and have elevated salinity. Activated sludge systems are usually protected from high salinity and pH by pretreatment of the wastewater entering the aeration tank; hence, these are most suited for treatment of organic wastes. However, pretreatment incurs cost; therefore, alternative methods employing organisms able to function under low pH and high salinity have to be adopted. A number of such reports have appeared in literature in recent times. Apart from the well known microbial degradations of aromatic, aliphatic, halogenated organics, PAHs, and dioxins (see subsequent chapters), microorganisms are known to degrade even hetero aromatic and hetero aliphatic compounds. Aniline and related hetero aromatic compounds have been found to degrade under aerobic fermentative, nitrate-reducing, and sulfate-reducing conditions at a variety of salt concentrations and pH values (Bromley-Challenor et al., 2000). Sulfur heterocycles, such as the benzothiozoles and their derivatives, are degraded both by anaerobic and aerobic means (Fig. 5-1) (Wever et al., 1997). More details are given in Chapter 25, Biodesulfurization. Thermophillic aerobic processes have also been reported to clean up effluents of organic industries.

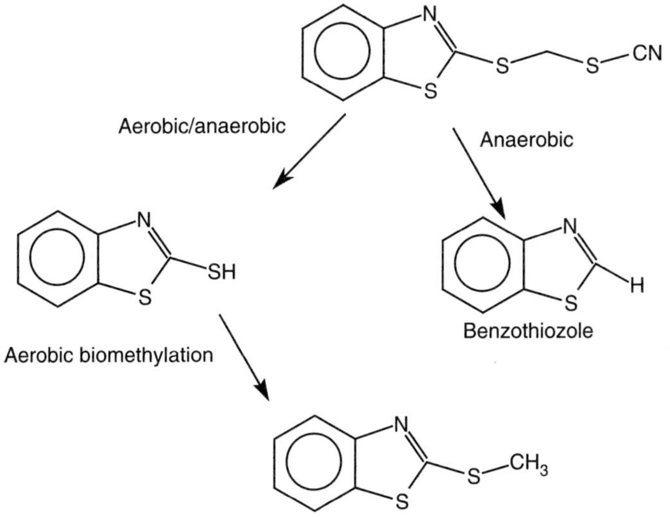

FIGURE 5-1. Biodegradation of benzothiazoles.

Depending on the type of organic or inorganic pollutant, appropriate biodegradation methods (aerobic/anaerobic) can be adopted. Suitable degradation strategies for toxic releases from the organic chemicals industry are given in Table 5-3. Complete mineralization of the pollutant is invariably brought about by a judicious combination of both processes. Anaerobic degradation usually provides intermediates that can be mineralized by subsequent aerobic processes. Excess salts and solid matter are ideally removed by pretreatment plants designed for the purpose. The effluent from the pretreatment is suitable for the biotreatment.

Another emerging application of bioremediation, the potential of which is yet to be fully realized, is biodegradation and/or removal of environmentally undesirable compounds through biofilter technology. Naturally occurring microorganisms are usually present in quantities adequate to handle easily biodegradable compounds like alcohols, ethers, and simple aromatics. More degradation-resistant chemicals, such as nitrogen- and sulfur-containing organics and especially chlorinated organics and aliphatics, may require inoculation with selected strains of microbes to achieve desired degradation efficiencies. Although every application must be evaluated individually, biofilter technology represents a volatile organic compound abatement option that is competitive in many cases on both efficiency and cost bases.

For purposes of bioremediation, aerobic microbial metabolism has traditionally been the focus of attention. Aerobic degradative pathways in microbes and in animals break down organic molecules oxidatively by using divalent oxygen or other active oxygen species, such as hydrogen peroxide, as electron acceptors. Aerobic catabolism of organics ultimately results in familiar mineral products—carbon dioxide and water. Aerobes are capable of degrading most organic wastes, provided enough oxygen is available. Some compounds, notably the organohalogens, are highly resistant to aerobic biodegradation (termed recalcitrant or persistent wastes). Resistance of most aromatic and aliphatic compounds to degradation is dramatically increased by halogenation (most commonly chlorination); further halogenation results in increased resistance.

Anaerobic microbes degrade organics reductively, eventually resulting in the mineral end product methane. In the case of carbohydrate compounds, carbon dioxide and free hydrogen also are produced. Although they are not usually utilized for routine waste degradation, some anaerobes are very adept at dechlorination of common recalcitrant organochlorine compounds, notably PCBs; organochlorine pesticides, such as DDT; and chlorinated aliphatics, such as the industrial solvent trichloroethylene (TCE). Thus anaerobic microbial catabolism (sometimes called fermentation) offers a bioremediation option to deal with persistent wastes. Complete anaerobic degradation of wastes, however, may be slow. The major problem with anaerobic digestion of organochlorine wastes is that biodegradation is often incomplete (at least on a practical time scale) and may result in

TABLE 5-3
Suitable Degradation Strategies for Organic Pollutants

Chemical name	*Aerobic degradation*	*Anaerobic degradation*	*Chemical/physical methods*
Ethylene	√	√	—
1,2-Dichloro-1,1,2-trifluoroethane	—	√	—
2,4-Dimethyl phenol	√	√	—
Acetamide	√	—	√
Acetonitrile	—	√	—
Acetophenone	√	√	—
Acrylamide	√	—	√
Acrylic acid	√	—	√
Acrylonitrile	√	√	—
Ammonia	√	—	√
Biphenyl	√	—	—
Bromine	—	—	√
Bromomethane	√	—	√
Carbon disulfide	√	—	√
Chlorobenzene	√	—	—
Chlorodifluoro methane	—	√	—
Cyanide compounds	—	√	√
Cyclohexanol	√	—	√
Dichloro fluoromethane	—	√	—
Diethyl sulfate	—	√	√
Ethylene glycol	—	—	√
Formaldehyde	—	—	√
Formic acid	—	—	√
Hydrogen cyanide	—	√	√
Malanonitrile	—	√	√
Manganese	√	—	√
m-Cresol	√	√	√
Methanol	—	√	√
Naphthalene	√	√	—
Nitrate compounds	—	√	√
Nitric acid	—	√	√
Nitro benzene	√	√	—
N-Methyl-2-pyrrolidine	√	√	—
o-Cresol	√	√	—
Propargyl alcohol	√	—	—
Propylene	√	—	√
Pthalic anhydride	√	—	√
Pyridine	√	—	√
Sodium nitrite	√	√	√
t-Butyl alcohol	√	—	√
Toluene	√	√	—
Vinyl acetate	√	√	√

toxic metabolites. The use of mixed cultures containing both aerobes and anaerobes facilitates mineralization of many organochlorines. In practice, a sequential bioreactor system utilizing both anaerobic and aerobic reactors could be employed. For example, PCBs or chlorinated aromatics could be dechlorinated anaerobically, then fed into an aerobic bioreactor to be fully mineralized to carbon dioxide and water. Similarly, TCE and perchloroethylene may be reductively metabolized to vinyl chloride (a toxic chemical), which can then be subjected to aerobic biodegradation. Commercial versions of such two-stage hybrid bioreactor systems are currently under development. Isolation and characterization of dehalogenases (dehalogenating bacterial enzymes) for possible development of immobilized enzyme reactors and biofilters are also being conducted (Janssen et al., 1990).

Appreciation of the potential of natural systems to regulate levels of aquatic toxicants has led to the development of constructed wetlands for bioremediation of complex wastes. It has been observed that wetlands have a buffering ability on surface waters with respect to circulating nutrient and pollutant levels. Wetlands have the capacity to store excess nutrients or wastes and to release stored excesses under the right environmental conditions (Hammer, 1989). A constructed wetland is an artificial habitat, most visibly made up of vascular plants and algal colonies, which also provide a structural and nutritional support for an associated, highly heterogeneous microbial community. One of the most promising applications of constructed wetlands is for in situ bioremediation of metal contamination. It is not always known to what extent the observed metal removal in natural wetlands is due to bacterial action and what is due to higher plant or algal activity. In any case, many of these organisms exist in a symbiotic arrangement, and multitrophic cultured systems are increasingly being viewed as an alternative to monocultures or even heterogeneous bacterial cultures. Field tests on acid mine drainage effluent have indicated that such systems are capable of removing metals via multiple pathway biological action (Batal et al., 1989). The use of both natural and constructed wetlands for heavy metal abatement is of great potential value, but questions remain about the eventual fates of the metals. Some means of extraction, such as removal of plant or sediment material, is necessary to prevent remobilization of metals from dead organic material or trophic transfer to grazing animals.

Phytoremediation

Plants can adapt to a wide range of environmental conditions and are capable of modifying conditions of the environment to some extent. The unique enzyme and protein systems of some plant species appear to be beneficial for phytoremediation. Additionally, since plants lack the ability to move, many plants have developed unique biochemical systems for nutrient acquisition, detoxification, and controlling local geochemical conditions (Sridhar Susarla et al., 2002). McFarlane et al. observed that the uptake and translocation of phenol, nitrobenzene and bromocil were directly related to transpiration rate in mature soyabean plants (McFarlane et al., 1987). Recently, the use of minced horseradish roots has been proposed for the decontamination of surface waters polluted with chlorinated phenols (Roper et al., 1996). Bruken and Schnoor used poplar trees for the uptake and metabolism of the pesticide atrazine. Results indicated that poplar trees can take-up, hydrolyze, and dealkylate atrazine to less toxic metabolites (Bruken et al., 1997). Thus, plants can contribute in many ways for environmental restoration of contaminated sites.

Bioremediation is an emerging field, the full potential of which is as yet unknown, especially in the cleanup of organic contaminants. There is a tremendous need for further basic research and development, especially in the areas of environmental site and waste diagnostics, waste-technology matching, and integration of multiple remediation techniques.

There is a clear need for improved methods of environmental surveillance for the prevention of adverse environmental conditions. Continued development of new methods, including lab-bench assays and gene-probe technologies and their utilization, may provide some of the desired information and early warning for environmental hazards. When required, bioremediative approaches need to be applied with the understanding that each local environment requires individual attention and detailed site evaluation. In bioremediation of a contaminated area, performance feedback to researchers with regard to the transport, fate, and possible toxicity of the metabolites produced is of tremendous value for method refinement. Moreover, the site evaluation processes must incorporate expertise from those knowledgeable in other remediation technologies as well as bioremediation experts. Coupled and integrated methods of containment, destruction, and biodegradation of pollutants are certain to yield more cost-effective cleanup solutions than procedures that focus on a single remediation technology. The primary limitation to the widespread use of many bioremediation approaches is often the extent to which the pollutant is available to the microbial population. The bioavailability of many chemicals diminishes with time as a result of weathering and aging phenomena, and the time window in which appropriate bioremediation technologies can be employed requires further definition. Many organic pollutants do not readily enter the bioactive, aqueous phase of soil and sediment environments. Their bioavailability to the microbial population might be appreciably increased by the use of appropriate surfactants, dispersants, chelators, or emulsifiers. The physical matrix in which pollutants are found largely determines the rate at which the pollutants become bioavailable. Improved bioremediation of complex mixtures might take advantage of the fact that microbes can be selected to mobilize, immobilize, or fix compounds or ions in such a way that they are rendered susceptible to further treatment. The first stage of the process may require the action of a biodegrading, surfactant-producing, or bioaccumulating organism.

References

Batal, W., L. S. Laudon, T. R. Wildeman. 1989. In: *Constructed Wetlands for Wastewater Treatment: Municipal, Industrial and Agricultural* (Hammer DA, ed). Chelsea, MI: Lewis Publishers, 550–557.

Bromley-Challenor, K. C. A., N. Caggiano, and J. S. Knapp. 2000. Bacterial growth on N, N-dimethyl formamide: implications for the biotreatment of industrial waste water. *J. Ind. Microbiol. Biotechnol.* 25(1):8–16.

Burken J. G. and J. L. Schnoor. 1997. *Environ. Sci. Technol.* 31:1399–1402.

De Wever, H. and H. Verachtert. 1997. *Wat. Res.*, 1(11):2673–2684.

Hammer, D. A., ed. 1989. Constructed Wetlands for Wastewater Treatment: Municipal, Industrial and Agricultural. Chelsea, MI: Lewis Publishers.

Janssen, D. B., M. Pentenga, J. Van der Ploeg, F. Pries, J. Van der Waarde, E. Wonink, A. J. Van den Wijngaard. 1990. *Biomolecular Study Center Annual Report.* Gröningen, The Netherlands: Groningen University, 6567.

McFarlane, J. C., C. Nolt, C. Wickliff, T. Pfleeger, R. Shimabuku and M. Mcdowell. 1987. *Environ. Toxicol. Chem.* 6:847–856.

National Research Council. 1985. *Oil in the Sea: Inputs, Fates, and Effects.* Washington: National Academy Press.

National Research Council. 1989. *Using Oil Spill Dispersants on the Sea.* Washington: National Academy Press.

O'Neill, F. J., K. C. A. Bromley-Challenor, R. J. Greenwood and J. S. Knapp. 2000. *Wat. Res.*, 34(18):4397–4409.

Roper, J. C., J. Dec, J. Bollag. 1996. *J. Environ. Qual.*, 25:1242–1247.

Susarla, S., V. F. Medina, S. C. McCutcheon. 2002. *Ecol. Eng.* 18:647–658.

Thayer, A. M. 1991. *Chem. Eng. News.* 69:23–44.

U.S. EPA. 2002. Office of Compliance Sector Notebook Project, Profile of the Organic Chemical Industry, 2nd edition, November.

Wever, H. D., K. Vereecken, A. Stolz, and H. Verachtert. 1998. Initial transformations in the biodegradation of Benzothiazoles by Rhodococcus isolates. *Appl. Environ. Microbiol.* 64(9):3270–3274.

CHAPTER 6

Chlorinated Hydrocarbons and Aromatics, and Dioxins

Introduction

Organic pollutants are among the most ubiquitous in our environment. They have accumulated because of a variety of anthropogenic causes and because of their greater hydrophobicity (i.e., their lack of solubility in water).

Occurrence

Organics, which include polycylic aromatic hydrocarbons (PAHs), chlorinated aromatic hydrocarbons, chlorinated aliphatic hydrocarbons, halogenated hydrocarbons, biphenyls, phenols, aniline derivatives, phenol ethoxylates, and benzoic acid derivatives, are ubiquitous in our environment. Both anthropogenic and natural causes are known for their accumulation. They are found in water, marine systems, soil, sewage, and air. These are the most common pollutants and are known to persist in the environment. Some of them, such as the PAHs, are potent carcinogens. All of them are reported to have adverse effects on human and animal health.

Some of these, like the polyhalogenated aromatics, are chemically inert and therefore can only be degraded by biological means. The degradation could be by either aerobic or anaerobic pathways. A brief outline of both these pathways is necessary to be able to design suitable degradation pathways for a given contaminant.

Aerobic Degradation

A number of bacteria and fungi are known to adopt the aerobic pathway. The enzymes involved in the fixing of oxygen (from air or water) into organic molecules are called "oxygenases." Most of the oxygen required by the microorganisms is used for oxidative phosphorylation, which generates energy for cellular processes. About 5 to 10% of the total oxygen requirement is normally used by these "oxygenases." The ability of oxygenases to incorporate oxygen into organic compounds is important because many of the hydrophobic pollutants such as PAHs are high in carbon and hydrogen but low in oxygen. Through the action of oxygenases, hydrophobic organic compounds become more water soluble and can be broken down by a large number of other microorganisms. The end result of the oxygenase reaction on these hydrocarbons is hydroxyl or carbonyl compounds, which are normally more water soluble than the parent compounds.

Two major classes of oxygenases are well known. They are monooxygenase and dioxygenase. Monooxygenases incorporate one atom of the oxygen molecule into the organic substrate while the second oxygen atom goes to form water. Dioxygenases incorporate both atoms of oxygen molecule into the substrates. (Note: The division is not absolute.) These enzymes participate in the oxidative metabolism of a wide variety of chemicals of pharmaceutical, agricultural, and environmental significance.

Dioxygenases

Dioxygenases are very important in initiating the biodegradation of a variety of chlorinated and nitrogenous aromatic compounds as well as nonsubstituted PAHs. There are two major types of dioxygenases. One type requires NADH or NADPH, and these enzymes hydroxylate the substrates (Cerniglia, 1992). The other type has no specific requirement for NAD(P)H, and it cleaves the aromatic ring (Eltis et al., 1993). The overall mechanism of degradation can be summarized as shown in Fig. 6-1. Dioxygenases are very important in initiating the biodegradation of a variety of chlorinated and nitrogenous aromatic compounds as well as nonsubstituted PAHs. Their main substrates seem to be derived from crude oil and lignin, as these are the major sources of aromatic compounds in the environment. Many of these compounds are first degraded to catechol or protocatechuate by oxygenases.

Monooxygenases

Monooxygenases are more abundant than the dioxygenases and are more commonly found in fungi and mammalian systems. They can catalyze several different types of oxygen insertion reactions. These classes of enzymes require two reductants (substrates). Since they oxidize two substrates, they are also called *mixed function oxidases*. Since one of the main substrates becomes hydroxylated, they are also called *hydroxylases* (Gibson, 1993). The overall mechanism of degradation can be summarized as shown in Fig. 6-2. The monooxygenases can initiate attack on aromatic compounds. They are more abundant than dioxygenases.

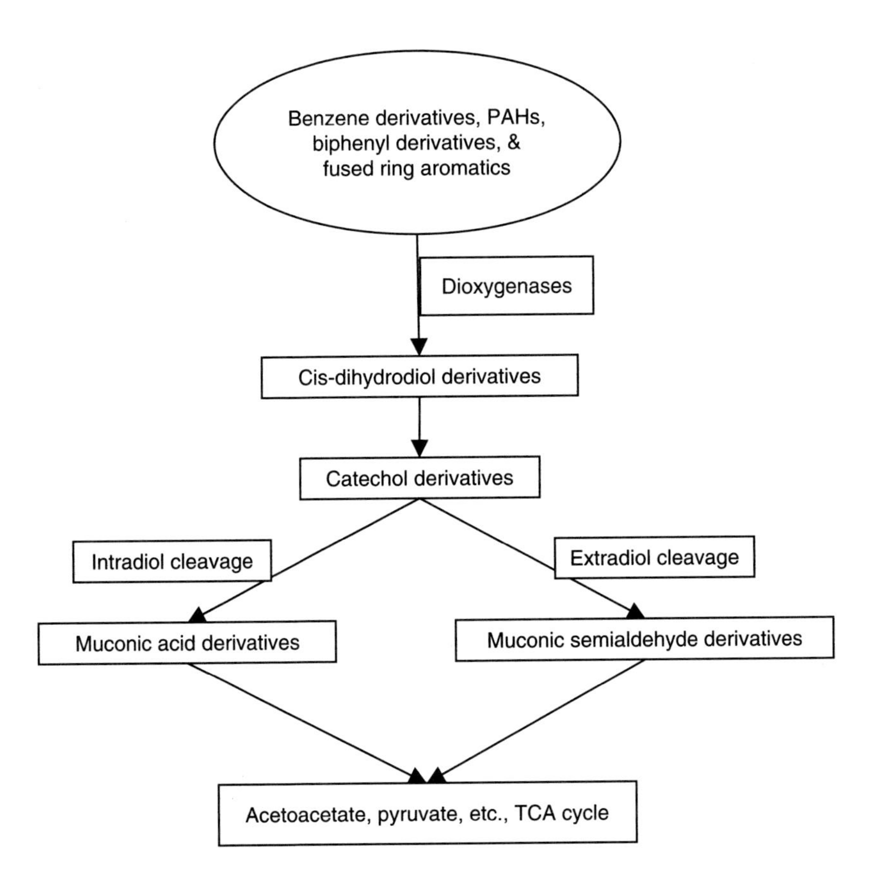

FIGURE 6-1. Dioxygenase degradation mechanism.

Anaerobic Degradation Pathways

The microbial mediated decomposition portion of the carbon cycle can be coupled with oxygen or can occur with no external electron acceptor. With oxygen, respiratory metabolism occurs and results in *higher energy yields* than fermentative metabolism (no external electron acceptor) does. Therefore, when oxygen is present, aerobic degradation predominates over anaerobic fermentation. Nonetheless, anaerobic decomposition still plays a key role in the carbon cycle in the ecosphere because of ecological effects. Since the late 1980s, an increasing number of novel microorganisms have been shown to utilize saturated and aromatic hydrocarbons as growth substrates under strictly anoxic conditions. In the absence of oxygen, a wide variety of alternative electron acceptors are used by the anaerobic bacteria for oxidation of organic compounds. Methanogenesis is predominant in freshwater sediments, while sulfate reduction is a dominant process in

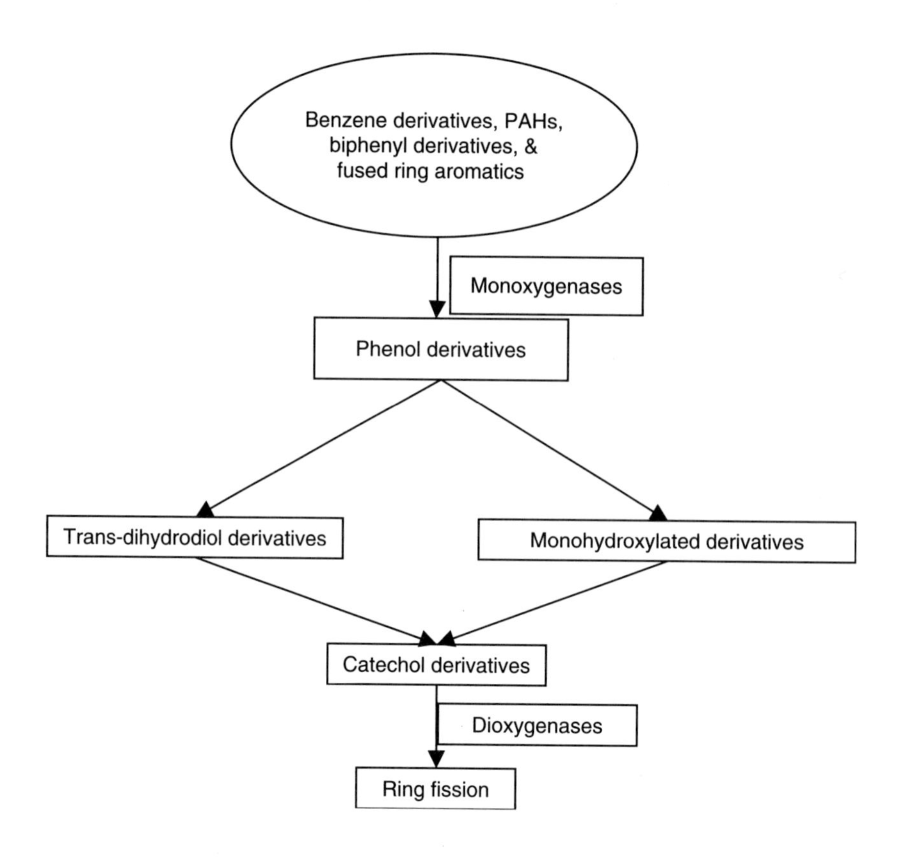

FIGURE 6-2. Monooxygenase degradation mechanism.

carbon metabolism in marine and estuarine sediments. Denitrification can be significant in regions of high nitrate input from agricultural runoff or sewage discharge. Fe(III) reduction is important in several sediments and anoxic soils. Halogenated aromatic compounds, phenols, benzoic acids, and PAHs are reported to have been degraded to carbon dioxide under a variety of redox conditions, with nitrate, iron, sulfate, and carbon dioxide as alternative electron acceptors. Oxidation of these substrates was coupled to reduction of the respective electron acceptor. Distinct anaerobic populations are enriched and responsible for metabolic patterns under different redox conditions. For the distinctive redox respiration to be effective, the anaerobic organisms grow in "syntrophic cocultures" with other anaerobes or grow by anoxygenic photosynthesis. The interactions and activities of diverse anaerobic communities need to be considered when evaluating the fate of anthropogenic contaminants in the environment and in developing bioremediation technologies. The overall picture of the bioremediation of organics is summarized in Fig. 6-3.

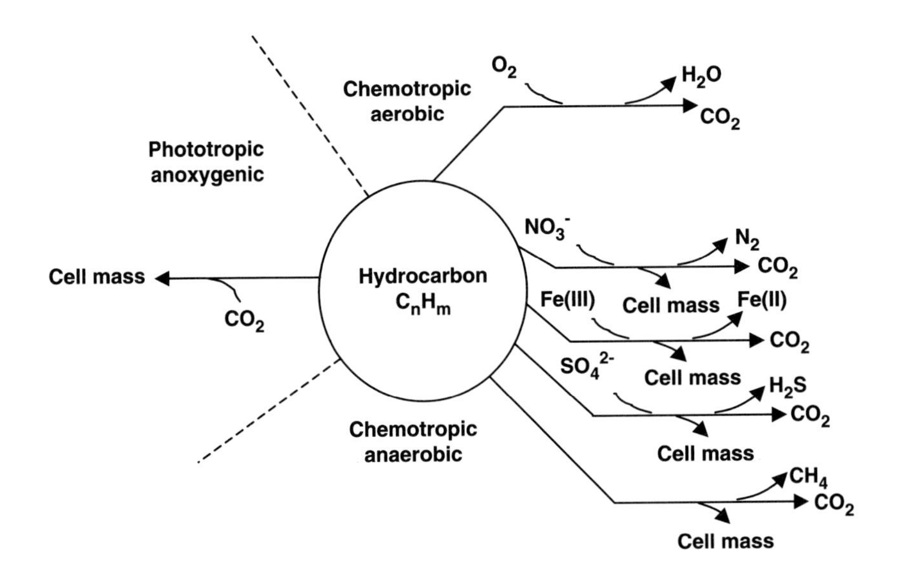

FIGURE 6-3. Bioremediation of organics.

There is no biochemical agent under anoxic conditions that exhibits the properties of the oxygen species involved in aerobic hydrocarbon activation; hence, the mechanisms of anaerobic hydrocarbon activation are completely different from oxygenase reactions. Indeed, all of the anaerobic (degradation) activation reactions of hydrocarbons are mechanistically unprecedented in biochemistry (Rabus et al., 2001). The number of electrons released and the free energy of some of these reactions are shown in Fig. 6-4.

Thus in designing anaerobic degradation technologies, importance should be given to the following:

- The type of inorganic substances present (nitrate, iron, sulfate, methane, etc.)
- The different communities of the microorganisms in the medium

Polynuclear Aromatic Hydrocarbons

Polynuclear aromatic hydrocarbons (PAHs) constitute one class of toxic environmental pollutant that has accumulated in the environment because of a variety of anthropogenic activities. Incomplete combustion of organic materials, in particular fossil fuels, is considered to be the source of these PAHs. Hence, domestic coal combustion, motor vehicle fuel combustion, and volatilization of the existing burden from contaminated soils are the primary sources. Of these, 71 to 80% is due to traffic emissions (Guerin and

$$CH_4 + 3H_2O \rightarrow HCO_3^- + 9H^+ + 8e^-$$

$$E_{in\ situ} = -0.289\ V\ (E^o = -0.239\ V)$$

$$SO_4^{2-} + 9H^+ + 8e^- \rightarrow HS^- + 4H_2O$$

$$E_{in\ situ} = -0.248\ V\ (E^o = -0.217\ V)$$

$$C_{16}H_{34} + 19.6\ NO^{3-} + 3.6H^+ \rightarrow 16HCO_3^- + 9.8\ N_2 + 10.8\ H_2O$$

$$\Delta G' = -983\ KJ / mole\ of\ N_2\ formed$$

$$C_{16}H_{34} + 12.25\ SO_4^{2-} \rightarrow 16HCO_3^- + 12.25\ HS^- + 3.75\ H^+ + H_2O$$

$$\Delta G' = -61\ KJ / mole\ of\ HS^-\ formed$$

$$C_{16}H_{34} + 11.25\ H_2O \rightarrow 12.25\ CH_4 + 3.75\ HCO_3^- + 3.75\ H^+$$

$$\Delta G' = -33\ KJ / mole\ of\ CH_4\ formed.$$

FIGURE 6-4. Anaerobic degradation of hydrocarbons.

Jones, 1988). Several PAHs have been shown to be acutely toxic. The most potent carcinogens of the PAH group in addition to benzo[a]pyrene include: the benzofluoranthenes, benzo[a]anthracene, dibenzo[ah]anthracene, and indenol[1,2,3-cd]pyrene (Fig. 6-5). Most of the PAHs are recognized by regulatory agencies such as the European Community (EC) and the U.S. Environmental Protection Agency as priority pollutants. Some of them are also classified as persistent organic pollutants (POPs).

General Aspects of PAH Degradation

The persistence of PAHs in the environment depends on the physical and chemical characteristics of the PAHs. The greater their lipophilic character (and corresponding hydrophobic character), the greater is their persistence. PAHs are degraded by photooxidation, chemical oxidation, and biological transformation. Microbial-mediated biological transformation is probably the most prevailing route of PAH cleanup (Mueller et al., 1989). The basic pathway of degradation is via the *cis* dihydrodiol formation as shown in Fig. 6-6 (Juhasz and Naidu, 2000). Although most bacteria possess the enzymes for the catabolism of PAHs, degradation of these compounds may not occur because these compounds are unable to pass through the bacterial

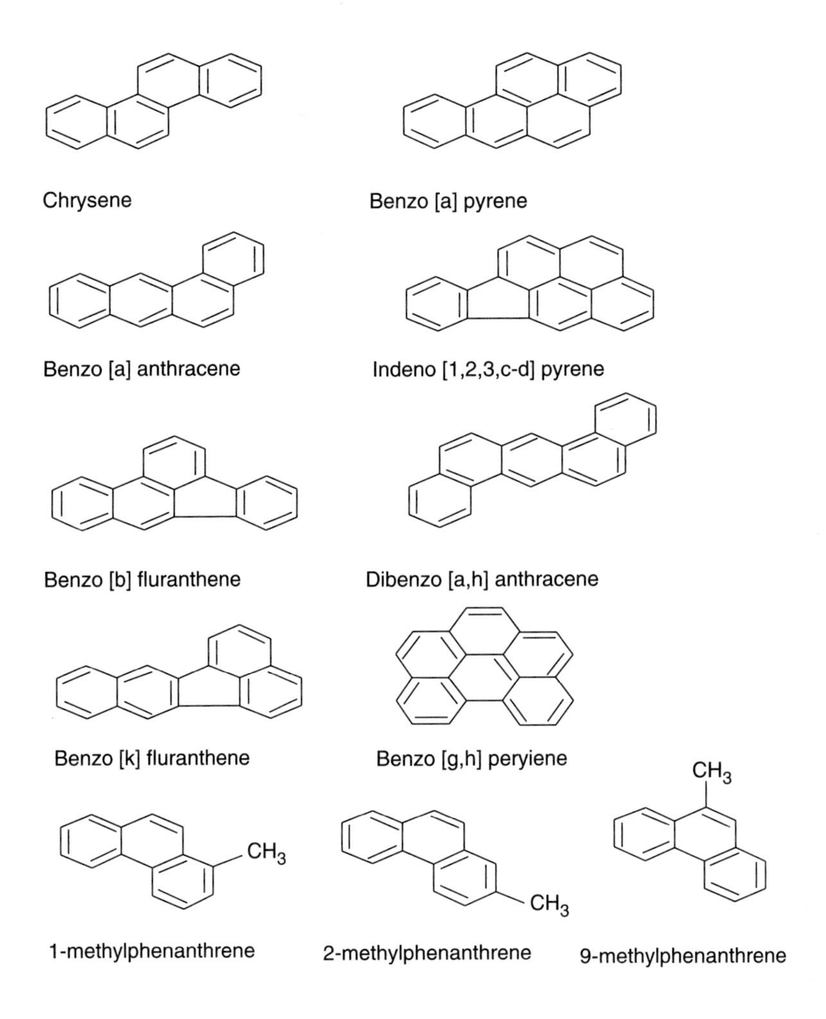

FIGURE 6-5. Structures of some PAHs.

cell walls. On the other hand, the ability of the fungi to produce extracellular enzymes such as lignin peroxidases (LiP) overcomes this problem (Duran and Espisito., 2000). *Thus fungal-bacterial cocultures are the choice for the complete degradation of PAHs.*

Halogenated Organic Compounds

Halogenated compounds constitute one of the largest groups of environmental pollutants. Contamination of marine and freshwater sediments by anthropogenic halogenated organic compounds, such as solvents (tetrachloro ethylene[PCE], trichloro ethylene [TCE], dichloro ethylene [DCE], chloroethylene [CE], trichloro methane[TCM], and dichloro methane

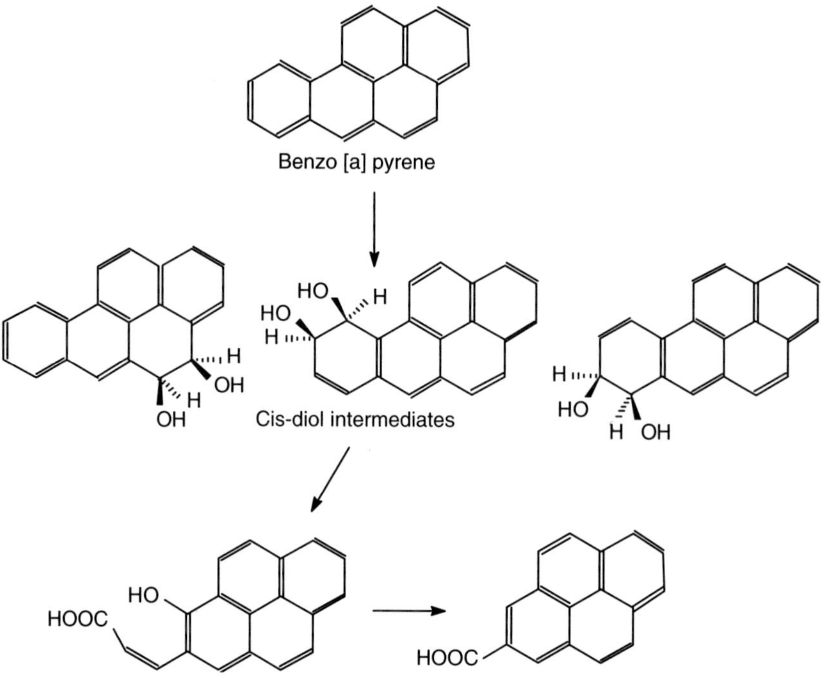

FIGURE 6-6. Biodegradation of PAHs.

[DCM]), pesticides (DDT, TCE), polychloro dibenzofurans (PCDFs), and dioxins, is a matter of increasing concern. The majority of these compounds are chlorinated, but brominated and fluorinated aromatic compounds are also in use. Microbial processes based on the metabolic activities of anaerobic bacteria are very effective in the degradation of these compounds. They are known to play important roles in nature by preparing many of these compounds for subsequent biodegradation, predominantly by the aerobic means. A critical step in degradation of organohalides is the cleavage of the carbon-halogen bond (Leisinger, 1996); two main strategies can be differentiated:

- The halogen substituent is removed as an initial step in degradation via reductive, hydrolytic, or oxygenolytic mechanisms, as shown in Fig. 6-7.
- Dehalogenation occurs after cleavage of the aromatic ring. This generally happens with lower chlorinated aromatics, wherein microorganisms may open the ring with dioxygenases before removal of chlorines, as shown in Fig. 6-8. This reaction is very similar to those acting on nonhalogenated substrates.

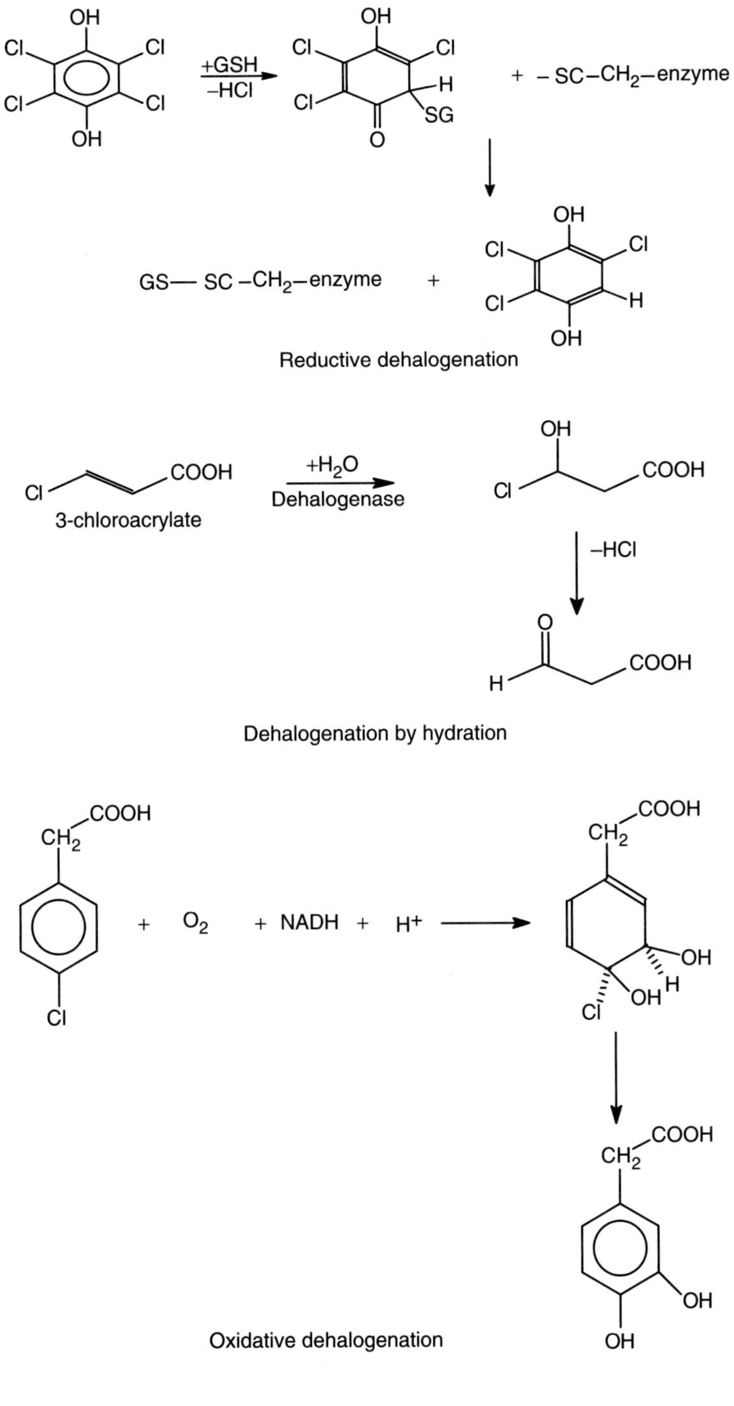
OH
Cl
Cl
Cl
Cl
OH
+GSH
−HCl
OH
Cl
Cl
Cl
H
SG
O
+ − SC−CH2−enzyme
OH
Cl
Cl
Cl
H
OH
GS— SC −CH2−enzyme
+
Reductive dehalogenation
Cl
COOH
3-chloroacrylate
+H2O
Dehalogenase
OH
Cl
COOH
−HCl
O
H
COOH
Dehalogenation by hydration
COOH
CH2
Cl
+ O2 + NADH + H+
COOH
CH2
OH
H
OH
Cl
COOH
CH2
OH
OH
Oxidative dehalogenation

FIGURE 6-7. Critical step in the degradation of organohalide.

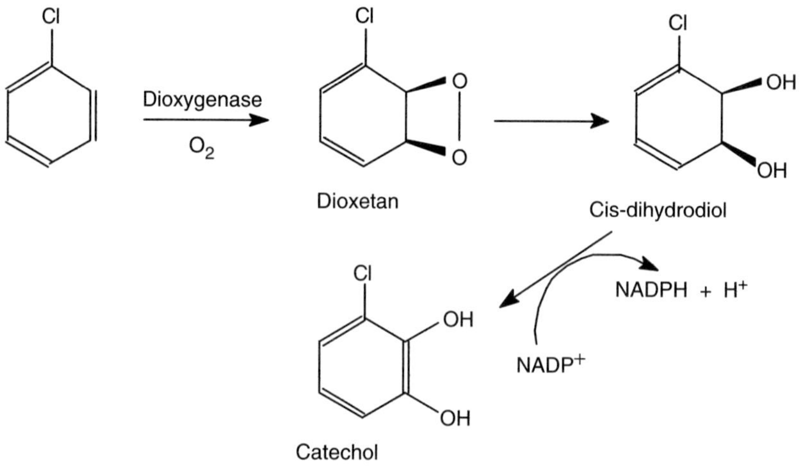

FIGURE 6-8. Degradation of organohalides by dioxygenases.

Reductive dehalogenation under aerobic conditions is brought about by conjugation with glutathione. A good example is the reaction catalyzed by tetrachlorohydroquinone reductive dehalogenase from *Sphingomonas chlorophenolicus*. This reaction requires 2 mol of reduced glutathione (GSH) per reaction. In the first part of the reaction, one molecule of GSH is oxidized and becomes attached to the substrate at the site of dechlorination. In the second part, another molecule of GSH extracts the first glutathione to form glutathione disulfide (GSSG) while replacing it with a hydrogen on the ring (Fig. 6-7). Under anaerobic conditions, reductive dehalogenation can yield energy for the microorganism through the process of halorespiration, where reductive dehalogenation is coupled to energy metabolism (Mohn and Tiedje, 1992). Here a halogenated compound like tetrachloroethene (PCE) serves as a terminal electron acceptor during oxidation of an electron-rich compound such as hydrogen, benzene, toluene, and similar organic substrates (Fig. 6-9). However, this process is often partly inhibited by other electron acceptors such as sulfate or nitrate. Several studies show that alternate electron acceptors, such as sulfate, iron (III), or nitrate, can support anaerobic degradation of halogenated phenols and benzoates. Mineralization of these compounds to CO_2 may be coupled to sulfate, iron(III), or nitrate reduction, as shown in Fig. 6-10.

In an interesting study on bioremediation of hazardous wastes, aromatic hydrocarbons such as benzene, toluene, ethyl benzene, xylenes, phenols, and cresols were used as electron donors to biologically reduce halogenated hydrocarbons (electron acceptors) such as tetrachloroethylene (PCE) and trichloroethylene (TCE), thereby achieving the degradation of both (U.S. Patent No. 5922204).

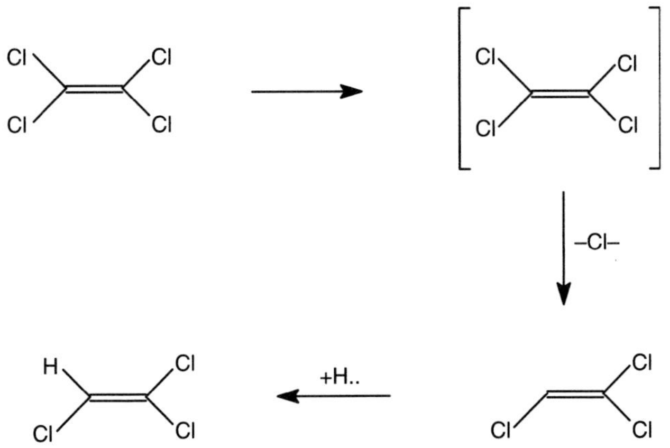

FIGURE 6-9. Halorespiration.

Oxidation:

$$C_6H_5O_2Cl + 12\,H_2O \longrightarrow 7\,CO_2 + Cl^- + 29\,H^+ + 28\,e^-$$

Reduction:

$$5.6\,NO_3^- + 33.6H^+ + 28\,e^- \longrightarrow 2.8\,N_2 + 16.8\,H_2O$$
$$28\,Fe^{3+} + 28\,e^- \longrightarrow 28\,Fe^{2+}$$
$$3.5\,SO_4^{2-} + 35\,H^+ + 28\,e^- \longrightarrow 3.5\,H_2S + 14\,H_2O$$
$$3.5\,HCO_3^- + 31.5\,H^+ + 28\,e^- \longrightarrow 3.5\,CH_4 + 10.5\,H_2O$$

FIGURE 6-10. Anaerobic degradation of halogenated phenols and benzoates.

Chlorinated Aliphatic Compounds

Chlorinated aliphatic compounds, a diverse group of industrial chemicals, play a significant role as environmental pollutants. Most prominent with respect to industrial use, environmental persistence, toxicity, and potential carcinogenicity are the chlorinated one-, two-, and three-carbon compounds. They are used as intermediates in the chemical industry, as solvents, and in some cases as pesticides (Kirechner, 1995). As mentioned earlier, dehalogenation is the critical step in the degradation of chlorinated aliphatics. Irrespective of whether it is based on hydrolytic, oxygenative, or reductive mechanisms, it commonly occurs via the first carbon-halogen–bond cleavage reaction. It will be of value to note that some of these chlorinated compounds serve as growth substrates for microbial cultures because of their electron acceptor ability. Mention must be made of chlorofluorocarbons (CFCs) and hydrofluorochlorocarbons (HCFCs), which are causing the depletion of the ozone layer. Since most of these exist as gases at room temperature, biodegradation is not observed.

Chlorinated Aromatic Compounds

Over the past few decades, the extensive use of chlorinated benzenes has led to considerable release of these compounds into the environment. It is estimated that about 15,000 chlorinated compounds are currently being used in various industries worldwide (McCarthy et al., 1996). The presence of these substituents on the aromatic rings hinders biodegradation. This is because these substituent groups deactivate the aromatic nucleus to electrophillic attack by oxygenases or other enzymes by withdrawing electrons from the ring. This deactivating effect increases with the number of halogen substituents, resulting in their greater persistence in the environment. However, since the enzymes of the anaerobic degradation do not attack the substrate in an electrophilic way, these deactivated rings are still suitable substrates for these enzymes. Thus, anaerobic biological degradation is the only process by which these heavily halogenated aromatics are transformed to a partly oxidized, partly dehalogenated state. These partly oxidized (hydroxylated), partly dehalogenated aromatics are then further degraded to carbon dioxide by an aerobic route. Thus, it is invariably a combination of anaerobic (first) and aerobic (subsequent) routes that finally mineralize these polychlorinated aromatics. Unlike the heavily chlorinated benzenes, mono-, di-, and trichloro-benzenes undergo aerobic degradation. For example, 1,4-dichloro benzene degradation by *Xanthobacter flavus 14p1* was initiated by dioxygenation and the ring opening proceeded via ortho cleavage, as shown in Fig. 6-11 (Gorisch et al., 1995). Some examples of the biodegradation of chlorinated organic compounds are given in Table 6-1.

Polychlorinated Biphenyls (PCBs)

Polychlorinated biphenyls (PCBs) are chemically inert liquids and are difficult to burn. They are excellent electric insulators; as a result, they have been used extensively as coolant fluids in power transformers and capacitors, as heat transfer fluids in machinery, as waterproofing agents etc. Because of their stability and extensive usage, together with inattentive disposal practices, PCBs became widespread and persistent environmental contaminants. Since the late 1950s, over 1 million metric tons of PCBs have been produced. Some of these compounds are reported to be carcinogenic. Apart from this, strong heating of PCBs in the presence of a source of oxygen can result in the production of small amounts of dibenzofurans (DFs) (Fig. 6-12). DFs like dioxins are highly toxic to humans and animals (Baird, 1999).

PCBs with relatively few chlorine atoms undergo oxidative aerobic biodegradation by a variety of microorganisms. Since the degradation occurs

TABLE 6-1
Biodegradation of Some Organic and Halogenated Organic Compounds

	Cosubstrate	*Organism*	*Enzyme/enzymes*	*Substrate*	*Metabolites*
1.	Pentachlorophenol	*Sphingomonas chlorophenolicus (ATCC 39723)*	Monooxygenases	Oxygen	2,3,5,6-tetrachlorohydroquinone
2.	Toulene	*Burkholderia cepecia G4*	Toulene-2-monooxygenase (TOM)	Oxygen	—
3.	Chlorinated aliphatic hydrocarbons	*Burkholderia cepecia G4*	—	Oxygen	—
4.	Toulene	*Pseudomonas putida F1*	Dioxygenases	Oxygen	—
5.	Aryl alkyl ethers	*Rhodococcus rhodochrous*	Monooxygenases	Oxygen	
6.	3-Chloroacrylic acid	*Coryneform bacteria*	Dehalogenases	—	3-oxopropionic acid
7.	Chlorinated aliphatic compounds (CAH)	*Autotrophic nitrifying bacteria*	Nonspecific monooxygenases	Ammonia	—
8.	Trichloro ethylene (TCE)	*Nitrosomonas europaea*	Nonspecific monooxygenases	Ammonia	
9.	Dichlorobenzenes (DCB)	*Pseudomonas* sp.	Dioxygenases	Oxygen	Chloro catechols
10.	Dichlorobenzenes (DCB)	*Alcaligenes* sp.	Dioxygenases	Oxygen	Chloro catechols
11.	Dichlorobenzenes (DCB)	*Xanthobacter flavus 14p1*	Dioxygenases	Oxygen	2,5-dichloro muconic acid
12.	Anthracene and fluoranthracene (PAHs)	*Absidia cylindrospora CS*			
		Asperigillus niger CS *Ulocladium chartorum CS*	Phenoloxidases	—	—
13.	Napthalene	*Pseudomonas putida*	Naphthalene-dioxygenase	Oxygen	Salicylate
14.	Toulene	*Desulfobacula toluolica*	—	Sulfate	—
15.	Chloro and fluoro derivatives of toulene	*Thauera* sp.	—	Nitrate	—

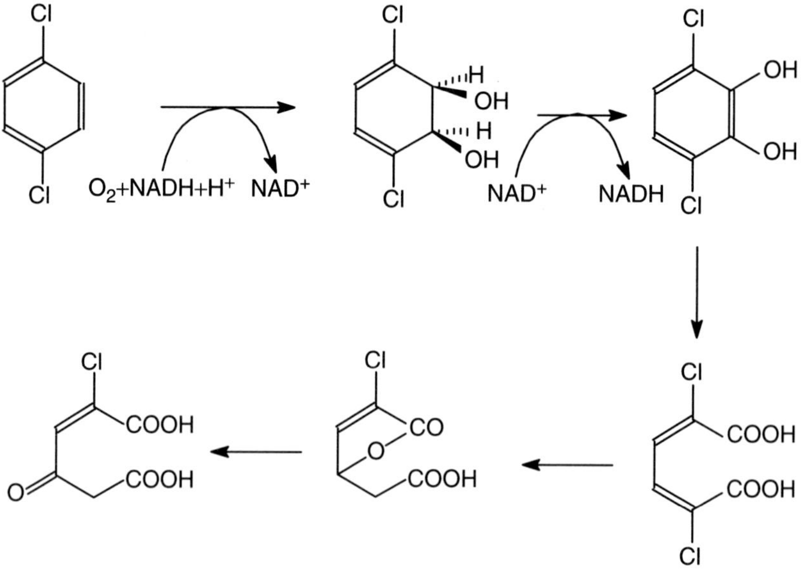

FIGURE 6-11. Degradation of 1,4-dichlorobenzene by *Xanthobacter flavus.*

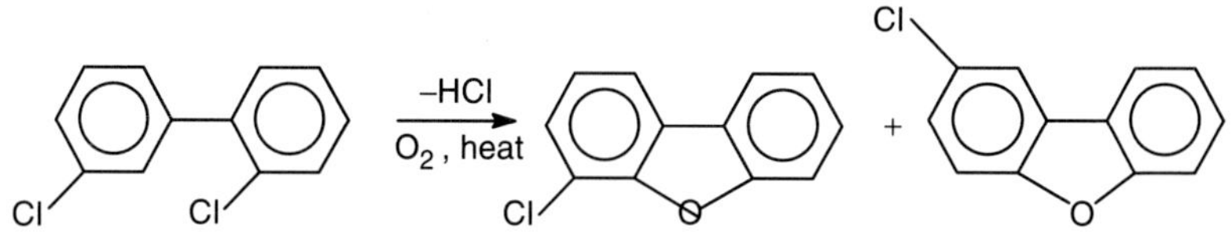

FIGURE 6-12. Formation of dibenzofurans from PCBs.

via *cis*-dihydrodiol formation followed by ring cleavage (Fig. 6-13), a pair of nonsubstituted sites one ortho to the point of connection between the rings and an adjacent one meta to the connection must be available on one of the benzene rings. PCB molecules that are heavily substituted with chlorines will undergo anaerobic degradation, as with polychlorinated organic compounds. Reductive dechlorination, as discussed earlier, occurs most readily with meta and para chlorines. Thus the products of anaerobic treatment of polychlorinated biphenyls are normally ortho-substituted congeners, that is, 2-chlorobiphenyl and/or 2,2′-dichlorobiphenyl (Fig. 6-14). Since dioxin-like toxicity of PCBs requires several meta and para chlorines, the anaerobic degradation process significantly reduces the health risk from PCB contamination. Anaerobic degradation also provides PCBs with free ortho and meta positions, thereby making them suitable substrates for aerobic microorganisms, leading to complete mineralization (Fig. 6-14).

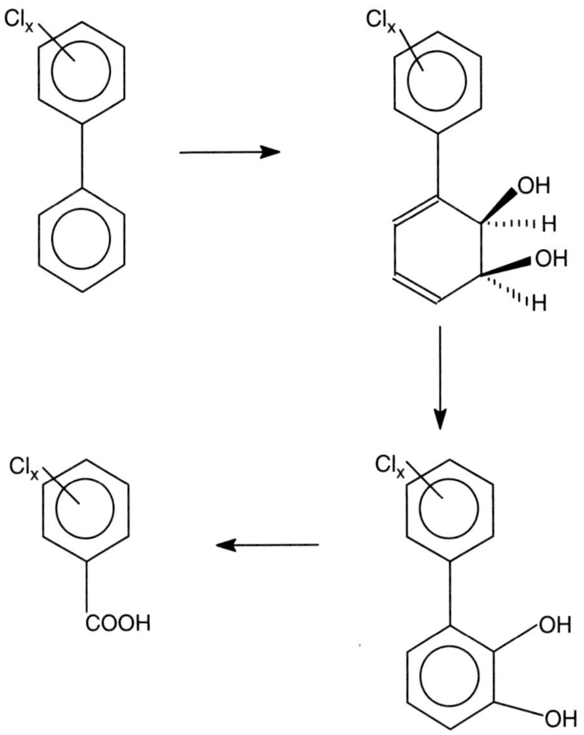

FIGURE 6-13. Aerobic degradation of PCBs.

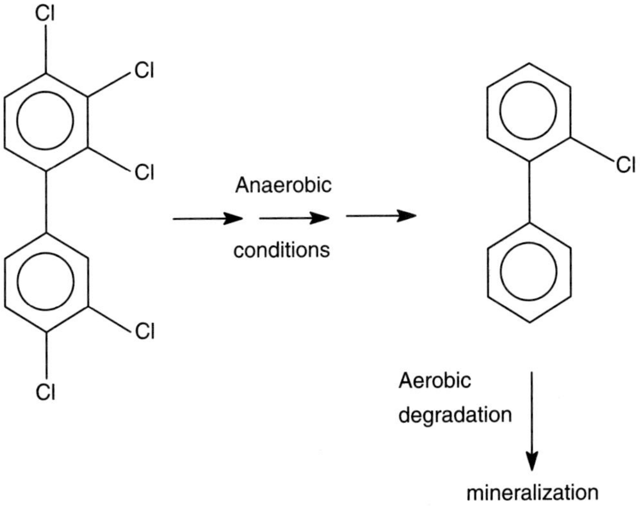

FIGURE 6-14. Anaerobic followed by aerobic degradation of PCBs.

Dioxins

Dioxin is the name given to group of compounds that have two chlorinated benzene rings connected through a central ring and two oxygen atoms located para to each other (Fig. 6-15). Ideally these compounds should be called polychlorodibenzo-*p*-dioxins. Dioxins along with dibenzofurans are the most widely known "*toxic byproducts.*" Polychlorinated dibenzofurans and polychlorodibenzo-*p*-dioxins are produced as by-products of a myriad of processes, such as:

- Synthesis of 2,4,5-trichlorophenoxy acetic acid (2,4,5-T), an herbicide
- Bleaching of pulp
- Incineration of garbage
- Recycling of metals
- Production of common solvents such as TCE and PCE.

Environmental contamination by dioxin also occurred as a result of an explosion in a chemical factory in Seveso, Italy, in 1976. The toxic health effects of dioxins are still debatable. All the same, researchers have indicated that ingestion of dioxins may lead to the following effects:

- Reproductive effects in the offspring
- Altered secretion of sex hormones
- Cancer

Studies on the structure-activity relationship of dioxins indicate that the toxicity of dioxins depends on the extent and pattern of chlorination. Toxic dioxins are those with three or four beta chlorine atoms and few if any alpha chlorines (carbon atoms that are bonded to those in the central dioxin rings are "alpha" carbons, and the outlying ones are "beta" carbons). Thus the most toxic is 2,3,7,8-tetrachlorodibenzo dioxin (2,3,7,8-TCDD) (Fig. 6-15). The toxicity equivalence factors for some important dioxins are mentioned in Table 6-2. (Data taken from Canadian Environmental Protection Act Priority Substance List, Assessment Report No. 1, 1990; available at http://www.ec.gc.ca/report_e.html).

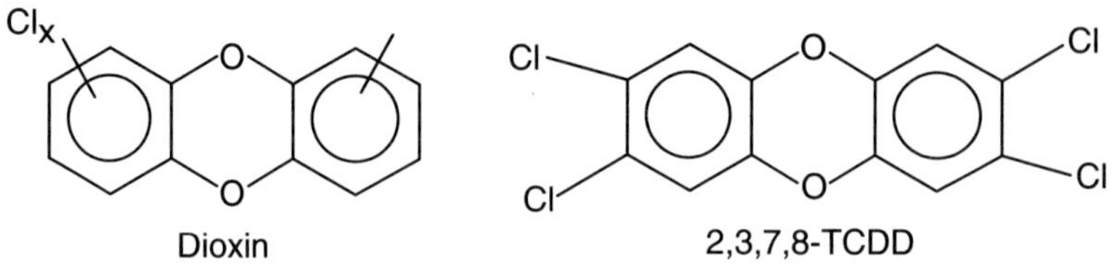

Cl_x O O

Dioxin

Cl O Cl Cl O Cl

2,3,7,8-TCDD

FIGURE 6-15. Structures of dioxins.

TABLE 6-2
Toxicity Equivalence Factor of Dioxins

Dioxins	*Toxicity equivalence factor*
2,3,7,8-Tetrachlorodibenzodioxin	1
1,2,3,7,8-Pentachlorodibenzodioxin	0.5
1,2,3,4,7,8-Hexachlorodibenzodioxin	0.1
1,2,3,4,6,7,8-Heptachlorodibenzodioxin	0.01
Octachlorodibenzodioxin	0.001

Biodegradation

Both aerobic and anaerobic organisms are known to degrade dioxins. The choice of the organism depends on the extent of chlorination. Similar to the degradation of PCBs, highly chlorinated dioxins (hexa, hepta, and octa) are dechlorinated by the action of anaerobic systems. These less-substituted dioxins are mineralized by aerobic systems (Fig. 6-14). Many reports have appeared in the literature on biodegradation of dioxins. Yeasts, fungi, and bacteria have been found to degrade dioxins. A few examples of detoxification of dioxins by various types of organisms and phytoremediation are listed in Table 6-3.

Microbes can be encouraged to biodegrade almost any organic chemical. Environmental chemists and microbial ecologists have extensively characterized the natural biodegradation pathways of a number of pollutant classes; recent reviews have been published for many, including polycyclic aromatic hydrocarbons (Cerniglia et al., 1989), polychlorinated biphenyls

TABLE 6-3
Some Examples of Biodegradation of Dioxins

	Compound name	*Organism*
1.	Polychlorinated dibenzo-*p*-dioxin	Recombinant yeast
2.	2,3-Dichloro dibenzo-*p*-dioxin	*Pseudomonas resinovorans*
3.	2,7-Dichloro dibenzo-*p*-dioxin	Wood rusting fungi
4.	Chlorinated dibenzo-*p*-dioxin	*Phlebia lindtneri*
5.	Chlorinated dibenzo-*p*-dioxin	*Terrabacter* sp.
6.	Chlorinated dibenzo-*p*-dioxin	*Sphingomonas* sp.
7.	TCDD	Ectomycorrhizal fungi
8.	Octachloro dibenzo-*p*-dioxin	Phytoremediation
9.	TCDD	*Pleurotus* sp.
10.	Dibenzo-*p*-dioxin	*Phanerochaete chrysosporium*

(PCBs) (Abramowicz, 1990), and pesticides (MacRae, 1989). Rapid screening assays are being developed by researchers to identify organisms capable of degrading specific wastes (Krieger, 1992). Molecular probes make it possible to test a small, mixed microbial population for specific degradative enzyme genes (Olson, 1991). Gene probing can also give an indication of the natural abundance of organisms with the potential to degrade specific pollutants at a given site.

References

Abramowicz, D. A. 1990. *Crit. Rev. Biotechnol.* 10:241–251.
Baird, C. 1999. *Environmental Chemistry*. New York: W. H. Freeman.
Cerniglia, C. E., 1992. Biodegradation of polycyclic aromatic hydrocarbons. *Biodegradation,* 3:351–368.
Cerniglia, C. E., and M. A. Heitkamp, 1989. *Metabolism of Polycyclic Aromatic Hydrocarbons in the Aquatic Environment.* Ed. U. Varanasi. 41–68 Boca Raton, FL: CRC Press.
Duran, N., and E. Esposito. 2000. *Applied Catalysis (B): Environmental*. 28:83–99.
Eltis, L. D., B. Hofmann, H. J. Hecht, H. Lunsdorf, and K. N. Timmis. 1993. *J. Biol. Chem.* 268:2727–2732.
Gibson, D. T., 1993. *J. Ind. Microbiol.* 12:1–12.
Gorisch, H., E. Spiess, and C. Sommer. 1995. *Appl. Environ. Microbiol.* 61:3884–3888.
Guerin, W. F., and G. E. Jones. 1988. *App. Env. Microbiol.* 54:929–936.
Juhasz, A. L. and R. Naidu. 2000. *Int. Biodeterioration & Biodegradation*. 45: 57–88.
Kirechner, E. M. 1995.*Chem. Eng. News.* 10:16–20.
Krieger, J. 1992. *Chem. Eng. News.* 70:36.
Leisinger, T. 1996.*Current Opinion in Biotechnology*. 7:295–300.
MacRae, I.C. 1989. *Rev. Environ. Contam. Toxicol.* 109:1–87.
McCarthy, D. L., S. Navarrete, W. S. Willet, P. C. Babbit, and S. D. Copley. 1996. *Biochem.* 35:14634–14642.
Mohn, W. W., and J. M. Tiedje. 1992. *Microbiol. Rev.* 56, 482–507.
Mueller, J. G., P. J. Chapman, and P. H. Pritchard. 1990. *Appl. Environ. Microbiol.* 55:3085–3090.
Olson, B. H. 1991. *Environ. Sci. Technol.* 25:604–611.
Rabus, R., H. Wilkes, A. Behrends, A. Armstroff, T. Fischer, A. Pierik, and F. Widdel. 2001. Anaerobic initial reaction of n-alkanes in a denitrifying bacterium: evidence for (1-methylpentyl) succinate as initial product and for involvement of an organic radical in n-hexane metabolism. *J. Bacteriol.* 183: 1707–1715.

Bibliography

Alkorta, I., and C. Garbisu. 2001. *Bioresource Tech.*, Sept.
Cerniglia, C. E. 1992. *Biodegradation* 3:351–368.
Dittmann, J., W. Heyser, and H. Bucking. 2002. *Chemosphere.* Oct.
Habe, H., K. Ide, M. Yotsumoto, H. Tsyi, T. Yoshida, H. Nojiri, and T. Omori. 2002. *Chemosphere*, July.
Habe, H., Y. Ashikawa, Y. Saiki, T. Yoshida, H. Nojiri, and T. Omori. 2002. *FEMS Microbiol.Letters*. May.
Marters, R., M. Wolter, M. Bahadir, and F. Zadrazil. 1999. *Soil Biology & Biochemistry.* Nov.
Michizoe, J., S. Y. Okazaki, M. Goto, and S. Funisakis. 2001.*Biochem. Eng. J.* Sept.
Mori, T. and R. Kondo. 2002. *FEMS Microbiol. Letters.* July.
Mori, T. and R. Kondo. 2002. *FEMS Microbiol. Letters*. Nov.
Sakaki, T, R. Shinkyo, T. Takita, M. Ohta and K. Inouye. 2002. *Arch. Biochem. Biophys.* May.
Widada, J., H. Nojiri, T. Yoshida, H. Habe and T. Omori. 2002. *Chemosphere.* Nov.

CHAPTER 7

Fluoride Removal

Introduction

Fluoride exists in the environment as a result of both natural and anthropogenic causes. The natural contamination of groundwater by fluoride ions is due to leaching of fluoride from rocks (soil) into the aquifers, while the wide use of fluorinated compounds by industry is the major anthropogenic cause. In the former situation, the fluoride is in an ionic form, while in the latter it may be present in a covalent form. This chapter deals with the removal of both these types of fluorine.

Organofluorine Compounds

The synthetic diversity of nature is also reflected in a large number of naturally produced halogenated compounds discovered in many different organisms. Until today, more than 3,500 halogenated metabolites have been isolated from bacteria, fungi, marine algae, lichens, higher plants, mammals, and insects. Whereas brominated metabolites are predominant in the marine environment, chlorine-containing metabolites are preferentially produced by terrestrial organisms. Although fluorine is the most abundant halogen in the earth's crust, biologically produced fluorinated metabolites are quite rare, as is the case of iodated metabolites. Hence, many fluorinated compounds in the environment are of anthropogenic origin, making them recalcitrant to degradation.

The chemicals of humanmade origin that are used as refrigerants, fire retardants, paints, solvents, herbicides, and pesticides are predominantly halogenated organic compounds and cause considerable environmental pollution and human health problems as a result of their persistence and toxicity (Mohn and Tiedje, 1992). They also transform into hazardous metabolites. As a general rule, the strength of resistance to enzymatic cleavage of carbon-halogen bonds is observed to increase with the electronegativity of the substituents, in the order F—C>Cl–C>Br–C>I—C.

Sodium monofluroacetate is highly toxic to most endothermic vertebrates and many invertebrates. Plant species belonging to the genus *Gastrolobium* are known to produce fluoroactates. Several genera of soil fungi (*Fusarium* and *Pencillium*) and bacteria (*Pseudomonas* and *Bacillus*) are known to degrade fluoroactates (Twigg and Socha, 2001).

Because of the concern over the depletion of stratospheric ozone by chlorofluorocarbons (CFCs), the use of hydrochlorofluorocarbons (HCFCs) such as HCFC-123 (2,2-dicloro-1,1,1-trifluoroethane) and HCFC-141 (B) (1,1-dichloro-1-fluoroethane) are banned (Fig. 7-1). Halothane (1-bromo-1-chloro-trifluoroethane) is used as an anesthetic gas. These compounds have many of the physical properties of the corresponding chloroflurocarbons (CFCs). The presence of C—H bond makes this group of compounds susceptible to hydroxylation by monooxygenases (Anders, 1991). There are at least two pathways by which these compounds are biodegraded—the reductive and the oxidative derivatives. The reductive pathway proceeds through a free radical (Fig. 7-2), while the oxidative pathway proceeds through the corresponding hydroxylated compound, which is further degraded to an acid derivative in the presence of water (Urban and Dekant, 1994) (Fig. 7-3).

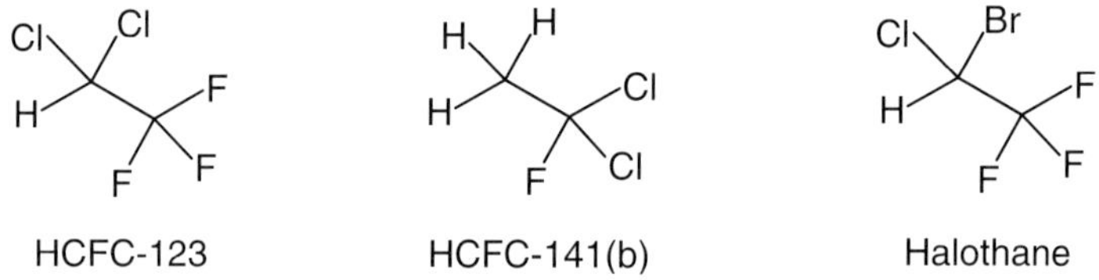

HCFC-123 HCFC-141(b) Halothane

FIGURE 7-1. Structures of common organofluorine compounds.

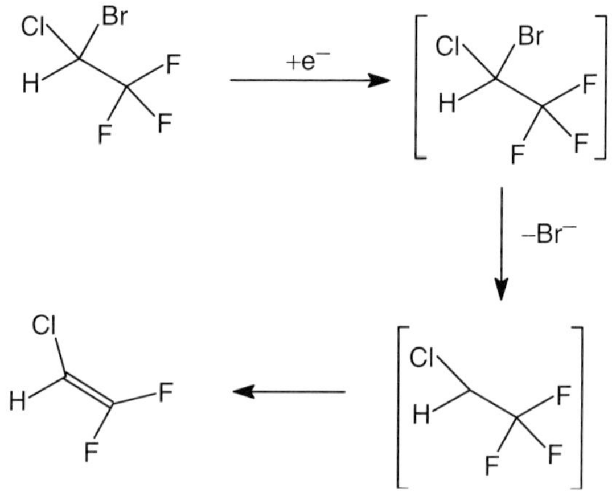

$+e^-$ $-Br^-$

FIGURE 7-2. Reductive degradation of halothane.

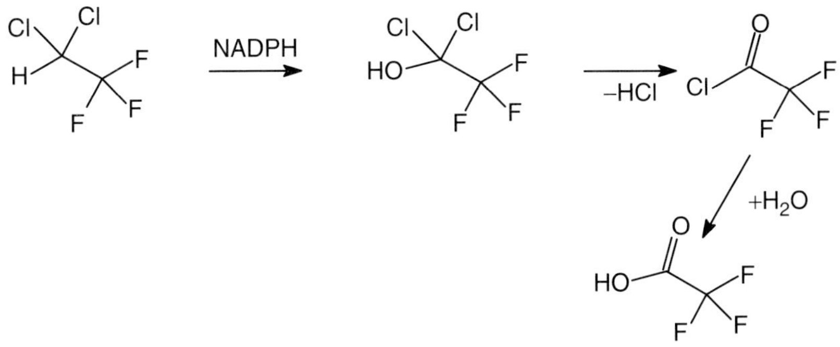

FIGURE 7-3. Oxidative degradation of HCFC-123.

New et al. (2000) reported the degradation of 4-fluorocinnamic acid (a common reagent in the synthesis of pharmaceuticals) to 4-fluorbenzoic acid using activated sludge. Fluorine is isosteric to hydrogen; hence most of the enzymes bringing about transformation of aromatic compounds will transform fluorinated aromatic compounds, too. In general, the *isosteric* replacement, even though it represents a subtle structural change, results in a modified profile: some properties of the parent molecule remain unaltered while others will be changed. The similar shape and polarity within a series of substrates of different reactivity (*bioisosteres*) eliminates effects due to differences in enzyme-substrate binding [ES] and *hence is a good method of extending the range of substrates that can be chosen for the transformation.*

A number of instances can be cited from the literature wherein the *isosteres* had similar transformations. The *isosteres*, 1,2-dihydro naphthalene, 2,3-dihydro benzothiophene, and 2,3-dihydro benzofuran, gave similar corresponding diol products on incubation with *Pseudomonas putida* UV4. Microbes, which possess the metabolic pathways to metabolize benzene, substituted benzenes, and phenols were found to metabolize fluorinated benzenes (*isosteres*) in a similar manner (Fig. 7-4).

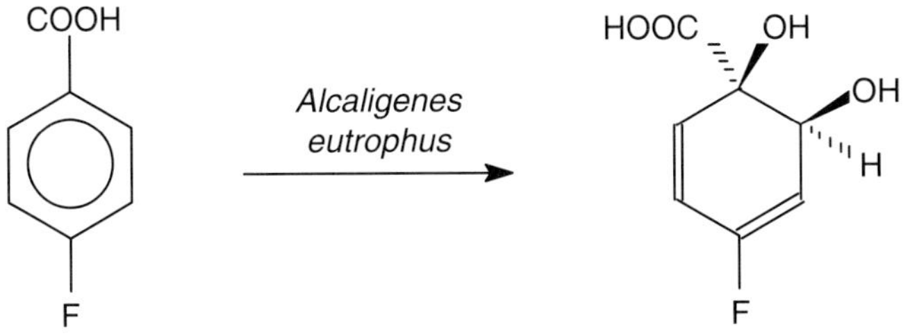

FIGURE 7-4. Degradation of benzoic acid isosteres.

Fluorobenzoic acids have been reported to degrade under both aerobic and anaerobic conditions (Vargas et al., 2000). Several pathways have been identified under aerobic conditions, but two are most widely reported. One pathway involves the degradation of fluorobenzoic acids into the corresponding fluorocatechol; in the other pathway, fluorobenzoic acid is transformed into hydroxyl benzoic acid.

Fluoride Contamination of Water and Treatment

Water gets contaminated by dissolving the pollutants in the lithosphere and also from anthropogenic causes. Fluorine is the thirteenth most abundant element in the earth's crust and is available in combined form as fluorspar (CaF_2), cryolite (Na_3AlF_6), fluoroapatite [$Ca_5F(PO_4)_3$], topaz [Al_2SiO_4 (OH, F)$_2$], sellaite (MgF_2), villiamite (NaF), bastnaesite (CoF_2), and fluorine hydrosilicates. Geological formation is the main source of fluoride in the groundwater. Fertilizers and pesticides, which contain about 1 to 3% fluoride, also contribute to its presence in the groundwater (Mariappan, 1996). The presence of fluoride ion in drinking water may be beneficial or detrimental to public health, depending on the concentration in which it is found. Fluoride intake beyond the limit of 1.5 mg/L causes dental and skeletal fluorosis and nonskeletal manifestations (Chen et al., 1995). Excess fluoride in drinking water is a major environmental problem in over 21 nations. About 15 million people are living in 3,500 endemic habitations of 16 states of India (Mariappan et al., 2000).

Because of the proven health danger of excess fluoride ion, governments routinely monitor the environment for its presence. In cases where control strategies have been implemented, there have been significant decreases in environmental metal levels.

Several methods have been advocated for defluoridation of drinking water. They can be broadly divided into two categories, viz., those based upon the addition of some material to the water during the softening or coagulation process and those based upon ion-exchange or adsorption processes. Adsorption or ion-exchange processes are recommended for low concentration treatment. These processes are performed by using lime and alum, bone char and synthetic bone, activated carbon and bauxite, ion-exchange, activated alumina, and reverse osmosis. Among these materials, activated alumina is supposed to be the most effective and economic adsorbent for removal from drinking water of fluoride in the lower concentration range. But so far most of the methods developed could not find any practical application because of high capital and operating cost and complexity of operating procedure. Even the Nalagonda Technique, involving the addition of aluminium salts, lime, and bleaching powder, has its shortcomings in the form of sludge disposal problems (Nawlakhe and Rao, 1990).

Defluoridation methods involving adsorption have been developed. Charred coconut shells or dry fibrous plant material have been used. These methods have the obvious problem of leachates that might alter the water quality, making it unsuitable for drinking purposes.

Pollutants that continue to be of enormous practical and economic importance are of heavy metals, such as lead, mercury, and cadmium; and inorganic anions such as fluoride, nitrate, and carbonate. These natural elements and compounds, found in the earth's crust, are utilized in many industrial processes and products, a use which has resulted in their release in higher concentrations and in more accessible form than is typical in natural systems. Incorporation of heavy metals into inorganic and organometallic complexes and anions into organic compounds (fluorocitrates) often alters their biological activity; such changes are just as likely to increase toxicity because of increased bioavailability as they are to decrease toxicity. Furthermore, depending on conditions of pH, increased temperature, etc., natural cycles may intervene to convert or mobilize relatively benign inorganic species to more toxic organic complexes (e.g., conversion of elemental mercury to methylmercury and fluoride to fluorocitrate).

Unlike organic pollutants, the toxicity of fluoride ion is inherent in its atomic structure, and it cannot be further transmuted or mineralized to a totally innocuous form. Its oxidation state, solubility, and association with other inorganic and organic molecules can vary, however; microbes as well as higher organisms may play a bioremediative role by transforming and concentrating these anions so that they are less available and less dangerous.

Many plants and bacteria have evolved various means of extracting essential nutrients, including anions, from their environment. Such organisms may provide the opportunity to make fluoride less available. However, a practical phytoremedial technology remains to be developed. Anion binding can be brought about by any of the following three methods, viz:

- Hydrogen bonding interaction
- Electrostatic interaction
- Hydrogen bonding with electrostatic interaction

Many microorganisms secrete high-affinity anion-binding compounds called ionophores (e.g., valinomycin, which binds to halides). The ionophores bind specific chemical forms of anions, and the anion-ionophore complex is then absorbed back into the organism for utilization. A bioremediation technology using native and chemically modified ionophores attached to inert support media would give good results.

In a recent study conducted on the ability of amino acids to bind and defluoridate water, the basic amino acids (lysine, arginine, and asparagine) were found to be effective (Kumar et al., unpublished work). Extension of this study led to the understanding that proteins are capable of selectively binding to fluoride and are therefore suitable for bringing about defluoridation of water. Microorganisms and plants exude a number of enzymes (proteins) that may have this ability of binding to fluoride, thus making it less available.

In the last decade, hairy roots have helped markedly in phytoremediation. The root zone is the part of the plant that is in intimate contact with the contaminant; hence this part should be targeted for the expression of foreign genes with a view toward enhancing the uptake, bioaccumulation, or biotransformation of specific compounds. Alongside roots of higher plants, the subterranean complex of mycelia associated with mushroom growth would appear to offer a number of possibilities in the field of remediation.

Existing technologies for defluoridation of drinking water are not practical; hence, a concerted effort to develop a bioremediation method is needed. Plants or microorganisms capable of transforming or accumulating fluoride ions are the only viable solutions to this vexing problem.

References

Anders, M. W., 1991. *Environ. Health Perspect.* 96:185–191

Chen, H. S., S. T. Huang, and H. R. Chen. 1995. *Bull. Environ. Contamin. Toxic.* 55:709–715.

Kumar, A. K., Ch. Janardhana, and S. Sateesh, unpublished work, Department of Chemistry, Sri Satya Sai Institute of Higher Learning, PN, A. P., India.

Maraippan, P., V. Yegnaraman, and T. Vasudevan. 2000. *Poll. Res.* 19(2): 165–177.

Mariappan, P. 1996. *J. IWWA.* XXVIII(3): 184.

Mohn, W. W., and J. M. Tiedje. 1992. *Microbiol. Revs.* 56:482–507.

Nawlakhe, W. G., and A. V. Jagannadha Roa. 1990. *J. IWWA.* XX(2): 287–291.

New, A. P., L. M. Freitas dos Santos, G. lo Biundo, and A. Spicq. 2000. *J. Chromat. (A).* 889: 177–184

Twigg, L. E., and L. V. Socha. 2001. *Soil Biology & Biochemistry* 33:227–234.

Urban, G., and W. Dekant. 1994. *Xenobiotica.* 24:881–892.

Vargas, C., B. Song, M. Camps, and M. M. Haggblom. 2000. *Appl. Microbiol. Biotechnol.* 53:342.

CHAPTER 8

Biodegradation of Pesticides

Introduction

Pesticides are a group of chemicals used for the control and prevention of pests such as fungi, insects, nematodes, weeds, bacteria, and viruses. Depending on the class of pests they act against, they are broadly classified as:

Fungicides	Kill fungi. Classes include dithiocarbamates, copper, mercurials, etc.
Herbicides	Kill weeds and other unwanted plants. Classes include carbamates, triazines, phenylureas, phenoxyacetic acids, etc.
Insecticides	Kill insects. Classes include organophosphates, carbamates, organochlorines, pyrethrins, pyrethroids, etc.
Nematocides	Kill nematodes.
Rodenticides	Kill rodents like mice.
Algicides	Control algae infestations in water channels and swimming pools.
Antifoulings	Control pests that affect underwater surfaces, like boats.
Biocides	Kill microorganisms.
Defoliants	Cause leaves or other foliage to drop from plants.
Desiccants	Cause drying of living tissues.
Plant growth regulators	Alter growth, blooming, or reproductive rates of plants.
Miticides or acaricides	Kill mites.

Pesticides can also be broadly classified on the basis of their significant chemical properties and reported behavior in soils and water as ionic and nonionic. The following list gives types of ionic pesticides and a few examples.

- Acidic—Dicamba, Ioxynil, 2,4,5-T, Dichlorprop, Mecoprop.
- Basic—Propazine, Cyanazine, Atrazine, Simazine
- Cationic—Diquat, Paraquat, Chlormequat
- Others—Isocil, Bromacil, cacodylic acid, MSMA

The categories of nonionic pesticides and some examples follow.

- Acetamides—CDAA
- Benzonitriles -Dichlobenil
- Esters—the methyl ester of Chloramben
- Thicarbamates—Diallate, Nabam, Metham
- Dinitro anilides—Nitralin, Isopropalin, Oryzalin
- Carbanilates—Swep, Barban, Prophan
- Chlorinated hydrocarbons—DDT, Heptachlor, Endrin, Methoxychlor
- Organophosphate—Ethion, Methyl parathion, Dementon
- Anilides—Diphenamide, Solan, Propanil
- Carbothioates—Molinate
- Ureas—Diuron, Buturon, Norea, Siduron
- Methyl carbamates—Zectran, Carbaryl, Terbutol

Pesticides are highly soluble in water and hence cannot be easily extracted. In addition they bind very strongly to soil. Several methods have been employed to degrade pesticides and to reduce their toxic nature. The main problem arises when the pesticides, which may be non-toxic, get degraded to toxic products. The methods used for degradation of pesticides are:

- Chemical treatment—A commonly used method is alkaline hydrolysis, where the pesticide is neutralized with an aqueous alkaline solution.
- Photodegradation—Process by which pesticides are broken down by the action of light, particularly sunlight.
- Electrochemistry—Effective degradation of chlorobenzoic and chlorophenoxy herbicides has been reported using electrochemical cells at a pH of 3.0.
- Incineration—This method is costly and requires long-distance transport to a central facility, which may not be approved by the general public.
- Bioremediation—This method uses microorganisms for degradation. Major problems encountered in the use of bioremediation are compound specificity, slow rates of degradation, incomplete metabolism, biofilm maintenance, and the survivability of engineered strains in the presence of natural populations.

Insecticides

Insecticides are among the most widely used compounds, finding application in agriculture and disease control. They may be broadly classified as follows:

Organochlorines—They contain organic carbon, hydrogen, and chlorine. They include compounds such as:

- Diphenyl aliphatics—Compounds like DDT (dichlorodiphenyl trichloro ethane) and hexchlorocyclohexane. They act primarily by blocking synaptic transmission in the nervous system of insects and are the most successful insecticides ever produced.
- Cyclodienes—Compounds like Aldrin, Deldrin, and Heptachlor. Used in the soil to control termites.

Organophosphates—All insecticides containing phosphorus. They are the most toxic of all pesticides to vertebrates; however, they are unstable or nonpersistent. They contain compounds like Malathion, Ethyl Parathion and Diazinon (Fig. 8-1).
Organosulphurs—Tetradifon.
Carbamates
Formamidines
Dinitrophenols
Organotins and several others.

Although pesticides like DDT may not be directly harmful to humans, even if the person is in close contact with the chemical, a phenomenon known as biomagnification or bioconcentration occurs, which is a very serious effect and affects organisms higher up in the food chain. The concept of biomagnification is shown in Fig. 8-2. Bioconcentration values of 58 to 5,100 were observed for Chlorpyrifos in fish. There were similar findings for Diazinon, where the numbers vary from 17.5 to 200 in fish. For carbofuran, it is ranges from 10 in snails to over 100 in fish. The World Health Organization estimates 3 million cases of acute severe poisonings, with 220,000 deaths, because of organochloro insecticides annually around the world.

DDT

This compound comes under the class of compounds known as diphenyl aliphatics. A gram-positive bacterium could degrade DDT, DDD (dichlorodiphenyl dichloro ethane), and DDE (dichlorodiphenyl ethane) in the presence of biphenyl. A consortium of the microorganisms (primarily *Serratia marcescens*) degraded 25 ppm of DDT in 144 h under aerobic conditions. DDT degradation was extensive in the presence of white rot fungus, *Phanerochaete chrysosporium,* which is a basidomycete.

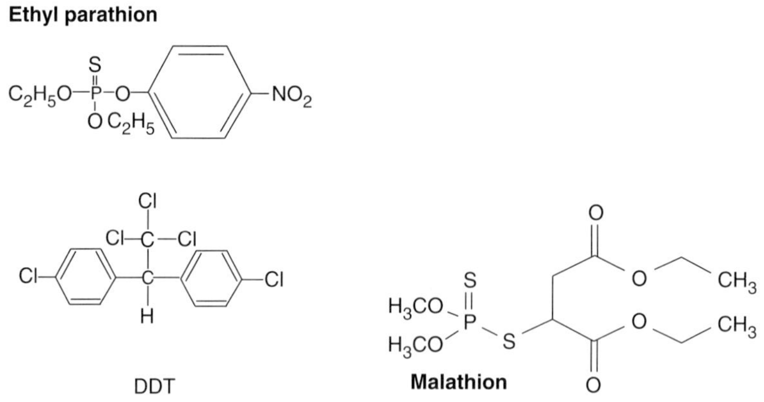

FIGURE 8-1. Structures of various insecticides.

TABLE 8-1
Strains and Reactor Systems Used for DDT Degradation

Strain	*Reactor system*
Various mutant strains of white rot fungi	Nylon web and polyurethane inserted into bioreactor
P. chrysosprium pellets	Pilot scale stirred tank reactor
P. chrysosprium BKM-F 1767 or ME 446	Air-lift
	Stirred tank
	Nylon sheet inserted into reactor
Various mutant strains of white rot fungi	Polyurethane inserted into stationary reactor
	Polyurethane inserted into agitated reactor
P. chrysosprium BKM-F 1767 and SC26	Rotating biological contactors
P. chrysosprium BKM-F 1767	Stirred tank reactor
	Hollow fiber reactor
	Silicon membrane reactor
P. chrysosprium ME 46, Inonotus dryophilus, and *Trametes versicolor*	Rotating tube bioreactors
P. chrysosprium I 1512	Nylon net inserted into reactor

Reactor Systems Different reactor systems have been employed under varying conditions, and they are summarized in Table 8-1. One of the most effective ways to treat DDT using the basidiomycete is the use of a packed bed reactor using wood chips as a carrier for the biomass. Optimum degradation occurred at 30°C and a pH of 4.5 when kept for 30 days at low glucose levels (0.1%) and without any nitrogen source. Intermediates isolated in the

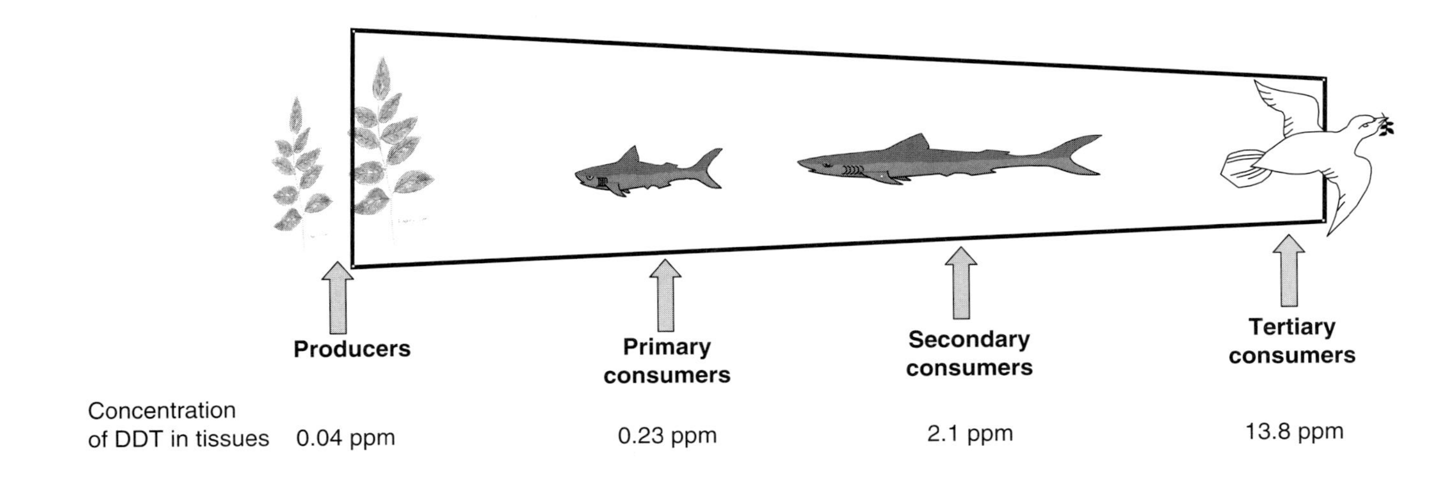

FIGURE 8-2. Biomagnification in nature.

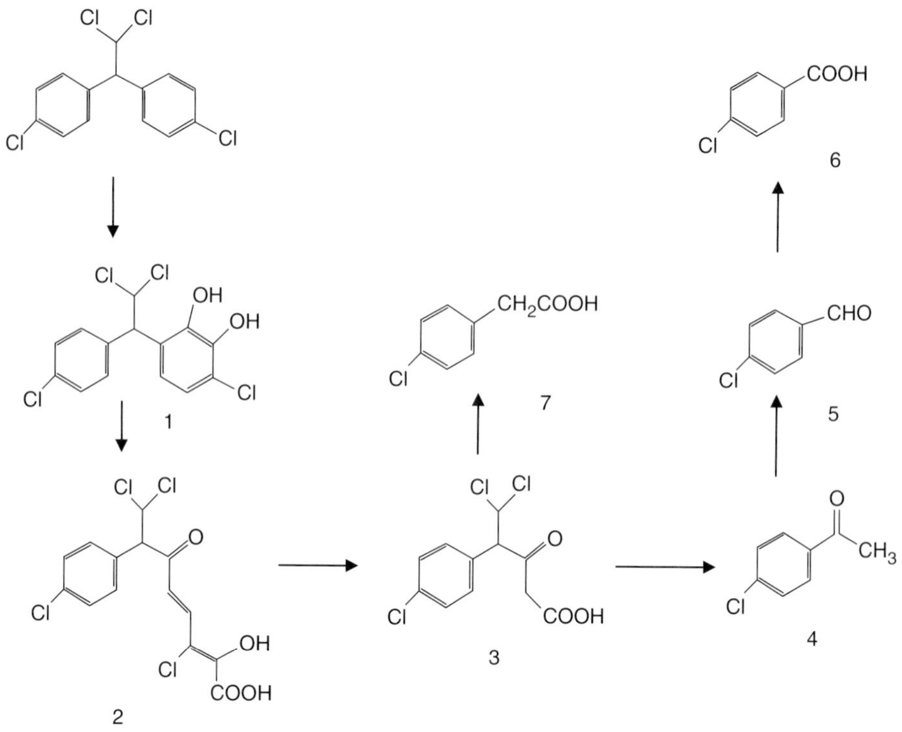

FIGURE 8-3. Degradation of DDD by dioxygenase.

pathway were DDD, 2,2,2-trichloro-1,1-bis(4-chlorophenyl)ethanol (dicofol), and 2,2-dichloro-1,1-bis(4-chlorophenyl)ethanol (FW-152). DDD, the first metabolite to be observed, was degraded in 3 days. The difocol formed was also mineralized. During the degradation of DDT, water-soluble polar compounds were observed.

Most microorganisms transform DDT to DDD under reducing conditions; however, further breakdown of DDD is generally not observed. It is a recalcitrant chlorinated hydrocarbon. *Ralstonia eutropha* is able to aerobically transform DDD by oxidative attack of the aromatic ring. The proposed pathway for the attack of DDD is an attack by a dioxygenase (Fig. 8-3), which inserts a molecular oxygen at the 2,3 positions forming a dihydrodiol. The further breakdown of the dihydrodiol is shown in the pathway, which yields para chloro benzoic acid as its final breakdown product.

A *Pseudomonas aeruginosa* strain degrades mono- and 1,4-dichlorobenzene but can only partially degrade 1,2,4-trichlorobenzene. The degradation ability of microorganisms is dependent on their long-term adaptation to contaminated habitats. Degradation of 1,2,4-TCB by an indigenous microbial population was very low (1% in 23 days), whereas in the soil from the contaminated site, the mineralization occurred very fast (62% within

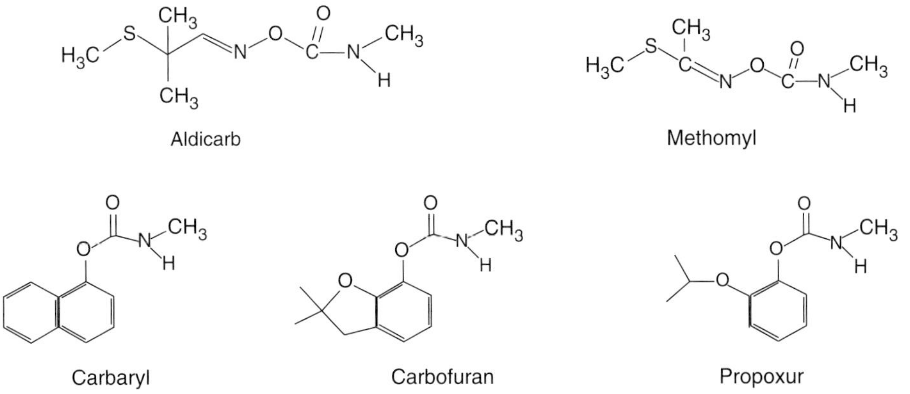

FIGURE 8-4. Structures of various carbamates.

23 days) (Schrolla et al., 2004). *Bacillus circulans* and *B. brevis* bacteria were found to degrade the four isomers of hexachlorocyclohexane insecticide under aerobic conditions. A bacterium, *Pseudomonas* spp., has also been reported to break down this insecticide (Gupta et al., 2000).

Phytoremediation has been used to treat chlorinated aromatics, including pesticides such as atrazine and DDT in contaminated soil and surface and groundwaters. In batch experiments, waterweed was found to degrade *p,p*-DDT and *o,p*-DDT to the corresponding isomers of DDD (Roper et al., 1996).

Carbamates

Carbamates are nerve poisons and act on a broad spectrum of insects. They are not as persistent as the organochlorines, but nevertheless their removal and degradation are essential. The main carbamates are Carbaryl, Carbofuran, and Propoxur (Fig. 8-4).

Carbofuran A *Pseudomonas* sp. was able to transform the highly toxic pesticide carbofuran (2,3-dihydro-2,2-dimethylbenzofuran-7-yl methylcarbamate) to 7-phenol (2,3-dihydro-2,2-dimethyl-7-hydroxy benzofuran) and several unknown metabolites. Among the metabolites assayed was 4-hydroxy carbofuran. It was reported that there was an oxidative transformation of carbofuran to 4-hydroxy carbofuran and a hydrolytic conversion by *Pseudomonas* sp. to 7-phenol, which was then broken down by other soil microorganisms (Fig. 8-5). Properly induced cells were able to degrade carbofuran to 7-phenol within 4 h, and to 4-hydroxycarbofuran in 8 h. Oxidative transformation of carbofuran is undertaken by many species of fungi, as well as *Spingomonas* sp. An *Achromobacter* strain transformed carbofuran by hydrolysis, and *Rhodococcus* transformed it to 5-hydroxycarbofuran.

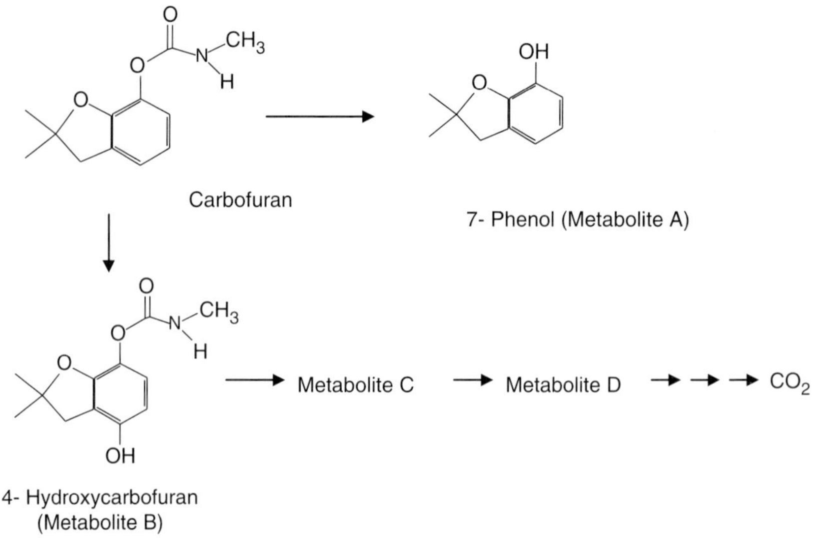

Carbofuran

7- Phenol (Metabolite A)

Metabolite C → Metabolite D → → → CO_2

4- Hydroxycarbofuran (Metabolite B)

FIGURE 8-5. Biodegradation of carbofuran.

Aldcarb Aldcarb is highly soluble in groundwater. When a *Methylosinus* sp. microorganism was immobilized on carboxymethyl cellulose microcarrier beads crosslinked with aluminum ions, it was found to be very effective in degrading this pesticide. Complete degradation was observed in 4 days when the reactions were carried out in a packed bed reactor with recycle.

Carbaryl It has been observed that a micrococcus species is able to utilize carbaryl as its sole source of carbon. The organism degraded carbaryl by hydrolysis to yield 1-naphthol and methylamine. 1-Naphthol was further metabolized via salicylate by a gentisate pathway. The organism also utilized carbofuran, naphthalene, 1-naphthol, and several other aromatic compounds as growth substrates. Carbamates are susceptible to hydrolysis by carboxylesterases (CbEs). The products of hydrolysis are an alcohol and carbamic acid, which instantaneously decomposes to carbon dioxide and methylamine. The *trans* isomers of the pyrethroids are degraded by CbEs faster than the *cis* isomers.

Organophosphates

Organophosphates include all insecticides containing phosphorus. They are the most toxic of all pesticides to vertebrates; however they are unstable or nonpersistent. They contain compounds like malathion, ethyl parathion, and diazinon. In 1989 almost 40% of the $6.2 billion global insecticide market was composed of organophosphates.

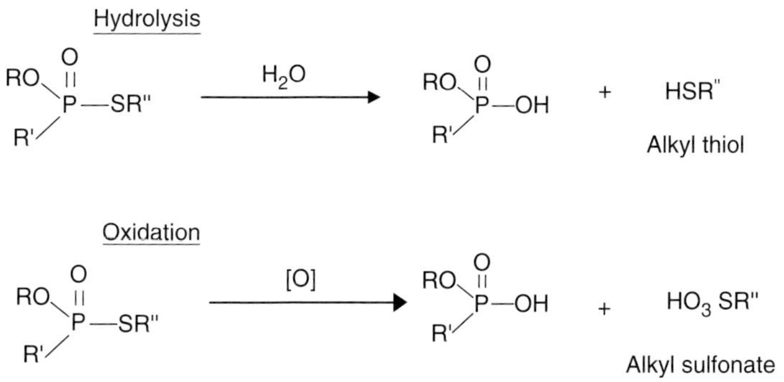

VX : R=C_2H_5, R' = CH_3 , R" = $CH_2CH_2N(iPr)_2$

RVX : R = iso-C_4H_9, R' = CH_3, R" = $CH_2CH_2N(Et)_2$

DiPr Amiton : R=C_2H_5, R' = OC_2H_5, R" = CH_2 $CH_2N(iPr)_2$

FIGURE 8-6. Organophosphate degradation.

Acetylcholine esterase is able to degrade malathion, methylparathion, and diazinon. During hydrolysis, the aromatic ring is used as a carbon source and the alkyl moiety (dithiomethyl phosphorothioate) is used as a source of phosphorus. Enzymatic degradation of organophosphates occurs either by the action of organophosphate acid hydrolases (OPH) or by organophosphate acid anhydrolases (OPAA). The OPH enzymes isolated from the organisms *Pseudomonas diminuta, Pseudomonas* sp., and *Flavobacterium* sp. ATCC 27551 showed high activity. The thermophilic bacteria *Altermonas* as well as the fungus *Pleurotus ostreatus* exhibit high activity of OPAA.

During the hydrolysis reaction, an alkyl thiol is liberated, whereas during oxidation, an alkyl sulfonate is produced (Fig. 8-6). The thiol product has an undesirable smell as well as a reasonable amount of toxicity; hence the oxidation reaction is preferred. The fungus *Pleurotus ostreatus* is able to degrade 75% of the pesticide in 16-20 h.

Fungicides

Fungicides are substances used to kill fungi. They can be of biological or chemical origin, and can be broadly classified into two major types:

- Preventive fungicides—These are substances that prevent fungal infections from occurring in a plant. They include compounds such as sulfur, dichlorocarbamates, organometallics, pthalimides, and benzimides.

- Curative fungicides—These are substances that move to the place where the infection has occurred and prevent further development of the pathogen. They include compounds such as acetimides, dicarboxymides, sterol inhibitors, and many others.

Pentachlorophenol (PCP) is one of the most commonly used fungicides. It acts as both a preventive and a curative fungicide. Many white rot fungi, including *Phanerochaete chrysosporium*, are effective in breaking down PCP as well as other compounds like DDT and phenanthrene. *Trametes versicolor* is another fungus that degrades PCP when it is in aerobic mode in a continuous fluidized bed; This fungus was also effective in batch reactors when the biomass was immobilized on foam cubes.

The fungicide mefenoxam was effectively degraded in 21 days (78%) by a *rhizosphere* system containing *Zinnia angustifolia* (Tropic Snow) in a bark and sand potting mix, whereas only 44% of the fungicide was degraded in the absence of the plant. Pure cultures of *Pseudomonas flurescens* and *Chyrsobacterium indologenes* isolated from the *rhizosphere* system could degrade the fungicide within 54 h (Pai et al., 2001).

Herbicides

Herbicides are a group of pesticides specifically designed to kill weeds. They have a high degree of toxicity and a long half-life, and they remain unaffected during treatment by regular wastewater treatment plants.

Pseudomonas sp. was able to completely degrade the Mecoprop (phenoxyalkyl carbonic acid) herbicide, but unable to degrade Isoproturon (phenylurea), Terbuthylazine (*s*-triazine), and Metamitron (triazine herbicide). Mecoprop biodegradation was not observed in the methanogenic (anaerobic) sulfate-reducing or iron-reducing microcosms. In the nitrate-reducing microcosm (*S*)-mecoprop did not degrade, but (*R*)-mecoprop degraded. In aerobic conditions (*S*)- and (*R*)-mecoprop degraded (Harrison et al., 2003). One hundred percent biodegradation of mecoprop was observed in activated sludge plants, but isoproturon, terbuthylazine, and metamitron herbicides did not biodegrade under the same conditions (Nitschke, 1999). Mecoprop is highly biodegradable in laboratory activated-sludge plants but requires long adaptation times (lag-phase).

Gramaxone and Matancha were degraded by *Pseudomonas putida* immobilized onto a calcium alginate gel in a batch reactor. Addition of activated carbon to the slurry increased the extent of degradation (from 48 to 95%).

Phenylurea herbicides are used for pre- or postemergence in cotton, fruit, or cereal production. *Sphingomonas* sp. strain SRS2 bacterium is found to mineralize the phenyl structure. Fungi that could degrade this herbicide include *Cunninghamella elegans, Mortierella isabellina, Talaromyces wortmanii, Rhizopus japonicus, Rhizoctonia solani,* and *Aspergillus niger.* (Sorensen et al., 2003). Two *A. globiformis* strains (D47 and N2) and one *B. sphaericus* strain (ATCC 12123) isolated from soils are capable of carrying out direct hydrolysis of a broad range of phenylurea herbicides and their aniline derivatives. *Bjerkandera adusta* and *Oxysporus sp* were able to degrade ~85% of chlortoluron diuron and isoproturon in 2 weeks (Khadrani et al., 1999).

Considerable variation was observed among the white rot fungi in their ability to degrade pesticides like Metalaxyl (phenylamide fungicide), Terbuthylazine (Triazine herbicide), Atrazine (Triazine herbicide), and Diuron (phenylurea herbicide). The fungus *Hypholoma fasciculare* was able to degrade 95% of Terbuthylazine in 42 days in a biofilm bed, whereas the other herbicides were only partially degraded. *Coriolus versicolor* was able to degrade more than 99% of diuron, while 80% of Atrazine was degraded during the same period and only 65% degradation was observed for the other two herbicides (Bending et al., 2002).

References

Bending, G. D., M. Friloux, and A. Walker. 2002. Degradation of contrasting pesticides by white rot fungi and its relationship with ligninolytic potential, *FEMS Microbiol. Lett.* 212:59–63.

Gupta, A., C. P. Kaushik, and A. Kaushik. 2000. Degradation of hexachlorocyclohexane (HCH; a, b, g and d) by *Bacillus circulans* and *Bacillus brevis* isolated from soil contaminated with HCH, *Soil Biol. Biochem.* 32:1803–1805.

Harrison, I., G. M. Williams, C. A. Carlick. 2003. Enantioselective biodegradation of mecoprop in aerobic and anaerobic microcosms. *Chemosphere* 53:539–549.

Khadrani, A., F. Seigle-Murandi, R. Steiman, and T. Vroumsia. 1999. Degradation of three phenylurea herbicides (chlortoluron, isoproturon and diuron) by micromycetes isolated from soil. *Chemosphere* 38(13):3041–3050.

Nitschke, L., A. Walk, W. Schossler, G. Metzner, and G. Lind, 1999. Biodegradation in laboratory activated sludge plants and aquatic toxicity of herbicides. *Chemosphere* 39(13):2313–2323.

Pai, S. G., M. B. Riley, and N. D. Camper. 2001. Microbial degradation of mefenoxam in rhizosphere of *Zinnia angustifolia*. *Chemosphere* 44:577–582.

Roper, J.C., J. Dec, and J. Bollag. 1996. Using minced horseradish roots for the treatment of polluted waters. *J. Environ. Qual.* 25:1242–1247.

Schrolla, R., F. Brahushia, U. Dorera, S. Kuhna, J. Feketeb, and J. C. Muncha. 2004. Biomineralisation of 1,2,4-trichlorobenzene in soils by an adapted microbial population. *Environ. Pollut.* 127:395–401.

Sorensen, S. R., G. D. Bending, C. S. Jacobsen, A. Walker, and J. Aamand. 2003. Microbial degradation of isoproturon and related phenylurea herbicides in and below agricultural fields. *FEMS Microbiol. Ecol.* 45:1–11.

Bibliography

Amitai, G., R. Adani, G. Sod-moriah, I. Rabinovitz, A. Vincze, H. Leader, B. Chefetz, L. Leibovitz-Persky, D. Friesem, and Y. Hadar. 1998. Oxidative Biodegradation of phosphorothiolates by fungal laccase. *FEBS Lett.* 438(3):195–200.

Bumpus, J.A., and S.D. Aust. 1987. Biodegradation of DDT [1,1,1-trichloro-2,2-bis(4-chlorophenyl)ethane] by the white rot fungus phanerochaete chrysosporium, *Appl. Environ. Microbiol.* 53(9):2001–2008.

Chaudhry, G. R., A. Mateen, B. Kaskar, M. Sardessai, M. Bloda, A. R. Bhatti, and S. K. Walia. 2002. Induction of carbofuran oxidation to 4-hydroxycarbofuran by *Pseudomonas* sp. 50432, *FEMS Microbiol Lett.* 214(2):171–176.

Hee-Sung, P., S.-J. Lim, Y. K. Chang, A. G. Livingston, and H. S. Kim. 1999. Degradation of chloronitrobenzenes by coculture of *Pseudomonas putida* and a *Rhodococcus* sp., *Appl. Environ. Microbiol.* 65(3):1083–1091.

Kopytko, M., G. Chalela, and F. Zauscher. 2002. Biodegradation of two commercial herbicides (Gramoxone and Matancha) by the bacteria *Pseudomonas putida. Electron. J. Biotech.* 5(2):August 15.

Pallerla, S., and R. P. Chambers. 1998. Reactor development for biodegradation of pentachlorophenol. *Catal. Today* 40(1):103–111.

Shimazu, M., A. Mulchandani, and W. Chen, 2001. Simultaneous degradation of organophosphorus pesticides and *p*-nitrophenol by a genetically engineered moraxella sp. with surface-expressed organophosphorus hydrolase. *Biotechnol Bioeng.* 76:318–324.

Sogorb, M. A., and E. Vilanova. 2002. Enzymes involved in the detoxification of organophosphorus, carbamate and pyrethroid insecticides through hydrolysis. *Toxicol. Lett.* 128:215–228.

CHAPTER 9

Degradation of Polymers

Introduction

Approximately 140 million tonnes of synthetic polymers are produced worldwide each year. Since polymers are extremely stable, their degradation cycles in the biosphere are limited. In Western Europe it is estimated that 7.4% of municipal solid waste is plastic; these plastics are classified as 65% polyethylene (PE)/polypropylene (PP), 15% polystyrene (PS), 10% polyvinyl chloride (PVC), 5% polyester terephthalate (PET), and miscellaneous others. Environmental pollution by synthetic polymers, such as waste plastics and water-soluble synthetic polymers in wastewater, has been recognized as a major problem. Degradation of polymers can be carried out by heat, radiation, or biochemical treatment. The radiant energy may be high-energy radiation from gamma rays, ion beams, and electrons or even low-energy radiation from ultraviolet (UV) light. UV stabilizers added to polymer products reduce the rate of degradation. Chemical degradation results from treatment with chemicals such as acids and alkalis. Biodegradation of polymers results from the use of microorganisms and enzymes.

Biodegradation

The biodegradability of a compound depends on its molecular weight, molecular form, and crystallinity. Biodegradability decreases with increase in molecular weight, while monomers, dimers, and repeating units degrade easily. Two types of depolymerases are involved in the process, namely, extracellular and intracellular. Microbial exoenzymes first break down the complex polymers in a process called depolymerization. The resulting short chains are small enough to permeate the cell walls, allowing them to be used as carbon and energy sources. When the end products are carbon dioxide, water, or methane, the process is called mineralization. Different end products are formed depends on the degradation pathway (Fig. 9-1).

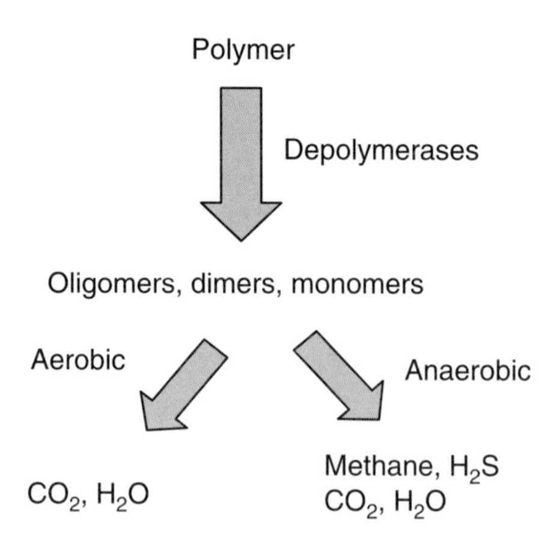

FIGURE 9-1. Reaction pathways during polymer biodegradation.

Polyethers (PE)

Polyethylene glycols (PEGs), polypropylene glycols (PPGs), and polytetramethylene glycol come under the class of polyethers and are used in pharmaceuticals, cosmetics, lubricants, inks, and surfactants. *Flavobacterium* sp. and *Pseudomonas* sp. together associate and mineralize PEG completely under aerobic conditions. During degradation, PEG molecules are reduced one glycol unit at a time after each oxidation cycle. *Pelobacter venetianus* was found to degrade PEG and ethylene glycol under anaerobic conditions (Kawai, 1987). High molecular weight PEGs (4,000 to 20,000) were degraded by *Sphingomonas macrogoltabidus* and *S. terrae,* while PPG was degraded by *Corynebacterium* sp.

Polyesters

Polyesters are polymers in which the component monomers are bonded via ester linkages. Many kinds of esters occur in nature, and the esterases that degrade them are ubiquitous in living organisms. Ester linkages are generally easy to hydrolyze, and hence a number of synthetic polyesters are biodegradable; bacterial polyesters (polyhydroxyalkanoates) have been used to make biodegradable plastics. Hydrolytic cleavage of the ester bond in low molecular weight polyesters by the lipase of *Pseudomonas* sp. has been reported.

Polyhydroxyalkanoates (PHA)

Polyhydroxybutyrate (PHB) is a naturally occurring polyester that accumulates in bacterial cells as a carbon and energy storage compound. PHB and copolymers containing hydroxyalkanoate PHA (e.g., 3-hydroxyvalerate) are being used for the manufacture of biodegradable plastics. Several PHA and PHB bacterial *depolymerases* are found to be capable of metabolizing PHB and other polyhydroxyalkanoate (PHA) polymers. The PHA *depolymerases* are serine hydrolases, usually having a single substrate-binding domain. Recently a PHB *depolymerase* with a two substrate-binding domain was reported. PHB depolymerases are able to degrade all-(*R*) chains, cyclic-(*R*) oligomers, oligolides, and racemic hydroxybutanoate polymers. The enzymes are generally obtained from microorganisms like *Alcaligenes faecalis* and *Pseudomonas stutzeri* (Shimao, 2001).

Atactic P(*R*, *S*-3HB) (atactic poly(*R*, *S*-3-hydroxybutyrate), which does not biodegrade in pure form, can undergo enzymatic hydrolysis in a P(*R*, *S*-3HB)/PMMA (polymethacrylate) blend, indicating that the enzymatic degradation can be induced by blending with an amorphous nonbiodegradable polymer. This is possibly because the blend gives P(*R*-3HB) *depolymerase* a more stable binding surface than that provided by the rubbery a-P(*R*,*S*-3HB). The *depolymerase* was purified from *Alcaligenes faecalis* (He et al., 2001). In order to modify their physical properties and retard enzymatic degradation of commercial microbial polyesters like PHA, they are blended with other degradable or nondegradable polymers such as PVA, PMMA, poly(ethylene oxide), PLA, cellulose, PCL, and polystyrene (PS).

Polylcaprolactone (PCL)

Polylcaprolactone (PCL) is a synthetic polyester that can be degraded by microorganisms and enzymes such as lipases and esterases. *Cutinases*, which are obtained from fungal phytopathogens, degrade cutin (the structural polymer of the plant cuticle) and act as PCL depolymerases. The biodegradability of polycaprolactone in the form of blend sheets (e.g., in polycarbonate-polycaprolactone blend sheets) is much reduced because the packed form of PCL in the blend sheets protects it from enzymatic digestion (Hirotsu et al., 2000). However, enzymatic degradation can be promoted by using oxygen plasma treatments to etch the surface. *Pencillium* spp. is known to utilize polyethylene adipate and polycaprolactone as its sole carbon and energy source, respectively.

Poly-L-Lactide Poly-L-lactide (PLLA) is a lactic acid–based aliphatic polyester that is used in medical and packaging applications. It can be degraded both aerobically and anaerobically. Several enzymes, including *proteinase* K, *pronase,* and bromelain, can degrade the polymer. Under thermophylic conditions, degradation with bromelain is faster than the others, probably because lactic acid is more favorable for anaerobic microorganisms than for aerobic organisms (Itavaara et al., 2002). PLLA is also found to degrade completely in 2 weeks in windrow composting.

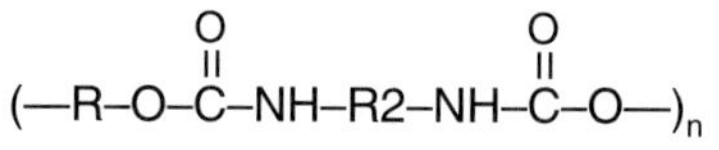

$$(\text{—R–O–}\overset{O}{\overset{\|}{C}}\text{–NH–R2–NH–}\overset{O}{\overset{\|}{C}}\text{–O—})_n$$

FIGURE 9-2. Structure of PUR.

Polylactic Acid (PLA) Polylactic acid (PLA) is absorbed easily in animals and humans, and hence has been extensively used in medicines. The degradation of the polymer in animals and humans is thought to proceed via nonenzymatic hydrolysis. Several enzymes, including *proteinase* K, *pronase,* and bromelain, can degrade the polymer (Shimao, 2001). PLA is also readily degraded in compost to CO_2 (about 90% degradation was achieved in 90 days). A PLA-degrading actinomycete strain reduced 100 mg of PLA film by 60% in the first 14 days in liquid culture at 303K. *Bacillus brevis* is also found to degrade 50 mg of PCL by around 20% in 20 days in liquid culture at 333K.

Poly(*p*-dioxanone) Poly(*p*-dioxanone) (PPDO) is known as a poly(ether-ester) and has good tensile strength and flexibility. It is used for bioabsorbable sutures in clinical applications. PPDO is degraded by strains that belong to the α and β subdivision of the class *Proteobacteria* and the class *Actinobacteria.* Degradation leads to the formation of monomeric acids (Nishida et al., 2000).

Polyurethane (PUR) Polyurethane (PUR) produced by the diisocyanate polyaddition process is the characteristic chain link of the urethane bond (Fig. 9-2). PUR degradation proceeds in a selective manner, with the amorphous regions being degraded before the crystalline regions. PUR synthesized from polyester polyol is termed "polyester PUR," and that synthesized from polyether polyol is termed "polyether PUR." Although most PUR used at present is polyether PUR, polyester PUR has recently become the focus of attention because of its biodegradability; therefore, it has advantages from the viewpoint of waste treatment. The PUR *depolymerases* of microorganisms have not been examined in detail, although because of the presence of the ester linkage, most degradation is carried out by *esterases. Comamonas acidovorans* TB-35 utilizes a polyester PUR that contains polydiethyleneglycol adipate as the sole source of carbon but not polyether PUR.

Phua et al (1987) found that two proteolytic enzymes, *papain* and *urease,* degraded medical polyester PUR. Bacteria like *Corynebacterium* sp. and *Pseudomonas aeruginosa* could degrade PUR in the presence of basal media (Howard, 2002). Several fungi are observed to grow on PUR surfaces, especially *Curvularia senegalensis,* which was observed to have a higher

TABLE 9-1
Polyurethane (PUR) Degrading Microorganisms

Microorganisms	*PUR degraded*
Fungi	
Aspergillus niger	PS, PE
A. flavus	PS, PE
A. fumigatus	PE
A. versicolor	PS, PE
Aureobasidium pullulans	PS, PE
Chaetomium globosum	PS, PE
Cladosporium sp	PS
Curvularia senegalensis	PS
Fusarium solani	PS
Gliocladium roseum	PS
Penicillium citrinum	PS
P. funiculosum	PS, PE
Trichoderma sp.	PS, PE
Bacteria	
Comamonas acidovorans	PS
Corynebacterium sp.	PS
Enterobacter agglomerans	PS
Serratia rubidaea	PS
Pseudomonas aeruginosa	PS
Staphylococcus epidermidis	PE

PE, polyether PUR; PS, polyester PUR

PU-degrading activity. Although cross-linking was considered to inhibit degradation, *papain* was found to diffuse through the film and break the structural integrity by hydrolyzing the urethane and urea linkage, producing free amine and hydroxyl group. *Porcine pancreatic elastase* degraded polyester PUR 10 times faster than its activity against polyether PUR. Table 9-1 lists the various microorganisms that degrade PU.

Polyvinyl alcohol (PVA) Polyvinyl alcohol (PVA) is a vinyl polymer joined by only carbon–carbon linkages. The linkage is the same as those of typical plastics such as polyethylene, polypropylene, and polystyrene, and of water-soluble polymers such as polyacrylamide and polyacrylic acid. Among the vinyl polymers produced industrially, PVA is the only one known to be mineralized by microorganisms. PVA is water soluble and biodegradable; hence it is used to make water-soluble and biodegradable carriers, which may be useful in the manufacture of delivery systems for chemicals such as fertilizers, pesticides, and herbicides.

PVA is completely degraded and utilized by a bacterial strain, *Pseudomonas* O-3, as a sole source of carbon and energy. However, PVA-degrading microorganisms are not ubiquitous within the environment. Almost all the degrading strains belong to the genus *Pseudomonas*, although some do belong to other genera (Chielliniet et al., 2003). Among the PVA-degrading bacteria reported so far, a few strains showed no requirement for pyrroloquinoline quinone (PQQ). From a PVA-utilizing mixed culture, *Pseudomonas* sp. VM15C and *P. putida* VM15A were isolated. Their symbiosis is based on a syntrophic interaction. VM15C is a PVA-degrading strain that degrades and metabolizes PVA, while VM15A excretes a growth factor that VM15C requires for PVA utilization.

Nylon High molecular weight nylon 66 membrane was degraded significantly by lignin-degrading white rot fungi grown under ligninolytic conditions with limited glucose or ammonium tartrate (Deguchi et al., 1997). The characteristics of a nylon-degrading enzyme purified from a culture supernatant of white rot fungal strain IZU-154 were identical to those of *manganese peroxidase*, but the reaction mechanism for nylon degradation differed significantly from manganese peroxidase. The enzyme could also degrade nylon-6 fibers. The nylon was degraded to soluble oligomers by drastic and regular erosion.

A thermophilic strain capable of degrading nylon 12 was isolated from 100 soil samples by enrichment culture technique at 60°C. At this temperature, the strain not only grew on nylon 12 but also reduced the molecular weight of the polymer. The strain was identified as a neighboring species to *Bacillus pallidus*. This strain had an optimum growth temperature of around 60°C. It was also found to degrade nylon 6 as well as nylon 12 but not nylon 66.

Polyvinyl Chloride Polyvinyl chloride (PVC) has become a universal polymer with many applications (e.g., for pipes, floor coverings, cable insulation, roofing sheets, packaging foils, bottles, and medical products) because of its low cost and physical, chemical, and weathering properties. PVC degrades at relatively low temperatures (~100°C) in the presence of light to release hydrogen chloride. Hence, to prevent degradation during processing, heat stabilizers are added, part of which are consumed during the processing. Degradation can also be achieved by exposure to molecular oxygen in the presence of alkali at higher temperature. The hydrogen chloride can be used for monomer production. Under regular anaerobic landfill conditions at 50°C, no changes were observed in the PVC, indicating that the polymer matrix is stable and no biodegradation has occurred.

Polyethylene (PE) Polyethylenes of low density are used widely as films in the packaging industry. They pose a serious problem because of their slow rate of degradation under natural conditions. They pose problems to the environment, freshwater, and animals. Extracellular *Streptomyces* sp. cultures were found to degrade starch-blended PE. *Phanerochaete chrysosporium* was also found to degrade starch-blended LDPE in soil (Orhan and Buyukgungor, 2000).

High molecular weight polyethylene is also degraded by lignin-degrading fungi under nitrogen-limited or carbon-limited conditions and by manganese peroxidase. Fungi such as *Mucor rouxii* NRRL 1835 and *Aspergillius flavus* and several strains of *Streptomyces* are capable of degrading polyethylene containing 6% starch. Degradation was monitored by observing changes in mechanical properties such as tensile strength and elongation (El-Shafei et al., 1998). The biodegradability of blends of LDPE and rice or potato starch was enhanced when the starch content exceeded 10% (w/w). No microorganism or bacteria has been found so far that could degrade PE that has no additives.

Polycarbonate (PC) Bisphenol-A polycarbonate (PC) is widely used because of its excellent physical properties such as transparency, high tensile strength, impact resistance, rigidity, and water resistance. Polycarbonate gets its name from the carbonate groups in its backbone chain. At 300°C in air, a 25% reduction in the molecular weight of PC was observed (Montaudo et. al., 2002).

PC is stable to bioorganism attack. PC sheets are known to degrade in lipase AK, but when they are blended with PC they become less biodegradable (Hirotsu et al., 2000). Because PC is hydrophobic, it probably suppresses biodegradation. Several authors have described enzymatic degradation of aliphatic polycarbonate (polyethylene carbonate, PEC). No degradation of PEC with a molecular weight of 300 to 450 kDa in hydrolytic enzymes (including lipase, esterase, lysozyme, chymotrypsin, trypsin, papin, pepsin, collagenase, pronase, and pronase E) was observed. This indicates that hydrolytic mechanisms based on hydrolases or aqueous conditions can be excluded for biodegradation of PEC (Dadsetan et al., 2003).

Polyimide Polyimides find application in the electronic and packaging industries. These polymers possess high strength and resistance to degradation. Fungi such as *Aspergillus versicolor, Cladosporium cladosporioides,* and *Chaetomium* sp. were found to degrade this polymer. Bacteria like *Acinetobacter johnsonii, Agrobacterium radiobacter, Alcaligenes denitricans, Comamonas acidovorans, Pseudomonas* sp., and *Vibrio anguillarum,* when tested, were not effective in biodegrading this polymer.

Fiber-Reinforced Polymeric Composite Fiber-reinforced polymeric composite materials (FRPCMs) are materials important in the aerospace and aviation industries. A fungal mixture consisting of *Aspergillus versicolor, Cladosporium cladosorioides,* and *Chaetomium* sp. and a mixed culture of bacteria including a sulfate-reducing bacterium were found to grow on this composite material. Only the fungi mixture could cause deterioration detectable over more than 350 days (Gu, 2003).

Polyacrylamide (PAA) Polyacrylamides are water-soluble synthetic linear polymers made of acrylamide or the combination of acrylamide and acrylic acid. Polyacrylamide finds applications in pulp and paper production, agriculture, food processing, mining, and as a flocculant in wastewater treatment. Polyacrylamide undergoes thermal degradation at 175 to 300°C (Smith et al., 1996) and can also undergo photodegradation.

Acrylamide is readily biodegraded under aerobic conditions by microorganisms in soil and water by deamination to acrylic acid and ammonia, which are utilized as carbon and nitrogen sources. *Pseudomonas stutzeri, Rhodococcus* spp., *Xanthomonas* spp., and mixed cultures have demonstrated degrading abilities under aerobic conditions in numerous studies (Haveroen, 2002).

Polyamide (PA) Polyamide-6 (PA-6) is a widely used engineering material. Oxidative degradation of PA-6 membranes was found using lignolytic white rot fungus IZU-154. *Aspergillus niger*–mediated degradation of polyamides based on tartaric acid and hexamethylenediamine and *Corynebacterium aurantiacum*–mediated degradation of ε-caprolactam, as well as its oligomers, have been reported (Marqué et al., 2000). Lignolytic fungus *Phanerochaete chrysosporium* is also found to degrade PA-6 (Klun et al., 2003). Degradation of the polymer was observed through a decrease in the average molecular mass (50% after 3 months), as well as in the physical damage to the fibers visible under a scanning electron microscope.

Rubber Biological attack of natural rubber latex is quite facile, but addition of sulfur and numerous other ingredients reduces biological attack. Straube et al. (1994) devulcanized scrap rubber by holding the comminuted scrap rubber in a bacterial suspension of chemolithotropic microorganisms with a supply of air until elemental sulfur or sulfuric acid was separated. This process can reclaim rubber and sulfur in a simplified manner. The biodegradation of the *cis*-1,4-polyisoprene chain was achieved by bacterium belonging to the genus *Nacardia* and led to considerable weight loss of different soft type NR-vulcanizates. Old tires with 1.6% sulfur were treated with different species of *Thiobacillus ferrooxidans, T. thiooxidans,* and *T. thioparus* in shake flasks and in a laboratory reactor. The best results were obtained with *T. thioparus* μ—4.7% of the total sulfur of the rubber powder was oxidized to sulfate within 40 days.

Conclusions

Since most of the polymers are resistant to degradation, research over the past couple of decades has focused on developing biodegradable polymers that are degraded and ultimately catabolized to carbon dioxide and water by bacteria and fungi under natural conditions. During the degradation process, they should not generate any substances that are harmful. These polymers can be classified into three major categories: (1) polyesters produced by microorganisms, (2) natural polysaccharides and other biopolymers like starch, and (3) synthetic polymers like aliphatic polymers (e.g., poly ε-caprolactone, poly L-lactide and poly butylenesuccinate, which are commercially produced).

Another approach toward achieving biodegradability has been through the addition of biodegradable groups into the main chain during the production of industrial polymers prepared by free radical copolymerization. Two such approaches are the use of ethylene bis(mercaptoacetate) as a chain transfer agent during the copolymerization of styrene and MMA, and the preparation of copolymers of vinylic monomers with cyclic comonomers containing the biodegradable functions such as ketene acetal and cyclic disulfides.

References

Chiellini, E., A. Corti, S. D'Antone, and R. Solaro. 2003. Biodegradation of poly (vinyl alcohol) based materials. *Prog. Polym. Sci.* 28:963–1014.

Dadsetan, M., E. M. Christenson, F. Unger, M. Ausborn, T. Kissel, A. Hiltner, and J.M. Anderson. 2003. In vivo biocompatibility and biodegradation of poly(ethylene carbonate), *J. Controlled Release* 93:259–270.

Deguchi, T., M. Kakezawa, and T. Nishida. 1997. Nylon biodegradation by lignin-degrading fungi. *Appl. Environ. Microbiol.* 63:329.

El-Shafei, H. A., N. H. Abd El-Nasser, A. L.Kansoh, and A. M. Ali. 1998. Biodegradation of disposable polyethylene by fungi and *Streptomyces* species. *Polym. Degrad. Stab.* 62:361–365.

Gu, J. D. 2003. Microbiological deterioration and degradation of synthetic polymeric materials: recent research advances. *Int. Biodeterior. Biodegrad.* 52:69–91.

He, Y., X. Shuai, A. Cao, K. Kasuya, Y. Doi, and Y. Inoue. 2001. Enzymatic biodegradation of synthetic atactic poly(R, S-3-hydroxybutyrate) enhanced by an amorphous nonbiodegradable polymer. *Polym. Degrad. Stab.* 3:193–199.

Hirotsu, T., A. A. J. Ketelaars, and K. Nakayama. 2000. Biodegradation of poly (e caprolactone)-polycarbonate blend sheets. *Polym. Degrad. Stab.* 68:311–316.

Howard, G. T. 2002. Biodegradation of polyurethane. *Int. Biodeterior. Biodegrad.* 49(4):213–216.

Itavaara, M., S. Karjomaa, and J.-F. Selin. 2002. Biodegradation of polylactide in aerobic and anaerobic thermophilic conditions. *Chemosphere* 46:879–885.

Kawai, F. 1987. The biochemistry of degradation of polyethers. *Crit. Rev. Biotechnol.* 6:273–307.

Klun, U., J. Friedrich, and A. Krzan. 2003. Polyamide-6 fibre degradation by a lignolytic fungus. *Polym. Degrad. Stab.* 79:99–104.

Marqués, M. S., C. Regano, J. Nyugen, L. Aidanpa, and S. Munoz-Guerra. 2000. Hydrolytic and fungal degradation of polyamides derived from tartaric acid and hexamethylenediamine. *Polymer* 41:2765–2768.

Montaudo, G., S. Carroccio, C. Puglisi. 2002. Mechanism of thermal oxidation of poly (bisphenol A carbonate). 2nd International conference on polymer modification, degradation, and stabilization. 30 June–4 July. Budapest, Hungary.

Nishida, H., M. Konno, A. Ikeda, and Y. Tokiwa. 2000. Microbial degradation of poly(p-dioxanone). I. Isolation of degrading microorganisms and microbial decomposition in pure culture, *Polym. Degrad. Stab.* 68:205–217.

Orhan, Y., and H. Buyukgungor. 2000. Enhancement of biodegradability of disposable polyethylene in controlled biological soil. *Int. Biodeterior. Biodegrad.* 45:49–55.

Phua, S. K., E. Castillo, J. M. Anderson, and A. Hiltner. 1987. Biodegradation of a polyurethane in vitro. *J. Biomed. Materials Res.* 21:231–246.

Shimao, M. 2001. Biodegradation of plastics. *Curr. Opin. Biotechnol.* 12(6):14–22.

Smith E. A., S. L. Prues, and F. W. Oehme. 1996. Environmental degradation of polyacrylamides, effects of artificial ecotoxicology and environmental safety. 35:121–135.

Straube G., E. Straube, W. Neumann, H. Ruckauf, R. Forkmann, M. Loffler, and H. Neustadt. 1994. Method for reprocessing scrap rubber. *Biotech. Adv.* 12(3):580–589.

Bibliography

Adhikari, B., D. De, and S. Maiti. 2000. Reclamation and recovery of waste rubber. *Prog. Polym. Sci.* 25(7).

Bockhorn, H., J. Hentschel, A. Hornung, and U. Hornung. 1999. Stepwise pyrolysis of plastic waste. *Chem. Eng. Sci.* 54:15–16.

Cacciari, I., P. Quatrini, G. Zirletta, E. Mincione, V. Vinciguerra, P. Lupattelli, and G. G. Sermanni. 1993. Isotactic polypropylene biodegradation by a microbial community: physicochemical characterization of metabolites produced. *Appl. Environ. Microbiol.* 59:3695–3700.

Fiddy, G. S., Y.-H. Lin, A. Ghanbari-Siakhali, P. N. Sharratt, and J. Dwyer. 1998. Catalytic degradation of high-density polyethylene. *Thermochim. Acta* 294(1):313–316.

Karlsson, S., O. Ljungquist, and A.-C. Albertsson. 1988. Biodegradation of polyethylene and the influence of surfactants. *Polym. Degrad. Stab.* 21:237–250.

Kawai, F., and H. Yamanaka. 1986. Biodegradation of polyethylene glycol by symbiotic mixed culture (obligate mutulism). *Arch. Microbiol.* 146:125–129.

Nawaz, M. S., S. M. Billedeau, C. E. Cerniglia. 1998. Influence of selected physical parameters on the biodegradation of acrylamide by immobilized cells of *Rhodococcus* sp. *Biodegradation* 9:381–387.

Okudaira, K. K., S. Hasegawa, P. T. Sprunger, E. Morikawa, V. Saile, K. Seki, Y. Harada, and N. Ueno. 1998. Photoemission study of pristine and photodegraded poly(methyl methacrylate). *J Appl. Phys.* 83(8):4292–4298.

Ramani, R., and C. Ranganathaiah. 2000. Degradation of acrylonitrile-butadiene-styrene and polycarbonate by UV irradiation. *Polym. Degrad. Stab.* 69:347–354.

Rangarajan, P. 1998. Conversion of waste plastics into transportation fuels. *Fuel and Energy Abs.* 38(2).

Rivaton, A., B. Mailhot, J. Soulestin, H. Varghese, J. L. Gardette. 2001. Comparison of the photochemical and thermal degradation of bisphenol-A polycarbonate and trimethylcyclohexane-polycarbonate, *Polym. Degrad. Stab.* 75(1):17-33.

Schink, E., P. H. Janssen, and J. Frings. 2002. Microbial degradation of natural and of new synthetic polymers, *FEMS Microbio. Lett.* 103(2–4):November.

Shama, G., and D. A. J. Wase. 1981. The biodegradation of caprolactam and some related compounds: a review. *Int. Biodeterior. Bull.* 17:1–16.

Troev, K., V. I. Atanasov, R. Tsev, G. Grancharov, and A. Tsekova. 1999. Chemical degradation of PUR. *Polym. Degrad. Stab.* 67(1): 14–22.

CHAPTER 10

Degradation of Dyes

Dyestuffs can be classified according to their origin, chemical and/or physical properties, or characteristics related to the application process. Another categorization is based on the applications sector (e.g., inks, disperse dyes, pigments, or vat dyes). A systematic classification of dyes according to chemical structure is the color index, namely, nitroso, nitro, monoazo, disazo, trisazo, polyazo, azoic, stilbene, carotenoid, diphenylmethane, triarylmethane, xanthene, acridine, quinoline, methine, thiazole, indamine/indophenol, azine, oxazine, thiazine, sulfur, lactone, aminoketone, hydroxyketone, anthraquinone, indigoid, phthalocyanine, natural, oxidation base, and inorganic. Synthetic dyes are also classified according to their most predominant chemical structures, namely, polyene and polymethine, diarylmethine, triarylmethine, nitro and nitroso, anthraquinone, and diazo (Fig. 10-1). Approximately 10,000 different dyes and pigments are manufactured worldwide with a total annual market of more than 7×10^5 tonnes per year. There are several structural varieties of dyes, such as acidic, reactive, basic, disperse, azo, diazo, anthraquinone-based, and metal-complex dyes. They all absorb light in the visible region. Untreated dye effluent is highly colored and hence reduces sunlight penetration, preventing photosynthesis. Many dyes are toxic to fish and mammalian life, inhibit growth of microorganisms, and affect flora and fauna. They are also carcinogenic in nature and hence can cause intestinal cancer and cerebral abnormalities in fetuses.

The physical and chemical methods for the treatment of dye-containing effluent includes physicochemical flocculation combined with flotation, electroflotation, flocculation with $Fe(II)/Ca(OH)_2$, membrane filtration, electrokinetic coagulation, electrochemical destruction, ion-exchange, irradiation, photochemical precipitation, oxidation, ozonation, adsorption with activated carbon, and the Katox treatment method, which involves the use of activated carbon and air mixtures. The chemical color removal process leads to 60 to 70% reduction in the color, while the decrease in biological oxygen demand (BOD) is only about 30 to 40% (Cooper, 1993; Nowak, 1992; Zamora et al., 1999).

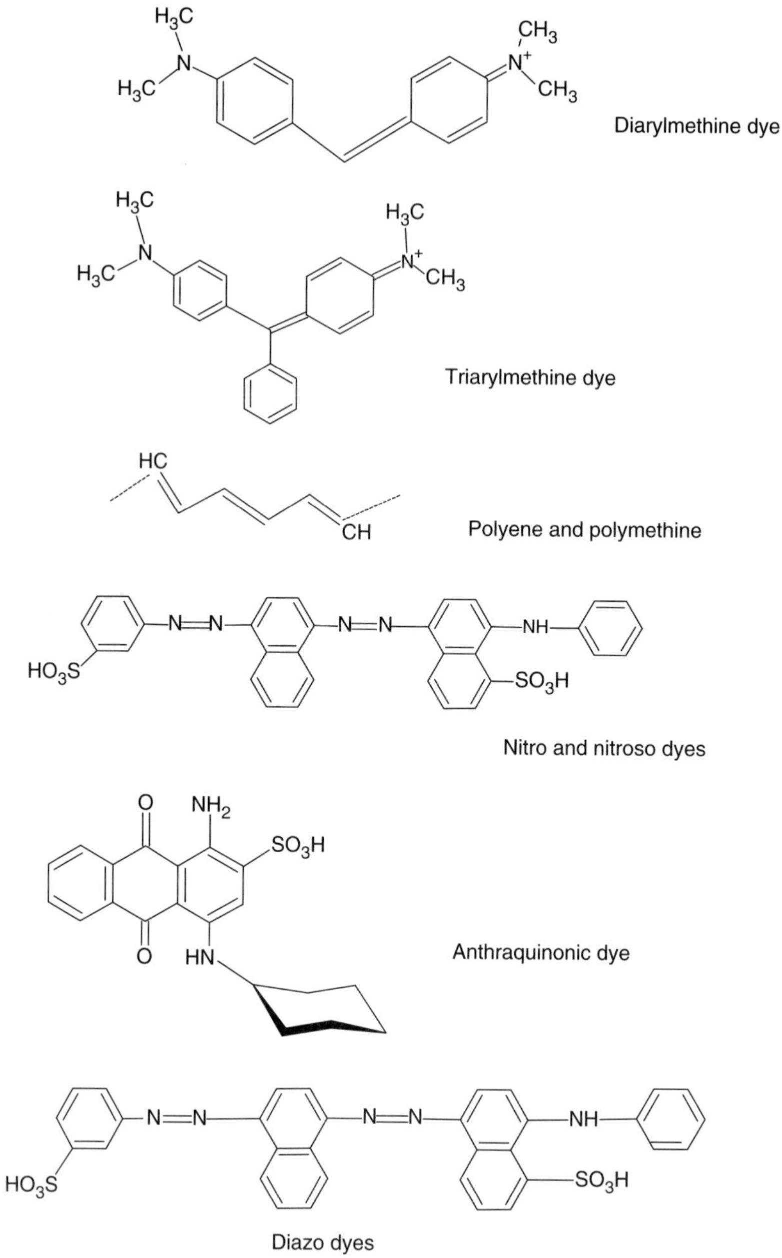

Diarylmethine dye

Triarylmethine dye

Polyene and polymethine

Nitro and nitroso dyes

Anthraquinonic dye

Diazo dyes

FIGURE 10-1. Structure of dyes based on predominant groups.

Textile Dyes

Textile industries consume two thirds of the dyes manufactured. The requirement for reactive dyes is high since cotton fabric with brilliant colors has a high demand. The reactive dyes bind to the cotton fibers by addition or substitution mechanisms under alkaline conditions and high temperature. Also, a significant fraction of the dye is hydrolyzed and released. Colored wastewater is a consequence of batch processes both in the dye-manufacturing and the dye-consuming industries. Two percent of dyes that are produced are discharged directly in the effluent, and a further 10% is lost during the textile coloration process. Generally the wastewater contains dye concentrations around 10 to 200 mg/L, as well as other organic and inorganic chemicals used in the dyeing process. The wastewater discharged from a dyeing process in the textile industry is highly colored and has low BOD and high chemical oxygen demand (COD) (because of the presence of grease, dirt, and/or sizing agents, as well as nutrients from dye bath additives). Alkali or acids from the bleaching, desizing, scouring, and mercerizing steps also end up in the effluent, resulting in extreme pH and high salt content (Chapter 11 deals with textile effluent).

Conventional biological processes have also been resorted to for the treatment of textile wastewater. This includes adsorption of dyestuff on activated sludge (Hu and Ko, 1992), decolorization of reactive azo dyes by transformation using *Pseudomanas luteola* (Hu, 1994), and biosorption of cationic dyes by dead macrofungus *Fomitopsis carnea* (Mittal and Gupta, 1996). Activated sludge has also been used as biomass in the adsorption of dyestuff, achieving about 90% of BOD, 40 to 50% of COD reduction, and 10 to 30% of color removal (Pagga and Taeger, 1994; Hitz et al., 1978).

Aerobic biological treatment alone generally cannot effectively decolorize wastewaters containing water-soluble dyes; hence a chemical treatment is a necessary primary stage. Effluent collected from a textile mill was chemically treated with sodium bisulfite and sodium borohydride as the catalyst and reduction agent, respectively, followed by aerobic biological oxidation leading to an 80% reduction in color, 98% reduction in BOD, 80% reduction in COD, and 95% reduction in TSS (Ghoreishi and Haghighi, 2003).

Reactive azo dyes, which are used for dyeing cellulose, produce the colored wastewater (Fig. 10-2). These dyes make up ~30% of the total dye market. Because of their stability and xenobiotic nature, reactive azo dyes are not totally degraded by conventional wastewater treatment processes that involve light, chemicals, or activated sludge. Azo dyes are not readily metabolized under aerobic conditions. Under anaerobic conditions, many bacteria reduce the electrophilic azo bond in the dye molecule to colorless amines. Although these amines are resistant to further anaerobic mineralization, they are good substrates for aerobic degradation through a hydroxylation pathway involving a ring-opening mechanism. Hence a combined anaerobic

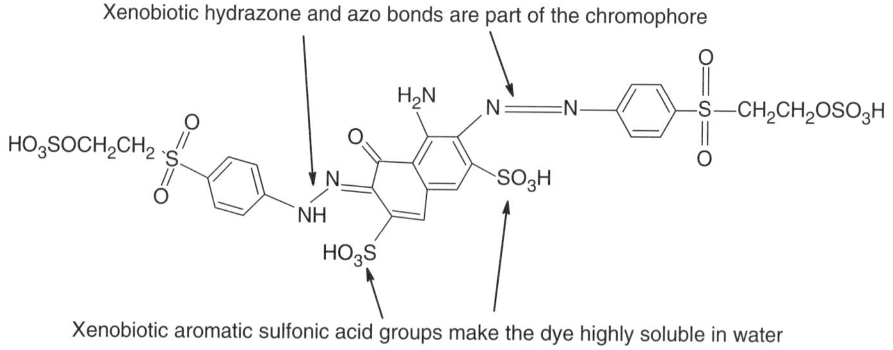

FIGURE 10-2. A typical reactive dye structure with its chromophore, containing azo/keto-hydrazone groups, the reactive centers and its solubilizing components (Remazol Black B—a reactive azo dye).

treatment followed by an aerobic one could be very effective. Microbial species, including bacteria, fungi, and algae, can remove the color of azo dye via biotransformation, biodegradation, or mineralization. Decolorization of azo dyes by bacteria is carried out by azoreductase-catalyzed reduction or by cleavage of azo bonds under anaerobic environment.

Pearce et al. (2003) have listed various literature examples dealing with mixed cultures and single bacterial strains that have been found to degrade these dyes effectively. A few examples of mixed culture include *Bacillus cereus, Sphaerotilus natans, Arthrobacter* sp., or activated sludge under anoxic conditions for reduction of azo dyes; *Alcaligenes faecalis* and *Commomonas acidovorans* for decolorization of reactive dyes, diazo dyes, azo dyes, disperse dyes, and phthalocyanine dyes under anaerobic conditions; *Alcaligenes faecalis and Commomonas acidovorans* have been used for the degradation of Remazol Black B. In addition, aerobic bacterial sludge and aerobic activated sludge have been used for degrading various azo, diazo, and reactive dyes.

A few examples of single bacterial strains include *Proteus vulgaris* under anaerobic conditions for treating azo food dyes; *Pseudomonas pseudomallei* for treating acid, direct, and basic dyes; immobilized *Pseudomonas* sp. for dyes having anthraquinone and metal-complex structures; *Streptococcus faeclis* for treating Red 2Gazo dye; *Pseudomonas luteola* for treating reactive azo dyes, direct azo dyes, and leather dyes, *Paenibacillus azoreducens* sp. nov. for treating Remazol Black B, and *Shewanella putrefaciens* for treating Remazol Black B and anthraquinone dyes.

Two mechanisms for the decoloration of azo dyes under anaerobic conditions in bacterial systems have been proposed (Keck et al., 1997; Pearce et al., 2003). The first one consists of direct electron transfer to azo dyes as

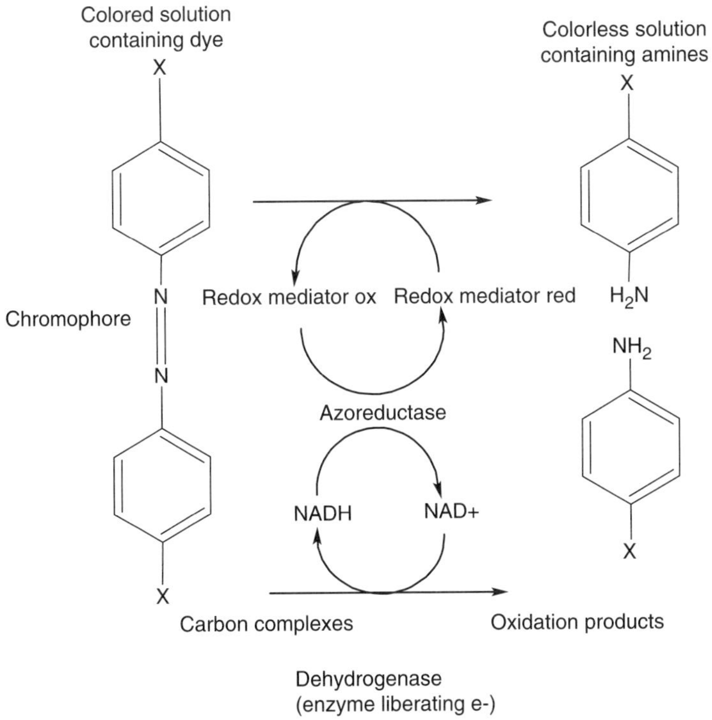

FIGURE 10-3. A proposed redox reaction for the degradation of azo dyes with whole bacterial cells.

terminal electron acceptors via enzymes during bacterial catabolism, connected to ATP-generation (energy conservation). The second one involves a free reduction of azo dyes by the end products of bacterial catabolism, not linked to ATP generation (e.g., reduction of the azo bond by reduced inorganic compounds, such as Fe^{2+} or H_2S, that are formed as the end product of certain anaerobic bacterial metabolic reactions). Figure 10-3 shows a possible pathway for the degradation of azo dyes under anaerobic conditions with whole bacterial cells.

During anaerobic degradation, a reduction of the azo bond in the dye molecule is observed. Then, aerobic conditions are required for the complete mineralization of the reactive azo dye molecule. The aromatic compounds produced by the initial reduction are degraded via hydroxylation and ring opening in the presence of oxygen. So for effective wastewater treatment, a two-stage process is necessary in which oxygen is introduced after the initial anaerobic reduction of the azo bond has taken place. The optimum pH for color removal is around pH 7 to 7.5. The rate of color removal tends to decrease rapidly under strongly acid or strongly alkaline conditions. The optimum cell culture growth temperature is between 35 and 45°C.

Operating Conditions

The efficiency of color removal depends on several factors, which include level of aeration, temperature, pH, and redox potential. The composition of textile wastewater is varied and can include in addition to the color, organics, nutrients, salts, sulfur compounds, and toxicants. The concentration of dye in the solution affects the rate of biodegradation; possible reasons include toxicity of the dye, toxicity of the metabolites formed during the degradation of the dye molecule, and ability of the enzyme to recognize the dye efficiently at very low concentrations. It has been found that during the decolorization of triphenylmethane dyes and textile dyestuff, effluent by *Kurthia* sp. was facile at low dye concentrations. But when the dye concentration was increased (~30 mM), the rate of color removal was reduced (Sani et al., 1999). If the dye reduction mechanism is nonenzymatic, then the reduction rate will be independent of the dye concentration. This type of behavior has been observed during the reduction of azo food dyes in cultures of *Proteus vulgaris* (Dubin and Wright, 1975).

It has been observed that simple structures and low molecular weight dyes degrade faster than highly substituted, high molecular weight dyes. If the dye reduction happens inside the cell, then the first step is diffusion of the molecule through the cell membrane. The presence of a sulfonate group could hinder this transfer rate, and the rate decrease could be proportional to the number of sulfonate groups. Of course, cultures could be adapted to produce azoreductase enzymes that have very high specificity toward particular dye structures. In addition, hydrogen bonding and electronegativity could affect the reduction rate.

Redox potential is a measure of the ease with which a molecule will accept electrons, which means that the more positive the redox potential, the more readily a molecule is reduced. The rate-controlling step in the dye reduction reaction involves a redox equilibrium between the dye and the extracellular reducing agent (see Fig. 10-3). The color removal process thus depends on the redox potential of the electron donors and acceptors. Different electron donors such as glucose, acetate, formate, etc., have different effects on the degradation reaction.

Enzymic reduction of azo groups is normally inhibited by dissolved oxygen; hence it is necessary that bacterial decolorization take place under nearly anaerobic conditions. Cell immobilization through entrapment with natural or synthetic materials is an ideal technique, which can create a local anaerobic environment favorable to oxygen-sensitive decolorization. Cell immobilization also enhances the stability, mechanical strength, and reusability of the biocatalyst.

Free and supported *Pseudomonas luteola* was able to reduce azo groups of C.I. Reactive Red 22 enzymatically (Chang et al., 2001). Immobilized cells exhibited lower activity because of mass transfer effects and were also less sensitive to dissolved oxygen levels and pH as compared with free suspended cells. The decolorization activity in all the cases increased as the temperature increased from 20 to 45°C (Mechsner and Wuhrmann, 1982).

Laccase immobilized on various supports decolorized textile reactive dyes (Dias et. al., 2004). The initial decolorization observed is due to the adsorption of the dye to the support, and the later decolorization is due to the enzymatic reaction. When the system is preirradiated, the reaction time is faster, probably because the small molecular fragments that are formed during the irradiation process are more compatible with the subsequent enzymatic process (Zamora et al., 2003). Use of enzyme for decolorization has several advantages over the use of fungi or bacteria. They include the absence of a lag phase, generation of a low amount of sludge, ease of controlling the process, and ability to operate the reactor at low or high contaminant concentration.

Fungi capable of decolorization include *Aspergillus sojae* B-10, *Myrothecum verrucaria*, *Myrothecum* sp., *Neurospora crassa*, and *Candida* sp. (Banat et al., 1996). A fungal strain ATCC 74414 isolated from a plant anise, *Pimpinella anisum*, aerobically decolorized two polymeric dyes, namely, Poly R-478 and Poly S-119 in liquid media; the process involved two steps: adsorption of the dye compound by fungal mycelia followed by biodegradation through microbial metabolism (Zheng et al., 1999).

Bacterial cultures capable of dye decolorization include *Aeromonas hydrophila* var 24B, *Pseudomonas luteola, P. cepacia,* and *Streptomycetes* BWI30 (Banat et al., 1996). Algal cultures *Chlorella* and *Oscillatoria* were able to degrade dyes to aromatic amines and subsequently to simpler compounds. *Geotrichum candidum* Dec 1 exhibits aerobic dye-decolorizing ability for 21 kinds of azo and anthraquinone dyes. It requires an external carbon source.

Reactors

White rot fungi have been used for the decomposition of several recalcitrant dyes in different reactor configurations, including fixed-film bioreactors (for the decolorization of dispersed dyes), packed bed reactors, rotating biological contactors, and pulsed flow reactors. Generally the operations were carried out either in batch or semibatch mode, although a few studies have reported using continuous mode. When carried out in rotating biological contactors, the degradation efficiency for decolorization of dispersed dyes was found to depend on the biofilm thickness, rotational speed, and carbon source concentration. Pulsed flow systems introduce oxygen in pulses. The white-rot fungus, *Pycnoporus cinnabarinus*, was found to decolorize high concentrations of dyes in a packed-bed reactor.

Selvam et al. (2003) have carried out treatment of dye industry effluent in batch and continuous modes using mycelia of *Thelephora* sp. Interestingly, they observed degradation rates that were higher in the batch mode (61% color removed in 3 days) as against continuous mode, where there was 50% color removal. Biodegradation of a simulated cotton textile effluent containing azo and diazo dyes was attempted in an anaerobic-aerobic sequencing batch reactor (SBR). A 24-h cycle with 10 h aeration time and 14 h of anaerobic time achieved 90% color removal. Fu et al. (2001) achieved 66% biodegradation of reactive dye in a similar reactor. When the same reaction was carried out in a two-reactor system, where one was anaerobic and the other aerobic, the degradation efficiency was 88%. Acid red 151 was aerobically biodegraded with an average efficiency of 88% in a sequencing batch biofilter using porous volcanic rock as packing (Buitron et al., 2004). The majority of the dye was transformed to CO_2. It was also found that 14 to 16% of the biotransformation was due to the anaerobic environments inside the porous support material.

Anaerobic degradation of black, red, and blue reactive dyes showed different results in a two-stage upward aerobic sludge blanket (UASB) system consisting of an acidification tank and in a reactor with and without the addition of an external carbon source such as tapioca starch (Chinwetkitvanich et al., 2000). Tapioca had no effect in the case of black dye, since the decolorization efficiency remained at ~ 70%. Degradation efficiency increased from 36 to 56% in the case of red dye and from 48 to 56% in the case of blue dye on the addition of tapioca. In these studies there was no correlation between the color removed and the amount of methane formed, indicating that methane-forming bacteria were not the only anaerobic microorganisms responsible for color removal. But Carliell et al. (1996) and Razo-Flores et al. (1997) suggested that during the methane production step of anaerobic decolorization, when the methanogenic bacteria used the azo bonds in the chromophores of the dye as electron acceptors, the azo bond was broken, resulting in the decolorization.

Anaerobic degradation on the order of 30 to 35% in COD was observed for six textile print dyes of various classes (azo, anthraquinone, cyanine, etc.) in an up-flow filter with milk whey as cosubstrate. It is thought that methanogenesis is inhibited by chemicals in the thickener, including surfactants and chelating agents, and by the high ammonia concentration in the filter due to hydrolysis of the urea present in the thickener. Eighty percent decolorization was observed in 24 hours and complete degradation in 4 days when textile dyes (Remazole Navy Blue and Red, Remazol Blue, Turquoise Blue, Black, Golden Yellow) were treated in a submerged anaerobic biofilm reactor with *Alcaligenes faecalis* and *Comomonas acidovorans* strains (Banat, 1996).

A fixed-bed reactor coupled with a pneumatic pulsation system has been found to increase the mass transfer rate and to enhance productivity for yeast and fungi systems. A similar design was found to be very effective for treating anthraquinone type (Poly R-478), azo type (Orange II), and phtalocyanine type (Reactive Blue 98) dyes. The decolorization efficiencies were on the order of 98% for several months at a dye loading of 0.2 g dye/m^3/day. Ninety-five percent continuous decolorization of Orange II dye using manganese peroxidase (MnP) in a continuous stirred tank reactor coupled with an external membrane unit as a filter was observed by López et al. (2002). The MnP, dye, and hydrogen peroxide were added continuously.

White Rot Fungi

White rot fungi are a heterogeneous group of organisms that have the capability of degrading lignin, several wood components, and many recalcitrant compounds. The enzymes are extracellular (limitations caused by substrate diffusion into the cell, generally encountered in bacteria, are not observed here), nonspecific (they can degrade a wide variety of recalcitrant compounds and even complex mixtures of pollutants), can tolerate a high concentration of pollutants, and are nonstereoselective. They also do not require any preconditioning since enzyme secretion depends on nutrient limitation, either nitrogen or carbon, and not on the presence of pollutant. Manganese peroxidases, lignin peroxidases, and laccases are the three lignin-modifying enzymes present that help to degrade lignin and various xenobiotic compounds including dyes. The main disadvantages are the low pH requirement for optimum activity of the enzymes, the complexity of the biodegradation mechanism of the ligninolytic system, and a requirement for some chemicals unlikely to be present in the wastewater.

Several white rot fungi studied for color removal include *Bjerkandera adusta* for degrading reactive Orange, Violet, Black and Blue; *Irpex lacteus* for degrading Methyl Red, Congo Red, and Naphtol Blue; *Phanerochaete chrysosporium* for degrading Remazol Turquoise Blue, azo dyes, Azure Blue, and Cresol Red; *Phlebia radiata* for degrading orange II and reactive blue; *Pleurotus ostreatus for* degrading Remazol Brilliant Blue and Poly R-478; and *Pycnoporus sanguineus* for degrading Orange G, Amaranth, Bromophenol Blue, and Malachite Green and several more. *Phanerochaete chrysosporium* has been found to degrade sulfonated azo dyes, heterocyclic, polymeric, anthraquinone, triphenylmethane, and azo dyes. The mechanism of color removal involves a lignin peroxidase and Mn-dependent peroxidase or laccase enzymes. Decolorization studies carried out by Selvam et al. (2003) of azo dyes Orange G, Congo Red, and Amido Black by a white rot fungus *Thelephora* sp. showed that the fungus was able to completely degrade (98%) Amido Black 10B in 24 h and Congo Red (> 97%) in 8 h. Only 33.3% of Orange G degraded in 9 days.

An activated sludge reactor containing white-rot fungus *Coriolus versicolor* could degrade 82% of a textile dye Everzol Turquoise Blue G (Kapdan and Kargi, 2002). Yang and Yu (1996) achieved 80% degradation of a dispersed dye in a continuous fixed-film bioreactor. Zhang et al. (1999) used a packed-bed reactor for the treatment of an azo dye, Orange II, and reached efficiencies on the order of 90%. The main problem in using white rot fungi for continuous effluent treatment is that they form a thick mycelial mat that can disrupt the reactor operation.

Conclusions

Several chemical and physical methods are available for the removal of color from the textile dye effluent. Decolorization by aerobic bacteria occurs mainly by adsorption of dyestuff on the cell surface rather than by biodegradation; therefore, low color removal efficiencies have been observed. However, anaerobic bacteria provide better COD and total organic carbon (TOC) removal than anaerobic bacteria do. The combination of anaerobic bacteria followed by aerobic bacteria is found to be very effective. Addition of adsorbent also provides several advantages including adsorption of toxic compounds, which reduces toxic effects on the microorganisms, and better sludge settling characteristics. The extracellular ligninolytic enzyme systems of the white-rot fungi *Phanerchaete chrysosporium* and *Coriolus versicolor* can degrade a wide variety of recalcitrant compounds, including xenobiotics, lignin, and dyestuffs. Several different reactors have been tried to achieve color removal, and a large number of bacteria and fungi have been identified as effective in this regard.

References

Banat, I. M., P. Nigam, D. Singh, and R. Marchant. 1996. Microbial decolorization of textile-dye-containing effluents: a review. *Bioresour. Technol.* 58:217–227.

Buitron, G., M. Quezada, and G. Moreno. 2004. Aerobic degradation of the azo dye acid red 151 in a sequencing batch biofilter. *Bioresour. Technol.* 92:143–149.

Carliell, C. M., S. J. Barclay, and C. A. Buckley. 1996. Treatment of exhausted reactive dyebath effluent using anaerobic digestion: laboratory and full-scale trials. *Water SA*. 22(3):225–233.

Chang, J. S., C. Chou, and S.-Y. Chen. 2001. Decolorization of azo dyes with immobilized *Pseudomonas luteola*. *Process Biochem.* 36:757–763.

Chinwetkitvanich, S., M. Tuntoolvest, and T. Panswad. 2000. Aerobic decolorization of reactive dyebath effluents by a two-stage UASB system with tapioca as a co-substrate. *Water Res.* 34(8):2223–2232.

Cooper, P. 1993. Removing color from dye house wastewaters—a critical review of technology available. *J. Soc. Dyers Colorists* 109:97–100.

Dias, A. A., R. M. Bezerra, and A. N. Pereira. 2004. Activity and elution profile of Laccase during biological decolorization and dephenolization of olive mill waste water. *Bioresour. Technol.* 92(1):7–13.

Dubin, P., and K. L. Wright. 1975. Reduction of azo food dyes in cultures of *Proteus vulgaris*. *Xenobiotica* 5(9):563–71.

Fu, L., and Q. L. Y. Qian. 2001. Treatment of dyeing waste water in two SBR systems. *Process Biochem.* 36(11):1111–1118.

Ghoreishi, S. M., and R. Haghighi. 2003. Chemical catalytic reaction and biological oxidation for treatment of non-biodegradable textile effluent. *Chem. Eng. J.* 95:163–169.

Hitz, H. R., W. Huber, and R. H. Reed. 1978. The adsorption of dyes on activated sludge. *J. Soc. Dyers Colorists* 94:71–76.

Hu, T. L. 1994. Decolorization of reactive azo dyes by transformation of *Pseudomanas luteola. Bioresour. Technol.* 49:47–51.

Hu, T. L., and W. L. Ko. 1992. Adsorption of reactive dyes by biomass, in: *Proceedings of the 17th Conference on Wastewater Treatment Technology, China*, pp. 105–116.

Kapdan, K. I., and F. Kargi. 2002. Simultaneous biodegradation and adsorption of textile dyestuff in an activated sludge unit. *Process Biochem.* 37:973–981.

Keck A., J. Klein, M. Kudlich, A. Stolz, H.-J. Knackmuss, and R. Mattes. 1997. Reduction of azo dyes by redox mediators originating in the naphthalenesulfonic acid degradation pathway of *Sphingomonas* sp. strain BN6. *Appl. Environ. Microbiol.* 63(9):3684–90.

López, C., I. Mielgo, M. T. Moreira, G. Feijoo, and J. M. Lema. 2002. Enzymatic membrane reactors for biodegradation of recalcitrant compounds. Application to dye decolourisation. *J. Biotech.* 99:249–257.

Mechsner K., and K. Wuhrmann. 1982. Cell permeability as a rate limiting factor in the microbial reduction of sulfonated azo dyes. *Europ. J. Appl. Microbiol. Biotechnol.* 15:123–126.

Mittal, A. K., and S. K. Gupta. 1996. Biosorption of cationic dyes by dead macro-fungus *Fomitopsis carnea*: batch studies. *Water Sci. Technol.* 34:81–87.

Nowak, K. M. 1992. Color removal by reverse osmosis. *J. Membr. Sci.* 68:307–315.

Pagga, U., and K. Taeger. 1994. Development of a method for adsorption of dyestuff on activated sludge. *Water Res.* 28:1051–1057.

Pearce, C. I., J. R. Lloyd, and J. T. Guthrie. 2003. The removal of colour from textile wastewater using whole bacterial cells: a review. *Dyes and Pigments* 58:179–196.

Razo-Flores, E., M. Luijten, B. Donlon, G. Lettinga, and J. Field. 1997. Biodegradation of selected azo dyes under methanogenic conditions. *Water Sci. Technol.* 36(67):65–72.

Sani, R. K., and U. C. Banerjee. 1999. Decolorization of triphenylmethane dyes and textile and dye-stuff effluent by *Kurthia* sp. *Enzyme Microbial Technol.* 24:433–437.

Selvam, K., K. Swaminathan, and K.-S. Chae. 2003. Decolourization of azo dyes and a dye industry effluent by a white rot fungus *Thelephora* sp. *Bioresour. Technol.* 88:115–119.

Yang, F. C., and J. T. Yu. 1996. Development of a bioreactor system using an immobilised white-rot fungus for decolorization. Part II: continuous decolorization tests. *Bioprocess Eng.* 16:9–11.

Zamora, P. P., A. Kunz, S. Gomes de Moraes, R. Pelegrini, P. de Campos Molelro, J. Reyes, and N. Duran. 1999. Degradation of reactive dyes in a comparative study of ozonation enzymatic and photochemical processes. *Chemosphere* 38(4):835–852.

Zhang, F., J. S. Knapp, and K.N. Tapley. 1999. Development of bioreactor systems for decolourisation of Orange II using white-rot fungus. *Enzyme Microbiol. Technol.* 24:48–53.

Zheng, Z., R. E. Levin, J. L. Pinkham, and K. Shetty. 1999. Decolorization of polymeric dyes by a novel *Penicillium* isolate. *Process Biochem.* 34:31–37.

Bibliography

Banas, J., E. Plaza, W. Styka, and J. Trela. 1999. SBR technology for advanced combined municipal and tannery wastewater treatment with high receiving water standards. *Water Sci. Technol.* 40(4–5):451–458.

Bernet, N., N. Delgenes, J. C. Akunna, J. P. Delgenes, and R. Molleta. 2000. Combined anaerobic-aerobic SBR for the treatment of piggery wastewater. *Water Res.* 34(2):611–619.

Helmreich, B., D. Schreff, and P. A. Wilderer. 2000. Full-scale experiments with small sequencing batch reactor plants in Bavaria. *Water Sci. Technol.* 41(1):89–96.

Kim, S. J., K. Ishikawa, M. Hirai, and M. Shoda. 1995. Characteristics of a newly isolated fungus, *Geotrichum candidurn* Dec 1, which decolorizes various dyes. I. *Ferment. Bioeng.* 79:601–607.

Lourenc, N. D., J. M. Novais, and H. M. Pinheiro. 2001. Effect of some operational parameters on textile dye biodegradation in a sequential batch reactor. *J. Biotechnol.* 89:163–174.
Malpei, F., V. Andreoni, D. Daffonchio, and A. Rozzi. 1999. Anaerobic digestion of print pastes: A preliminary screening of inhibition of dyes and biodegradability of thickeners. *Bioresour. Technol.* 63:49–56.
Mielgo, I., M. T. Moreira, G. Feijoo, and J. M. Lema. 2001. A packed-bed fungal bioreactor for the continuous decolourisation of azo-dyes (Orange II) *J. Biotechnol.* 89:99–106.
Shaul, G. M., T. J. Holdsworth, C. R. Dempsey, and K. A. Dostall. 1991. Fate of water soluble azo dyes in the activated sludge process. *Chemosphere* 22:107–119.
Zamora, P. P., C. M. Pereira, E. R. L. Tiburtius, S. G. Moraes, M. A. Rosa, R. C. Minussi, and N. Durán. 2003. Decolorization of reactive dyes by immobilized laccase. *Appl. Catal., B: Environ.* 42:131–144.

CHAPTER 11

Textile Effluent

Textile industries receive and prepare fibers; transform them into yarn, thread, or webbing; convert the yarn into fabric or related products; and dye and finish these materials to the final product. Textile manufacturing consumes a considerable amount of water. The principal pollutants in the textile effluent are recalcitrant organics, color, toxicants and inhibitory compounds, surfactants, soaps, detergents, chlorinated compounds, and salts. Dye is the most difficult constituent of the textile wastewater to treat. The type of dye in the effluent could vary daily, or even hourly, depending upon the campaign.

The various processes in a textile mill that generate effluent are desizing, scouring, bleaching, dyeing, and printing. Effluent from the desizing operation produces the most chemical oxygen demand (COD) (40%) and effluent from the final rinsing operation the least (7.5%). The suspended solids load is highest in the effluent generated from desizing (47.7%) and the least from the scouring and bleaching operation (7.3%). The volume of effluent produced is highest from the combined scouring and bleaching operation (53.9%) and the least from the final washing (7.5%) (see Figs. 11-1 to 11-3). The typical cotton textile effluent stream will have a pH of 8.9 to 9.9, a biological oxygen demand (BOD) of 90 to 170 mg/L, a COD of 1,018 to 1,062 mg/L, suspended solids of 110 to 180 mg/L, a total kjeldahl Nitrogen (TKN) of 20 to 57 mg/L, a phosphate concentration of 0.9 to 2.4 mg/L, sulfate concentration of 1.9 to 6.4 mg/L, Cl^- of 557 to 610 mg/L, and a color reading (at 669 nm using UV-visible spectrophotometer) of 0.15 to 0.21 (Sen and Demirer, 2003).

Physical Treatment

Chemical, physical, and biological methods have been tried for the treatment of textile effluent containing dyes (Robinson et al., 2001). The physical treatment methods include (1) adsorption with activated carbon, peat, wood chips, fly ash and coal (mixture), silica gel, and other materials such as natural clay, corn, cobs, rice hulls, etc.; (2) membrane filtration; (3) ion

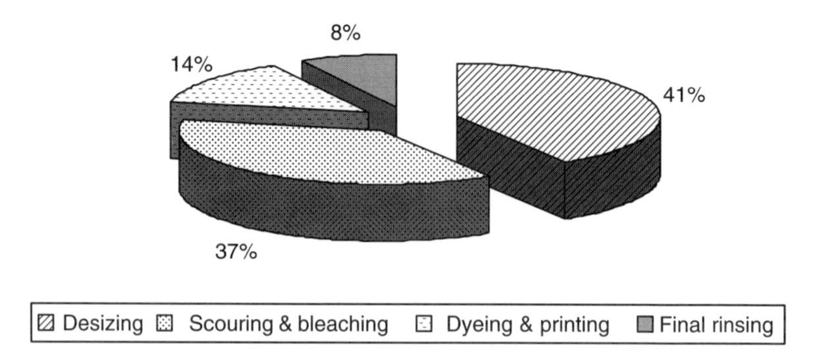

FIGURE 11-1. Chemical oxygen demand contribution to the final effluent from various operations from a cotton textile mill.

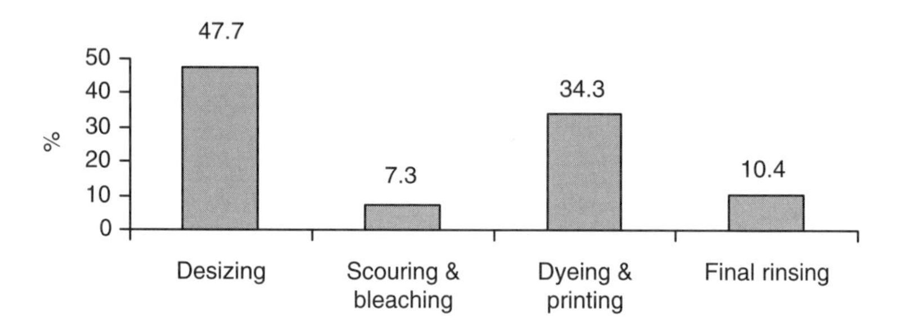

FIGURE 11-2. Contribution of suspended solids to the final effluent from various operations from a cotton textile mill.

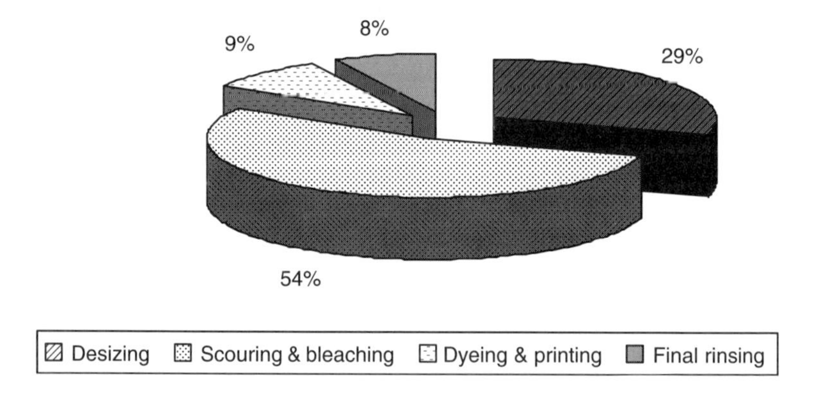

FIGURE 11-3. Volume contribution to the final effluent from various operations of a cotton textile mill.

exchange; (4) irradiation; and (5) electrokinetic coagulation. The chemical methods include oxidative processes using hydrogen peroxide, H_2O_2-Fe(II) salts (Fenton's reagent), ozonation, photochemistry, sodium hypochlorite, cucurbituril, and electrochemical destruction. Alum treatment is well suited for disperse, vat, and sulfur dyes, while activated carbon treatment is ideal for azoic, reactive, acid, and basic dyes. Ozone has been found effective for azoic, reactive, acid, basic vat, and sulfur dyes. The biological treatment methods include (1) decolorization by white rot fungi; (2) use of other microbial cultures like mixed bacterial cultures, *Pseudomonas* strains, *Sphingomonas* under aerobic, anaerobic, or mixed conditions; and (3) adsorption by living or dead microbial biomass. Physical and chemical methods are expensive and are very effective only if the effluent volume is small.

Biodegradation

It is generally difficult to degrade wastewater from the textile industry by conventional biological treatment processes because the BOD/COD ratio is less than 0.3. Decolorization of dyes by bacteria can be brought about by either adsorption to microbial cells or biodegradation. Since dyes of different types are generally used in a textile industry, the microorganisms should be nonspecific and secrete extracellular enzymes.

The worldwide annual production of wool is about 1.6 million tonnes (clean weight, 2001 figures). The scouring (contaminant extraction) or washing of this wool removes an approximately equal weight of wool wax and uses about 10 m^3 of water per tonne of raw wool. The wastewater consists of a stable emulsion of wool wax (~75%) in an aqueous solution containing dissolved organic and inorganic compounds with very high COD (~45,000 mg/L). A chemical flocculation process known as Sirolan CF removes over 95% of wool wax. The sludge produced from the process is used in composting. The effluent left from this process contains the water-soluble organic and inorganic compounds washed from the raw wool, known as "suint," which is a highly colored mixture of mainly potassium salts of fatty acids, as well as peptides with a COD of 5,000 to 15,000 mg/L. The effluent contains nonionic surfactants like nonylphenol polyethoxylates, which are used in the scouring process. The treatment of this effluent for 1 to 7 days with anaerobic bacteria resulted in partial grease flocculation; the efficiency of this process varied widely (from 30 to 80%), depending on the concentration of free surfactant, rather than total surfactant, in the effluent. Anaerobic microbes (taken from the sludge of a municipal wastewater treatment plant) could partially degrade this ethoxylate within 7 days by shortening the hydrophilic chain, causing coagulation and subsequent flocculation of wool grease from the liquor (Charlesa et al., 1996).

Wastewaters resulting from printing matrix washing and dyestuff leakage from the silk and Lycra printing industry contain a high concentration of ammonia nitrogen (urea, ammonium sulfate, and tartrate used for the dyestuff preparation) together with organic compounds. The effluent quality was improved by a single-sludge treatment process with an externally added carbon source in the form of a mixture of methyl and ethyl alcohols from pharmaceutical industry wastes. This treated effluent when subsequently filtered and sterilized using UV radiation was found to be suitable for reuse for washing purposes, achieving a 30 to 40% reduction in water usage. The treated water did not accumulate any recalcitrant or toxic compounds. When the water was treated prior to filtration with a dose of 20 mg/L of ozone, the residual color and nonionic surfactants were degraded, allowing the water to be used for preparing dyes (Sterna et al., 1996).

Aeromonas hydrophila could decolorize 24 azo, anthraquinone, and indigo textile dyes containing various substituents such as nitro and sulfonic groups with 2 days of incubation (Chen et al., 2003). The decolorization efficiency was enhanced by the addition of an extra nitrogen source such as yeast extract or peptone. Glucose inhibited the decolorization activity, probably because the acids formed during glucose conversion decreased the pH of the culture medium, thus inhibiting cell growth and decolorization activity. The presence of oxygen inhibited the bacterial degradation of azo dyes, because the oxygen was probably competing with the azo group as the electron receptor in the oxidation of the reduced electron carrier.

When solid textile mill sludge mixed with cow dung was vermicomposted (with *Eisenia foetida*) (30% on dry weight basis) for 90 days, there was a significant reduction in the C:N ratio and an increase in TKN. Dehydrogenase activity increased up to 75 days and decreased on further incubation (Kaushik and Garg, 2003).

Biosorption

The uptake and subsequent accumulation of chemicals and toxins by biomass is termed "biosorption." Dead bacteria, yeast, and fungi have been found to biosorb dyes from textile effluents and decolorize them (Robinson et al., 2001). The interaction involves adsorption, deposition, and ion-exchange. For living cells, however, the major mechanism for mineralizing dyes is biodegradation through the lignin-modifying enzymes they produce, such as laccase, manganese peroxidase, and lignin peroxidase. Biosorption can be used when the dye is very toxic to the growth of the microorganism. It had been observed that the biomass derived from the thermotolerant ethanol-producing yeast strain *Kluyveromyces marxianus* IMB3 had a relatively high affinity for heavy metals and biosorbed dyes from textile effluents. Actinomyces have been used as an adsorbent for decolorization of effluents containing anthroquinone, phalocyanine, and azo dyes. Biosorption is generally quick—a few minutes in algae to a few hours in bacteria. *Aspergillus niger* can decolorize acidic and basic blue dyes by biosorption in 30 to 48 h.

Dead fungal cells such as *Botrytis cinerea, Cryptococcuss, Candida rugosa,* and *Endothiella aggregata* have been able to decolorize several reactive dyes through an adsorption process (Fu and Viraraghavan, 2001). Autoclaved *Phanerochaete chrysosporium* could decolorize a solution containing Congo Red better than living cells (90 and 70%, respectively). Several authors have found a similar pattern in the performance of dead and live cells, and it was concluded that because of the rupture of dead cells there was an increase in the surface area for adsorption. Degradation using living cells has several limitations and disadvantages. Living cells require a nutrient if the influent lacks it, and their performance depends on the appropriate operating conditions. Waste products from fermentation could be a source for dead biomass. Fungal biomass can be regenerated using organic solvents like methanol and ethanol and nonionic surfactants such as Tween and alkali solutions. It was suggested that alcohols such as methanol modified the hydrophilic-hydrophobic interaction between the dye molecule and the biomass.

Pretreatment of the biomass increased its adsorption capacity. There are two basic methods: (1) autoclaving or (2) washing with solvents such as formaldehyde or with inorganic chemicals such as sulfuric acid, sodium hydroxide, and sodium bicarbonate. Autoclaving increased the biosorption of Basic Blue 9 dye by a factor of 15, probably by disrupting the fungal structure and exposing the binding sites for the dye. Sulfuric acid treatment doubled the adsorption capacity by changing the negatively charged surface of the biomass to a positively charged one, thereby increasing the attraction between the fungus and the dye. The factors that affected the growth of the fungi were: characteristics of the medium (pure nutrient medium), type of carbon source (glucose, starch, maltose, and cellobiose were good carbon sources), type of nitrogen source, nutrient concentration, pH (between 4 to 5) and temperature of the medium (20 to 35°C), incubation time (1 to 2 days), and concentration of oxygen in the medium (Zhang et al., 1999). The factors that pertain to the wastewater are:

- Structure of the dye.
- Metal ions. Metal ions could neutralize the surface charges and thus reduce the repulsive forces, bringing them closer and making the biosorption process more favorable.
- Surfactants. The presence of detergent in the wastewater could reduce the binding efficiency of the cells.
- Temperature. Lower temperatures favored biosorption if the process is physical absorption.
- pH. Affected the solubility of some dyes as well as the biosorption capacity.
- Ionic strength.

So far no clear conclusions can be drawn about how the structure of the dye affects its degradation. Spadaro et al. (1992) reported hydroxyl-, amino-, acetamido-, or nitro-substituted aromatic dyes degraded faster than unsubstituted ones, whereas a few authors have not seen any difference in the degradation of substituted and unsubstituted sulfonated dyes.

Combined Treatments

Because of large variability of the composition of textile wastewaters, most of the traditional methods for treatment are inadequate. Hence current research is focused on a combination of physical, chemical, and biological methods for treating this effluent in a cost-effective way.

The principal enzymes, lignin peroxidase and manganese peroxidase, present in the white rot fungus *Phanerochaete chrysosporium* can oxidize substrates by an electron transfer process or by radicals generated during the enzyme catalytic cycle. Other white rot fungi capable of decolorizing dyes include *Trametes versicolor, Coriolus versicolor*, and *Funalia trogii*. A combined biotreatment with *Phanerochaete chrysosporium* fungus followed by ozonization could lead to better degradation. Ozone attacks nucleophilic centers like carbon–carbon or nitrogen–nitrogen double or triple bonds or acts through hydrogen abstraction, electron transfer, or radical addition, which degrade recalcitrant compounds like dye much more quickly. Kunz et al. (2001) observed that enzyme treatment led to a 30 to 40% decolorization, and the subsequent ozone treatment led to 55% phenol degradation and 40% color reduction. Chloride ion present in the effluent slows down the activity of the enzyme because it competitively binds to the active site, preventing the binding of hydrogen peroxide and thereby inhibiting the electron transfer in the first step of the cycle. A typical dye house effluent containing an anthraquinone dye, an anionic detergent, a softening agent, and NaCl was treated aerobically using an activated sludge process followed by UV radiation; this treatment decreased the inhibition of microbial growth from 47 to 30%. The addition of 2 mL H_2O_2 to 1 L of UV-irradiated sample decreased the inhibitory effect further to 26%.

A combination of aerobic treatment and the addition of an adsorbent-type flocculent such as powdered activated carbon, bentonite, activated clay, or commercial synthetic inorganic clay has been found to degrade color more effectively. The addition of activated carbon into the aeration tank increased the removal efficiency to 90% (Marquez and Costa, 1996). Pala and Tokat (2002) observed a 94% COD reduction and an 80% color removal efficiency when an organic flocculant was added to the activated sludge used to treat a textile dye effluent. The color removal efficiency was only 35% when the adsorbent was not added. Although activated carbon removed dyes effectively from a waste stream, they were present in a more concentrated and toxic form in the liquid, and their safe disposal increased the treatment cost. The regeneration of activated carbon also added to the operating cost. Addition of low cost adsorbents such as peat, wood, silica, and fly ash during the activated sludge process was found to be very effective in color removal (Ramakrishnan and Viraraghavan, 1997). Removal of textile reactive dyes by renewable biosorbents like apple pomace and wheat straw have shown promising results (Robinson et al., 2002).

A continuous process of combined chemical coagulation using poly aluminum chloride (PAC) and a polymer, followed by electrochemical oxidation and finally aerobic activated sludge treatment for 1 h was able to reduce the COD of the textile effluent by 85%. The treatment cost of this combined process was estimated to be about $ 0.34 per ton of wastewater treated (Lin and Peng, 1996). The current cost figure of the conventional treatment process was about $0.45 per ton of wastewater.

Reactors

A baffled anaerobic reactor was used to degrade a cotton textile mill effluent using sucrose, peptone, and nutrient as cosubstrates. When a hydraulic retention time of 20 h was used, 70% COD (influent = 1000 mg/L) and 90% color removal rates were achieved. A two-stage upflow anaerobic sludge blanket (UASB) reactor operated with tapioca as the cosubstrate (1,500 mg/L) achieved a 70% color removal and an almost 90% COD reduction. When only one UASB reactor was used without any cosubstrate, a 60% COD reduction and 80% color removal were achieved.

Anaerobic treatment of textile wastewater and decolorization using a fluidized bed reactor with an external carbon source in the form of glucose led to an 82, 94, and 59% COD, BOD, and color removal, respectively, at a hydraulic retention time (HRT) of 24 h (Sen and Demirer, 2003). Pumice was used as the support material for the growth of the microorganisms. Various types of reactors (including a sequencing batch reactor, UASB, anaerobic filter, batch, and fluidized bed) have been used to treat effluents containing synthetic phthalocyanine, diazo, azo, reactive, anthraquinone, and basic dyes under anaerobic conditions (Sen and Demirer, 2003).

Synthetic wastewater from the desizing and dyeing sections of a textile mill was treated in an anaerobic/aerobic sequential batch reactor (SBR) with a hydraulic retention time of 2.6 days (Shaw et al., 2002) The desizing section effluent contained starch, polyvinyl alcohol (PVOH), and carboxymethyl cellulose (CMC), and the dyeing section effluent contained alkali, sodium salts, and an azo dye (Remazol Black). The reactor cycle (~5 h and 50 min) consisted of fill, anaerobic reaction, aerobic reaction, settle, decant, and idle (Fig. 11-4). Color and dye reduction was greater than 90%, and partial degradation of PVOH was observed. Decolorization improved with aeration despite the reduction in total organic carbon (TOC) removal, indicating that the decolorization process was independent of the methanogenic activity as long as the redox potential remained low. It has been reported that PVOH is more easily degraded under aerobic conditions.

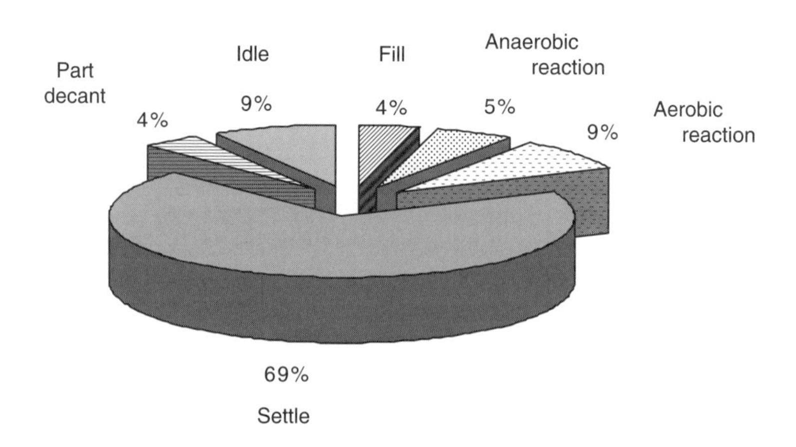

FIGURE 11-4. Various operational cycles (time) in a typical anaerobic-aerobic sequential batch reactor (SBR).

Yu et al. (1996) achieved a 92% degradation of a 75,000 MW PVOH in two reactors—one was an anaerobic SBR, the other was aerobic. A simulated cotton mill effluent was treated using a UASB and an aerobic reactor placed in series; the COD, BOD, and color removal efficiencies achieved were 88, 99, and 77%, respectively. Most of the color removal took place in the UASB (O'Neill et al., 2000).

Effluent from the Sirolan CF process was aerobically treated using a laboratory-agitated fermenter and 100- and 3000-L pilot-scale airlift fermenters with external recycle for 24-, 45-, and 94-h retention times, respectively, to achieve 65, 55, and 95% COD, TOC, and BOD removal efficiencies, respectively (Poole et al., 1999). The biomass used to treat this effluent was identified as from the genus *Epistylis.* A combination of the Sirolan CF process followed by aerobic degradation could remove 90% of the COD, much more than any other process, but this process requires high capital and operating expenditures.

Textile effluent was processed in a pressurized bioreactor (~3 bar of pressure) coupled with an ultrafiltration membrane unit for sludge retention. The treated effluent was then processed in a nanofiltration unit and was good enough to be reused in the plant. The bioreactor was operated at very high biomass concentration, but the excess sludge production was very small (Krautha, 1996). A pilot-scale dynamic up-flow sand filter was used as a biofilm reactor for decolorization and denitrification as well as for filtration of suspended solids from a pretreated textile effluent. Biomass growth and the sloughing of biological film did not prevent the removal of high concentrations of influent suspended solids. At low nitrate-loading rates, the filter followed the ideal plug-flow hydrodynamics. In the lower part of the filter,

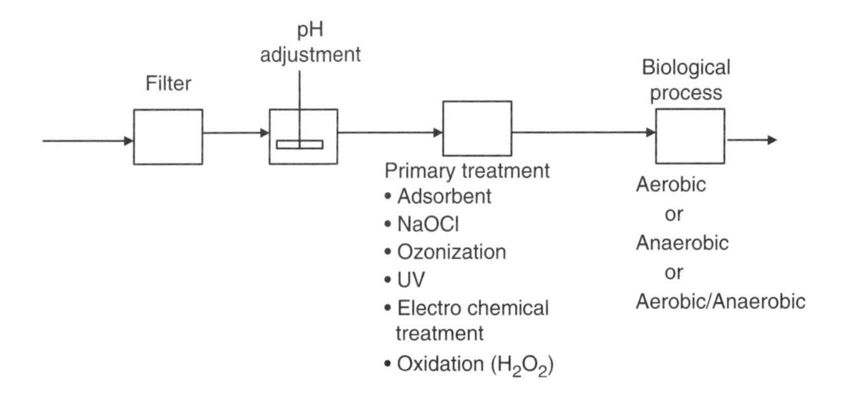

FIGURE 11-5. A typical flow sheet for treatment of textile dye effluent.

denitrification removal rates followed zero-order kinetics, while in the upper part of the filter, denitrification followed half-order kinetics (Canziania and Bonomoa, 1998).

Conclusions

Textile industries produce considerable amounts of effluent characterized by large amounts of suspended solids, high COD, fluctuating pH, high temperature, and a mixture of dyes. A combination of biochemical, chemical, and physical processes appears to be promising in degrading such an effluent as shown in Fig. 11-5. The presence of dyes in the effluent poses the biggest problem since they are recalcitrant and toxic. Both aerobic and anaerobic processes have been successfully used for degrading the dyes, but the best appears to be a combination of both. Adsorption of dyes by dead cells appears to be a better alternative than treatment with live cells. Biological activity in liquid state fermentation is slow and hence is inefficient on a continuous basis. Solid-state fermentation appears to be a good alternative for handling an enriched biomass. Chapter 10, Degradation of Dyes, deals exclusively with the decolorization of dyes.

References

Canziania, R., and L. Bonomoa. 1998. Biological denitrification of a textile effluent in a dynamic sand filter. *Water Sci. Technol.* 38(1):123–132.

Charlesa, W., H. Goena, and R. Cord-Ruwischa. 1996. Anaerobic bioflocculation of wool scouring effluent: the influence of non-ionic surfactant on efficiency. *Water Sci. Technol.* 34(11):1–8.

Chen, K. C., J.-Y. Wua, D.-J. Liou, and S.-C. J. Hwang. 2003. Decolorization of the textile dyes by newly isolated bacterial strains. *J. Biotechnol.* 101:57–68.

Fu, Y., and T. Viraraghavan. 2001. Fungal decolorization of dye waste waters: a review. *Bioresource Technol.* 79:251–262.

Kaushik, P., and V. K. Garg. 2003. Vermicomposting of mixed solid textile mill sludge and cow dung with the epigeic earthworm *Eisenia foetida. Bioresour. Technol.* 90:311–316.

Krautha, K. 1996. Sustainable sewage treatment plants—application of nanofiltration and ultrafiltration to a pressurized bioreactor. *Water Sci. Technol.* 34(3–4):389–394.

Kunz, A., V. Reginatto, and N. Duran. 2001. Combined treatment of textile effluent using the sequence *Phanerochaete chrysosporium. Chemosphere* 44:281–287.

Lin, S. H., and C. F. Peng. 1996. Continuous treatment of textile wastewater by combined coagulation, electrochemical oxidation and activated sludge. *Water Res.* 30(3):587–592.

Marquez, M. C., and C. Costa. 1996. Biomass concentration in PACT process. *Water Res.* 30(9):2079–2085.

O'Neill, C. O., F. R. Hawkes, D. L. Hawkes, S. Esteves, and S. J. Wilcox. 2000. Anaerobic-aerobic biotreatment of simulated textile effluent containing varied ratios of starch and azo dye. *Water Res.* 34(8):2355–2361.

Pala, A. and E. Tokat. 2002. Color removal from cotton textile industry wastewater in an activated sludge system with various additives. *Water Res.* 36:2920–2925.

Poole, A. J., R. Cord-Ruwisch, and F. W. Jones. 1999. Biological treatment of chemically flocculated agro-industrial waste from the wool scouring industry by an aerobic process without sludge recycle. *Water Res.* 33(9):1981–1988.

Ramakrishnan, K. R., and T. Viraraghavan. 1997. Dye removal using low cost adsorbents, *Water Sci. Technol.* 26(2–3):189–196.

Robinson, T., B. Chandran, and P. Nigam. 2002. Removal of dyes from a synthetic textile dye effluent by biosorption on apple pomace and wheat straw. *Water Res.* 36:2824–2830.

Robinson, T., G. McMullan, R. Marchant, and P. Nigam. 2001. Remediation of dyes in textile effluent: a critical review on current treatment technologies with a proposed alternative, *Bioresour. Technol.* 77:247–255.

Sen, S., and G. N. Demirer. 2003. Anaerobic treatment of real textile wastewater with a fluidized bed reactor, *Water Res.* 37:1868–1878.

Shaw, C. B., C. M. Carliell, and A. D. Wheatley. 2002. Anaerobic/aerobic treatment of coloured textile effluents using sequencing batch reactors. *Water Res.* 36:1993–2001.

Spadaro, J. T., M. H. Gold, and V. Renganathan. 1992. Degradation of azo dyes by the lignin degrading fungus *Phanerochaete chrysosporium. Appl. Environ. Microbiol.* 58(8):2397–2401.

Sterna, S. R., L. Szpyrkowiczb, and F. Zilio-Grandib. 1996. Treatment of silk and lycra printing wastewaters with the objective of water reuse. *Water Sci. Technol.* 33(8):95–104.

Yu, H., G. Gu, and L. Song. 1996. Degradation of polyvinyl alcohol in sequencing batch reactors. *Environ. Technol.* 17:1261–1267.

Zhang, F., J. S. Knapp, and K. N. Tapley. 1999. Decolorization of cotton bleaching effluents with wood rotting fungus. *Water Res.* 33(4):919-928.

Bibliography

Ledakowicz , S., M. Solecka, and R. Zylla. 2001. Biodegradation, decolourisation and detoxification of textile wastewater enhanced by advanced oxidation processes. *J. Biotechnol.* 89:175–184.

Marmagne, O., and C. Costa. 1996. Color removal from textile plant effluents. *Am. Dyest Rept.* 85(4):15–21.

CHAPTER 12

Tannery Effluent

The global leather industry produces about 18 billion square feet of leather a year (2003 data) with an estimated value of about $40 billion. Developing countries now produce over 60% of the world's leather needs. About 65% of the world production of leather goes into leather footwear, and the global production of footwear is estimated at around 11 billion pairs (worth an estimated $150 billion at wholesale prices). Conversion of rawhide into leather (an unalterable and imputrescible product) requires several mechanical and chemical operations involving many chemicals in an aqueous medium, including acids, alkalis, chromium salts, tannins, solvents, auxiliaries, surfactants, acids, and metallorganic dyes; natural or synthetic tanning agents; sulfonated oils, and salts. The quantity of effluent generated is about 30 L for every kilogram of hide or skin processed. The total quantity of effluent discharged by Indian tanneries is about 50,000 m^3/day and contains high concentrations of organic pollutants. Tannery effluents can be divided into three types based on the operations carried out in the three different sections (Fig. 12-1):

- Unhairing and liming wastewater with high sulfide and lime content and high pH [accounts for 45% of the effluent volume and contributes to 30% of the overall biological oxygen demand (BOD) and chemical oxygen demand (COD)].
- Tanning wastewater with high salinity and high chrome levels.
- Retanning, dyeing, and fat liquoring wastewater (accounts for ~20% of the total COD).

The first area of operations is the beam house in which the raw hides and skins are cleaned and prepared so that the hides are more pliable, attractive, and useful. The operations include siding, trimming, washing, soaking, fleshing, and unhairing. The last operation involves treatment with lime and sodium sulfide as the primary chemicals to dissolve the hair. These wastewaters are highly alkaline (pH of 10 to 12). Raw leather is made up of three main layers. The upper layer (epidermis) contains hair, glands, muscles, etc.

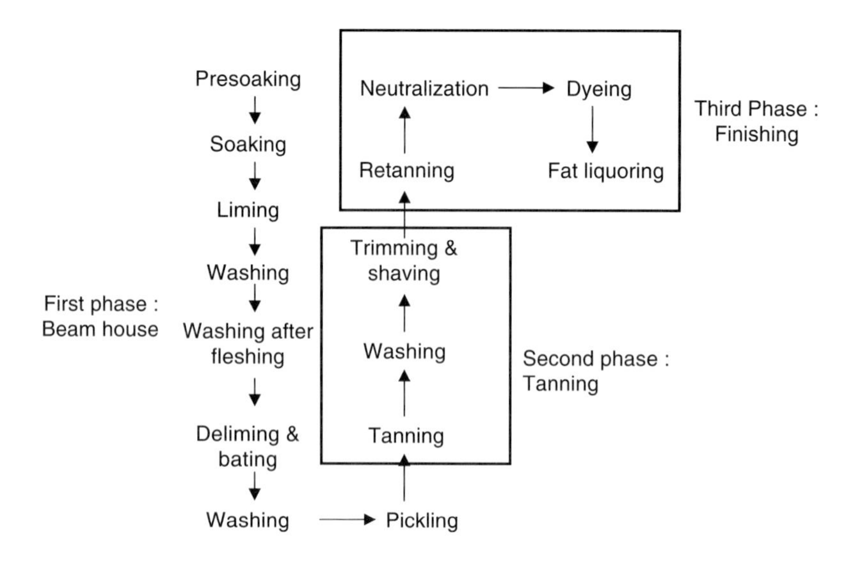

FIGURE 12-1. Various steps in conversion of hide to leather.

The middle layer (corium), which is useful and constitutes the leather product, is made to react with the tanning agent, while the bottom layer is removed by mechanical means. Removal of the upper layer is carried out through liming and bating processes. The skin is rehydrated by soaking and removing the globular proteins. Hair is removed from the skin by the reductive liming process. Two important processes take place during this step—hydrolysis and chemical reduction of keratin. Hydrolysis occurs in an alkaline medium, and lime acts as a buffer, maintaining the pH around 12.5. While globular proteins are easily solubilized, keratin and collagen are resistant to hydrolysis. Reduction of keratin is achieved by the use of alkali sulfide salts. Elastin, another strong fibrous protein that is not soluble under the conditions of liming, is removed using enzymes (the bating process). Pickling is another pretanning step where some globular proteins are removed with the aid of pretanning agents.

The second area of operations is the tanyard in which a durable material is produced from the animal hides or skins. The proteinaceous matter in the hides is made to react with the tanning agent for stabilization. This is accomplished by using synthetic tanning agents containing trivalent chromium salts, aluminum, zirconium, etc., or using vegetable tannins extracted from the bark of certain trees; 90% of tanning is done using the former. These operations are carried out in an acidic medium, and the wastewater generated usually has a pH in the range of 2.5 to 3.5. The tanning process stabilizes the skin structure by forming transverse bonds among its fibers. In the case of mineral tanning agents, the tanning agent blocks the carboxylic groups (or in the case of vegetable tanning agents, the amine groups) and joins the proteinaceous colloid, thus increasing the crosslinking of the collagen fibers. The resulting stabilized leather material cannot be degraded by physical or biological means. Of the tanning agents added, 15% does not get fixed to the leather and is discharged with the effluent. The third set of process operations involve retanning and wet finishing, which gives the tanned hides special or desired features such as bleached appearance, added coloring, lubrication, or further tanning for finished leather properties. These operations usually do not alter the pH of the wastewaters.

Solid waste is generated, including trimmings, degraded hide, and hair from the beam house, amounting to 70% of the wet weight of the original hide. The tannery wastewater has a very high salinity. The main contributors are 30% chlorides from the pickling bath (where skins are prepared with salts and acids prior to adding tanning agent) and 60% sulfates from the tanning bath. Hydrogen sulfide is released during dehairing, and ammonia is released during deliming. Characteristics of the effluent released from a tannery producing full chrome upper leather from dry salted bovine hides is given in Table 12-1; the effluent from an industrial district housing several tanneries is given in Table 12-2.

TABLE 12-1
Characteristics of Effluent Released by a Tannery Treating Bovine Hide

pH	7.5–9.0
SS	1,500–4,000 mg/L
TS	29,000–45,000 mg/L
Chromium	100 mg/L
COD	5,000–10,000 mg/L
BOD	1,500–2,000 mg/L

TABLE 12-2
Wastewater Characteristics of Tannery Effluent (Turkey)

Characteristic, mg/L	*Raw wastewater*	*Clarifier effluent*
Total COD	5,094	2,216
Soluble COD	2,336	1,187
BOD_5	1,760	958
SS	2,229	794
VSS	—	506
Total nitrogen	358	226
Organic nitrogen	223	62
NH_3—N	135	164
Total phosphorus	—	5.1
Total chromium (Cr^{3+})	116	41
Sulfur	51	27

Nitrogen exists in leather tanning wastewaters as ammonia and organic nitrogen and is present in both particulate and soluble forms. All nitrogen originating from the bovine leather processing plant is from soaking, liming, deliming, bating, pickling, and tanning, and the washings from these processes (Zengin et al., 2002), and the total nitrogen in the effluent is below 1%. The main source of nitrogen is from the liming step (~7390 mg/L), followed by deliming, bating, washing, and pickling steps. The main source for protein in the effluent comes from the liming step (~13,660 mg/L), followed by the washing, deliming, and bating steps (Kabdasli et al., 2003).

Physical and Chemical Treatment

Mixing all three effluents and then treating is found to be very inefficient. The unhairing wastewater is treated with oxygen in the presence of a manganese (II) salt, which acts as a catalyst to transform the sulfide to sulfate. The wastewater of the tanning process, which contains large quantities of chromium, is directly recycled or treated with alkali to precipitate chromium hydroxide, which is subsequently reused. The retanning, dyeing, and fat liquoring wastewater is treated with iron (II) sulfate to precipitate a large amount of protein and organic contaminants. Still, some contaminants such as surface active agents and fat liquors remain unaltered by a normal physicochemical treatment. The recovery of chromium from tannery waste is carried out in three ways: namely, Cr (III) extraction with sulfuric acid solution at a pH of 1, Cr(III) oxidation to Cr(VI) with H_2O_2, and Cr(VI) separation from other cations and its subsequent reduction to Cr(III). The overall Cr (III) recovery yield from sludges is about 80% (Macchi et al., 1991).

Adsorption using clays such as bentonite, sepiolite, or activated carbon, or expensive tertiary treatments like the Fenton process have also been found effective in treating the final effluent (Espantaleo et al, 2003). Coagulation followed by removal of sludge is an old technique that has been effectively adapted for treating effluents of different types. The most widely used coagulants are aluminum (III) and iron (III) salts. Although vegetable tannage produces small amounts of wastewater, it has a negative impact on the primary treatments like flocculation or coagulation, and leads to increased use of chemicals (Scholz and Lucas, 2003).

The wastewater from leather tanning contains organic nitrogen that comes from proteins and amino acids. These compounds have been recovered by several physical and chemical means, including addition of miscible solvents such as alcohol or acetone; addition of metal cations such as Zn^{2+}, Cd^{2+}, Cu^{2+}, or bulky anions such as perchlorate, trichloroacetate; salting out using ammonium sulfate or other salts; isoelectric pH precipitation; and polyelectrolyte aggregation precipitation. Many of these techniques are not economical and cannot be used on an industrial scale. For protein removal, 50% is achieved by isoelectric pH precipitation carried out between the optimum pH interval of 2.1 to 3.8, and 60% is achieved with $FeCl_3$; 85%

ammonia removal and 50% protein removal is achieved with magnesium ammonium phosphate precipitation followed by acid precipitation of the protein (Kabdasli et al., 2003).

Biochemical Treatment

Aerobic

A combination of biochemical oxidation and chemical ozonation of tannery effluent has been found to yield excellent results, with the first part performed in an upflow sequencing batch biofilm reactor provided with external recycle (Iaconi et al., 2002). COD, $NH_4—N$, and total suspended solids (TSS) removals were 95, 98, and 99.9%, respectively. The combined process produced very low sludge, about 0.03 kg TSS/kg COD removed, which is much lower than the values reported in the literature for conventional biological systems. Ozone helped in the mineralization of some organic substances and the partial oxidation of some others, leading to enhancement of the biodegradability of the effluent. The aerobic treatment of the beam house and tanyard wastewater substreams, followed by an oxidative treatment using ozone, and a second aerobic treatment improved the aerobic biodegradability of refractory organic compounds. Also, full nitrification was achieved during the subsequent aerobic degradation, and the remaining ammonia was completely removed (Jochimsena et al., 1997).

A membrane sequencing batch reactor performed better than a sequencing batch reactor in treating beam house wastewater that was collected after the oxidation of sulfide compounds. The former reactor achieved a removal efficiency of 100% for ammonium ion and 90% for COD, and the latter reactor achieved low ammonium removal and 90% for COD. The sequential batch reactor exhibited a washout in 90 days, whereas the membrane bioreactor was very stable for more than 120 days of operation (Martinez et al., 2003; Goltara et al., 2003).

A settled vegetable-tanning process effluent was treated successfully in a mixed continuous-flow laboratory-scale plant. The BOD and COD removal efficiencies were 85 to 96% and 86 to 97%, respectively, under steady-state conditions (Murugesan and Elangovan, 1989). Activated-sludge treatment of chrome-tanning waste mixed with sewage was able to remove 87 to 96% of BOD under steady state conditions (Murugesan et al., 1996). About 84 to 92% of the influent BOD was removed from a tannery effluent when it was treated in an activated sludge well mixed reactor (Ram et al., 1999; Orhon et al., 1999).

A comparative study of tannery waste treatment was done using an upflow anaerobic sludge blanket reactor (UASB) and activated sludge (AS) reactor; interestingly, the latter was found to have more advantages than the former with respect to the capital and operating costs as well as the quality of performance (Tare et al., 2003). Total annualized costs, including capital, operating, and maintenance costs, for the UASB and AS plants were Indian Rs. 4.24 million/million liters per day (MLD) and Indian Rs. 3.36 million/MLD, respectively. The treated UASB effluent had higher BOD and COD and considerable amounts of chromium and sulfide when compared with the AS reactor effluent.

Anaerobic Treatment

Although the anaerobic treatment of tannery beam house wastewater looks attractive, the presence of sulfide, namely hydrogen sulfide, inhibits methanogenic bacteria. Undissociated H_2S is toxic, since it diffuses "freely" through the cell membrane, denatures proteins and enzymes (sulfide cross-linking), and affects internal cell pH. In a continuous-flow fixed-film reactor, a concentration of 100 mg/L of undissociated sulfide inhibited the efficiency and degree of degradation, which was eliminated by incorporating a sulfide-stripping system that reduced the concentration of undissociated sulfide to 30 mg/L. This modification improved the efficiency of the degradation by 15%. Acidogenic bacteria were not inhibited by hydrogen sulfide (Schenka et al., 1999).

A combined anaerobic and aerobic treatment of tannery wastewater effluent containing 900 mg/L organic carbon content (DOC) gave a removal efficiency of 85% (Reemtsma, and Jekel, 1997). Under anaerobic conditions, sulfate is converted by sulfate-reducing bacteria (SRB) into sulfide, which is not only a toxic compound but also a strong inhibitor of methanogenesis. Also, SRB compete with methane producing bacteria (MPB) for substrates such as hydrogen and acetate and with syntrophic acetogenic bacteria (SAB) for intermediate substrates such as short-chain volatile fatty acids (VFA) and alcohols. This results in a reduction of organics available for conversion to methane. So it is advisable to remove the sulfide either before the anaerobic treatment or as part of the biotreatment cycle as shown in Fig. 12-2. A combination of the sulfur recovery unit integrated with a USAB reactor for treating tannery effluent led to improved biogas production and also recovery of elemental sulfur (Suthanthararajan et al., 2004). The sulfur removal unit consisted of a stripper column, absorber column, regeneration unit, and sulfur separator. A stripper efficiency of about 65 to 95% in terms of sulfide removal was achieved. The efficiency of degradation in a continuous flow fixed film reactor improved by 15% when the concentration of undissociated sulfide was reduced from 100 to 30 mg/L with the help of a continuous sulfide removal system (Wiemanna et al., 1998).

Chromium

Chromium is an important heavy metal used in the leather, electroplating and metallurgical industries. More than 170,000 tonnes of chromium wastes are discharged into the environment. In India 700,000 tons of wet salted hides

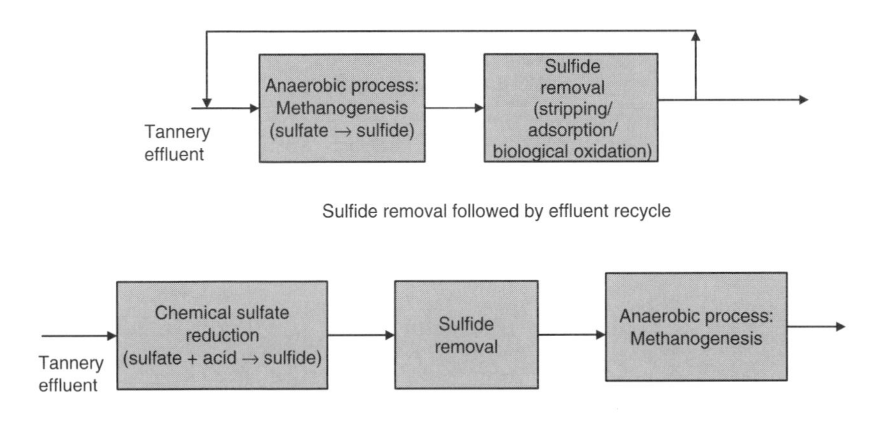

FIGURE 12-2. Sulfide handling.

and skins are processed annually in about 3,000 tanneries that discharge 3×10^7L of wastewater with suspended solids in the range of 3,000 to 5,000 mg/L and chromium as Cr in the range of 100 to 200 mg/L (1999 data). Some 200,000 tons of partly dried (40% dry matter) chromium-containing sludge is generated and is dewatered mostly in pen sludge drying beds using sand and gravel, a chamber filter press, or a belt press. The moisture content of the dewatered sludge from sludge drying beds ranges from 50 to 70% and the concentration of chromium as Cr ranges from 1 to 3% on a dry solid weight basis. The disposal of such large quantities of hazardous solid chromium waste poses serious environmental and health problems (Rajamani et al., 2000).

The earliest technique practiced for the disposal of chromium sludge consisted of solidification of the waste with cement and organic clay. The acidic ion exchange resins Amberlite IRC 76 and Amberlite IRC 718 retained 95% of the chromium at pH 5 (Kocaoba and Akcin, 2002). Extraction using supported liquid membranes (Djane et al., 1999), chemical methods such as reduction by sodium metabisulfite, ferrous ions, zero valent iron, and a mixture of dimethyl dithiocarbamate, ferrous sulfate, and aluminum chloride have also been tested. The last method is also practiced commercially (Chang, 2003).

Several heterotrophs and coliforms were tolerant to a chromate level of greater than 50 g/mL, and many coliforms were resistant to higher levels of chromate too, whereas only a few heterotrophs were resistant to Cr^{6+} at a level of greater than 150 g/mL (Vermaa et al., 2001). A few important microbes involved in the reduction of chromium are *Pseudomonas aeruginosa, Enterobacter cloacae,* and *P. fluorescens. Desulfovibrio desulfuricans* immobilized on a polyacrylamide gel reduced 80% of 0.5 *M* Cr (VI) with lactate or H_2 as the electron donor and Cr (VI) as the electron acceptor (Kamaludeen et al., 2003a).

NCIM 5080 and NCIM 5109 actinomycetes strains have been found to reduce chromium levels by 99% within 24 h and at the same time reducing 70 to 80% of the COD in 74 to 96 h (Kumar, 2003). Two strains, *Bacillus circulans* and *B. megaterium*, were able to bioaccumulate 34.5 and 32.0 mg Cr per gram of dry weight, respectively, and decrease Cr(VI) concentration from 50 mg/L to less than 0.1 mg/L in 24 h. Living and dead cells of *B. coagulans* biosorbed 23.8 and 39.9 mg Cr per gram of dry weight, respectively, and living and dead cells of *B. megaterium* biosorbed 15.7 and 30.7 mg Cr per gram of dry weight, respectively (Srinatha et al., 2002). Biosorption by the dead cells was higher than that of the living cells due to pH conditioning of the dead cells. Microbes that were able to biosorb chromium include *Oscillatoria* sp., *Arthrobacter* sp., *Agrobacterium* sp., *Pseudomonas aeruginosa* 5128, and sulfate-reducing bacteria. Pretreatment enhances the biosorption capacity as seen in the case of dead *Rhizopus nigricans.* Chromium uptake varied, depending on the type of pretreatment at pH of 2.0. Biosorbent efficiency decreased when the microorganisms were treated with mild alkali, while treatment with acids, alcohols, and acetone improved the chromium uptake capacity (Bai and Abraham, 2002). The waste *Mucor meihi* biomass was found to be an effective biosorbent for the removal of chromium from industrial tanning effluents, reaching sorption levels of 1.15 mmol/g (Tobin and Roux, 1998).

Dried and classified *Pinus sylvestris* bark was able to remove 90% of trivalent chromium. Pretreatment of the bark helped to increase its chromium sorption capacity (Alves et al., 1993). A column packed with calcium alginate (CA) beads with humic acid could adsorb 54% of the chromium from a tannery effluent and also reduce its ecotoxicity in 72 h of operation (Pandeya, 2003). *Dunaliella,* a unicellular biflagellate halophilic green algae, biosorbed 45 to 58 mg/g Cr(VI). The green algae *Carlina vulgaris, Scenedesmus obliquus, Synechocystis* sp., *Cladonia crispata,* and *Spirogyra* sp., had maximum uptakes of 33.8, 30.2, 39.0, 39.5, and 14.7 mg/g, respectively. Fungal species of *Mucor meihi, Rhizopus nigricans, R. arrhizus,* and, and *Aspergillus niger* biosorbed 59.8, 119.7, 58.1, and 15.6 mg Cr/g, respectively (Donmez and Aksu, 2002). A strain of *Streptomyces griseus* was found to grow in glucose/sodium acetate medium and reduce Cr^{6+} to Cr^{3+} (Laxman and More, 2002).

Conclusions

Figure 12-3 lists the various physical, chemical, and biochemical methods that have been tried for treating tannery effluents. Several issues such as the cost of physical and chemical methods, the toxicity of chromium on

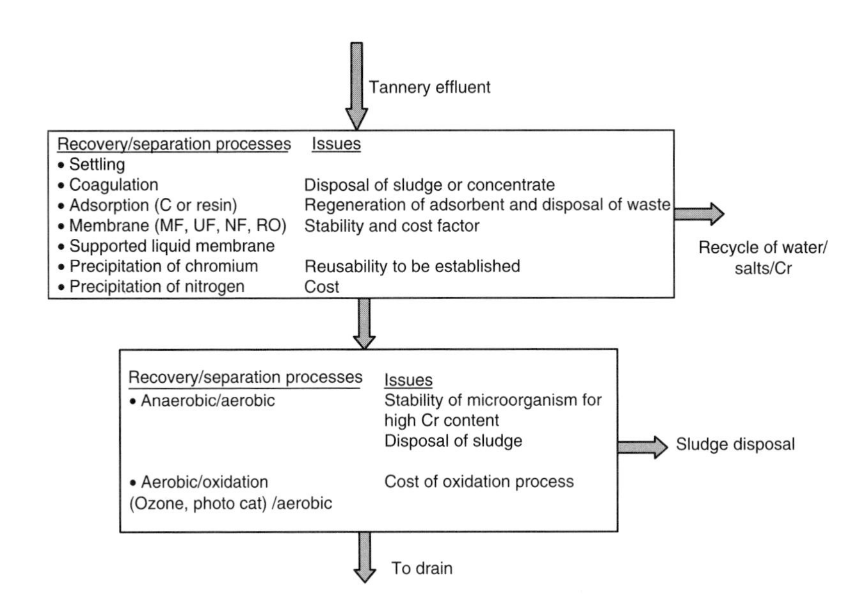

FIGURE 12-3. Physical, chemical, and biochemical treatment techniques for tannery effluent.

biochemical methods, and the inhibitory nature of sulfides in the anaerobic degradation process have not been fully resolved. The reduction and recycle of the various streams at the source appears to be a good approach to dramatically decrease the present quality and quantity of effluent generated by this industry. Disposal of the sludge after biochemical treatment, however, has not been satisfactorily solved. Biochemical treatment of chromium is also discussed in Chapter 13, Treatment of Waste from Metal Processing and Electrochemical Industries.

References

Alves, M. M., C. G. González Beça, R. Guedes de Carvalho, J. M. Castanheiral, M. C. Sol Pereira, and L. A. T. Vasconcelos. 1993. Chromium removal in tannery wastewaters "polishing" by *Pinus sylvestris* bark. *Water Res.* 27(8):1333–1338.

Bai, S. R., and T. E. Abraham. 2002. Studies on enhancement of Cr(VI) biosorption by chemically modified biomass of *Rhizopus nigricans*. *Water Res.* 36:1224–1236.

Cassano, A., J. Adzet, R. Molinari, M. G. Buonomenna, J. Roig, and E. Drioli. 2003. Membrane treatment by nanofiltration of exhausted vegetable tannin liquors from the leather industry. *Water Res.* 37:2426–2434.

Chang, L. Y. 2003. Alternate chromium reduction and heavy metal precipitation methods for industrial wastewater. *Environ. Prog.* 22(3):174–182.

Djane, N. K., K. Ndung'u, C. Johnsson, H. Sartz, T. Tornstrom, and L. Mathiasson. 1999. Chromium speciation in natural waters using serially connected supported liquid membranes. *Talanta* 48:1121–1132.

Donmez, G., and Z. Aksu. 2002. Removal of chromium (VI) from saline wastewaters by *Dunaliella* species. *Process Biochem.* 38:751–762.

Goltara, A., J. Martinez, and R. Mendez, 2003. Carbon and nitrogen removal from tannery wastewater with a membrane bioreactor. *Water Sci. Technol.* 48(1):207–214.

Iaconi, C. D., A. Lopeza, R. Ramadoria, A. C. Di Pintob, and R. Passino. 2002. Combined chemical and biological degradation of tannery wastewater by a periodic submerged filter (SBBR). *Water Res.* 36:2205–2214.

Jochimsena, J. C., H. Schenkb, M. R. Jekela, and W. Hegemann. 1997. Combined oxidative and biological treatment for separated streams of tannery wastewater. *Water Sci. Technol.* 36(2–3):209–216.

Kabdasli, I., T. Ölmez, and O. Tünay. 2003. Nitrogen removal from tannery wastewater by protein recovery. *Water Sci. Technol.* 48(1):215–223.

Kamaludeen, S. P. B., K R A. Kumar, S. Avudainayagam, and K. Ramasamy. 2003a. Bioremediation of chromium contaminated effluents. *Indian J. Exper. Biol.* 41(9):972–985.

Kamaludeen, S. P. B., M. Megharaj, R. Naidu, I. Singleton, A. L. Juhasz, B. G. Hawke, and N. Sethunathan. 2003b. Microbial activity and phospholipid fatty acid pattern in long-term tannery waste-contaminated soil. *Ecotoxicol. Environ. Safe.* 56:302–310.

Kocaoba, S., and G. Akcin. 2002. Removal and recovery of chromium and chromium speciation with MINTEQA2. *Talanta* 57:23–30.

Kumar, A. 2003. Chromium-chomping bacteria to clean tannery waste. *Terra Green* 36:May 15.

Laxman, R. S., and S. More. 2002. Reduction of hexavalent chromium by *Streptomyces griseus. Minerals Eng.* 15:831–837.

Macchi, G., M. Pagano, M. Pettine, M. Santori, and G. Tiravanti. 1991. A bench study on chromium recovery from tannery sludge. *Water Res.* 25(8):1019–1026.

Martinez, J. M., A. Goltara, and R. Mendez. 2003. Tannery wastewater treatment: comparison between SBR and MSBR. *Water Supply* 3(5):275–282.

Murugesan, V., B. Arabindoo, and R. Elangovan. 1996. Treatability studies and evaluation of biokinetic parameters for chrome tanning wastewater admixtured with sewage. *J. Ind. Pollution Control* 12(1):41–53.

Murugesan, V., and R. Elangovan. 1989. Biokinetic parameters for activated sludge treating vegetable tannery waste. *Indian J. Environ. Protection* 14:511–515.

Orhon, D., O. Karahan, and S. Sozen. 1999. The effect of residual microbial products on the experimental assessment of the particulate inert cod in wastewaters. *Water Res.* 33(14):3191–3203.

Pandeya, K., S. D. Pandeya, V. Misra, and A. K. Srimal. 2003. Removal of chromium and reduction of toxicity to Microtox system from tannery effluent by the use of calcium alginate beads containing humic acid. *Chemosphere* 51(4):329–333.

Rajamani, S., E. Ravindranath, R. S. Rajan, K. Chitra, B. U. Maheswari, and T. Ramasami. 2000. Generation of hazardous sludge from tannery effluent treatment plants and disposal problems in India, in 2000 Pacific Basin Conference, Manila, Philippines, April 12–14. University of the Philippines: Pacific Basin Consortium for Hazardous Waste Research and Management.

Ram, B., P. K. Bajpai, and H. K. Parwana. 1999. Kinetics of chrome-tannery effluent treatment by the activated-sludge system. *Process Biochem.* 35:255–265.

Reemtsma, T., and M. Jekel. 1997. Dissolved organics in tannery wastewaters and their alteration by a combined anaerobic and aerobic treatment. *Water Res.* 31(5):1035–1046.

Schenka, H., M. Wiemannb, and W. Hegemannc. 1999. Improvement of anaerobic treatment of tannery beamhouse wastewater by an integrated sulphide elimination process. *Water Sci. Technol.* 40(1):245–252.

Scholz, W., and M. Lucas. 2003. Techno-economic evaluation of membrane filtration for the recovery and re-use of tanning chemicals. *Water Res.* 37:1859–1867.

Srinatha, T., T. Vermaa, P. W. Ramteke, and S. K. Garg. 2002. Chromium (VI) biosorption and bioaccumulation by chromate resistant bacteria. *Chemosphere* 48(4):427–435.

Suthanthararajan, R., K. Chitra, E. Ravindranath, B. Umamaheswari, S. Rajamani, and T. Ramesh. 2004. Anaerobic treatment of tannery wastewater with sulfide removal and recovery of sulfur from wastewater and biogas. *J. Amer. Leather Chemists* 99(2):67–72.

Tare, V., S. Gupta, and P. Bose. 2003. Case studies on biological treatment of tannery effluents in India. *J. Air & Waste Manage. Assoc.* 53:976–982.
Tobin, J. M., and J. C. Roux. 1998. Mucor biosorbent for chromium removal from tanning effluent. *Water Res.* 32(5):1407–1416.
Vermaa, T., T. Srinatha, R. U. Gadpaylea, P. W. Ramteke, R. K. Hansa, and S. K. Garg. 2001. Chromate tolerant bacteria isolated from tannery effluent. *Bioresource Technol.* 78(1):31–35.
Wiemanna, M., H. Schenka and W. Hegemanna. 1998. Anaerobic treatment of tannery wastewater with simultaneous sulphide elimination. *Water Res.* 32(3):774–780.
Zengin, G., T. Ölmez, S. Dogçruel, I. Kabdaşlý, and O. Tünay. 2002. Assessment of source-based nitrogen removal alternatives in leather tanning industry wastewater. *Wat. Sci. Tech.* 45(12):205–215.

Bibliography

Espantaleon, A. G., J. A. Nieto, M. Fernandez, and A. Marsal. 2003. Use of activated clays in the removal of dyes and surfactants from tannery waste waters. *Appl. Clay Sci.* 24:105–110.
Gokcay, C. F., and U. Yetis. 1991. Effect of chromium (VI) on activated sludge. *Water Res.* 25(1):65–73.
Lawrence, A. W., and P. L. McCarty. 1970. Unified basis for biological treatment design and operation. *J. Sanit. Eng. Div. Am. Soc. Civil Engr.* 96(SA3):757–778.
Orhon, D., E. A. Genceli, and S. Sözen. 2000. Experimental evaluation of the nitrification kinetics for tannery wastewaters. *Water SA* 26(1):43–49.
Rossini, M., J. Garcia Garrido and M. Galluzzo. 1999. Optimization of the coagulation-flocculation treatment: influence of rapid mix parameters. *Water Res.* 33(8):1817–1826.
Schrank, S. G., H.J. José, and R. F. P. M. Moreira. 2002. Simultaneous photocatalytic Cr(VI) reduction and dye oxidation in a TiO_2 slurry reactor. *J. Photochem. Photobiol. A: Chemistry* 147:71–76.
Song, Z., C. J. Williams, and R. G. J. Edyvean. 2000. Sedimentation of tannery wastewater, *Water Res.* 34(7):2171–2176.

CHAPTER 13

Treatment of Waste from Metal Processing and Electrochemical Industries

The mining, electroplating, tannery, steel works, automobile, battery, and semiconductor industries are faced with the problem of heavy metals in their effluent streams, which harm the soil and the waterways. In the United States, approximately 217,000 sites are polluted and 31,000 sites are contaminated by only heavy metals. The National Priority (Superfund) List includes 1,200 sites that are contaminated by heavy metals. The metals most often encountered include lead, chromium, copper, zinc, arsenic, and cadmium (Meunier et al., 2004). Hence treating, neutralizing, and remediating these heavy-metal–polluted sites have become an utmost priority to these industries. Unlike organic contaminants that can be degraded to harmless products, metals cannot be further transmuted or mineralized to a totally innocuous form. Their oxidation state, solubility, and association with other inorganic and organic molecules can be varied so that they are made harmless. Although this does not solve the problem of pollution, it makes the environment less harmful and also aids in the recycling of the metals. Many enzymes have divalent or transition elements in their active center. For normal cell metabolism, minute quantities of these metals are required. But when these metals are present in excess, they could be toxic to the same enzymes. Many other metals cause damage to the cells by blocking and inactivating the sulfhydryl groups of proteins. Metals can be divided into three groups based on their effect on cells and microorganisms: (1) essential and nontoxic metals such as Ca or Mg; (2) essential but could be toxic at high concentrations like Fe, Mn, Zn, Cu, Co, Ni, and Mo; and (3) toxic at all levels, for example, Hg or Cd.

Chemical methods for treatment of this wastewater include neutralization, precipitation and filtration, electrochemistry, reverse osmosis, encapsulation, ion exchange, adsorption, or solvent extraction. These methods are effective and well established, but require large quantities of expensive chemicals or are capital intensive, and they also generate large quantities of sludge that must be recycled or disposed of effectively. Few of these methods are very effective at high metal concentrations, and they become uneconomical under dilute conditions. Recently, biological methods have attracted interest because of their simplicity. Other advantages of biological techniques over the physical and chemical methods include their higher specificity, suitability to in situ methodologies, and avoidance of high energy and toxic chemical addition. Different methods have been studied for metal extraction from contaminated soil, including leaching by inorganic acids (H_2SO_4, HCl, HNO_3, etc.), leaching by organic acids (citric acid, acetic acid, etc.), bioleaching, and use of chelating agents (EDTA, ADA, DTPA, NTA, etc.), surfactants, and biosurfactants (Meunier et al., 2004).

Mechanisms of Metal-Microorganism Interaction

The microorganisms convert the metal contaminants to forms that are precipitated or volatilized from the solution, alter the redox state so that they become more soluble leading to its leaching from soil, or allow its biosorption on microbial mass, thereby preventing its migration. Different types of reactions, which take place in various parts of the cell, are shown in the following list (the movement of the substrate is from 1 to 5) (Valls and de Lorenzo, 2002).

1. Extracellular reactions
    - Precipitation with excreted products
    - Complexation and chelation
    - Siderophores
2. Cell-associated materials
    - Ion-exchange
    - Particulate entrapment
    - Nonspecific binding
    - Precipitation
3. Cell wall
    - Adsorption or ion-exchange, covalent binding
    - Entrapment of particles
    - Redox reactions
    - Precipitation
4. Cell membrane/periplasmic space
    - Adsorption/ion-exchange
    - Redox reactions/transformations
    - Precipitation
    - Diffusion and transport (influx and efflux)
5. Intracellular
    - Metallothionein

Metal y-glutamyl peptides
Nonspecific binding/sequestration
Organellar compartmentation
Redox reactions or transformations

Eukaryotes are more sensitive to metal toxicity than bacteria. *Cyanobacterium synechococcus* is resistant to Zn^{2+} and Cd^{2+} because of the production of the metallo thioneins (MT) gene, which is known to bind these metals. Sulfate-reducing bacteria (SRB) are anaerobes that produce sulfide and immobilize toxic ions (Cu, Fe, Zn, Ni, Cd, Pb, etc.) as metal sulfides; hence they exhibit metal tolerance as a secondary outcome of their metabolism. Hydrogen sulfide produced during the sulfate ion reduction reacts to form the precipitate. Ferric iron is also precipitated as its hydroxide. Its elimination probably occurs through two steps—the first being the reduction to ferrous iron and the second to divalent metallic sulfide as a precipitate (Eger, 1994).

Thiobacilli and *Thermophilic archaea,* iron- and sulfur-oxidizing bacteria, grow at the highest metal concentrations. *Thiobacillus ferrooxidans* is dependent on Fe(II), but it is also resistant to Al, Cu, Co, Ni, Mn, and Zn at a concentration of 0.1 to 0.3 *M*. Reduction of Cr (VI) to insoluble Cr(III) by SRB may be due to bacterial respiration or indirect reduction by Fe^{2+} and sulfide. Under iron-limiting conditions, microorganisms such as bacteria, fungi, cyanobacteria, and algae excrete siderophores, which are low molecular weight Fe(III) coordination compounds, to capture iron from the environment. In addition to iron, siderophores and analogous compounds can complex other metals including Ga(III), Cr(III), Sc, In, Ni, U, Th, Pu(IV), Fe(III), Pu(VI), and Th(IV) (White et al., 1995).

In dissimilatory processes, the transformation of the target metal is unrelated to its intake by the microbe. The chemical species that result from the cognate biological activity generally end up in the extracellular medium. An example is Cr (VI) reduction under both aerobic and anaerobic conditions, with NADH and electron transport systems serving as the respective electron donors. *Desulfovibrio desulfuricans* couples the oxidation of a variety of electron donors to the reduction of Tc (VII) by a periplasmic hydrogenase. Membrane-bound enzymes catalyze dissimilatory metal-reducing activities.

Dissimilatory Fe^{3+}-reducers, *Geobacter metallireducens* and *Shewanella putrefaciens,* can reduce highly soluble Tc (VII) to less soluble, reduced forms of technetium and Co^{3+}-EDTA to Co^{2+}-EDTA. *Thauera selenatis* can reduce highly soluble Se (VI) to insoluble Se^0. A phosphatase-containing *Citrobacter* sp. can precipitate uranium (U^{6+}) when supplied with an organic phosphate donor. Dissimilatory iron-reducing bacteria such as *G. metallireducens* and *S. putrefaciens* couple the oxidation of H_2 or organic substrates to the reduction of ferric iron (Fredrickson and Gorby, 1996). *S. putrefaciens* could also grow by coupling the oxidation of formate or lactate to the reduction of magnetite Fe(III). *G. metallireducens* or

S. putrefaciens has been shown to reduce uranium from its relatively mobile oxidized state U(VI) to insoluble U(IV), which precipitates as the insoluble mineral uraninite. Fe(III)-reducing *Bacillus* strains were able to solubilize up to 90% of PuO_2 over a period of 6 to 7 days in the presence of nitrilotriacetic acid.

Metal ions exported from the cytoplasm will form metal-bicarbonates and carbonates around the cell surface, and at supersaturated concentration will crystallize on the cell-bound metal ions, serving as crystallization centers. This crystallization process leads to very high metal to biomass ratios (between 0.5 and 5.0 on a weight basis). Once the bioprecipitation process has reached a certain level, nucleation proteins and polysaccharides are released from the cells and bioprecipitation continues on these released foci.

Biosorption and Bioaccumulation

Biosorbents are natural ion-exchange materials that primarily contain weakly acidic and basic groups. They have advantages over their chemical counterpart since they can remove ions at very low concentrations (on the order of 2 to 10 mg/L). Biosorbents are more specific and hence prevent the binding of alkaline earth material. Also, they have the potential of genetic modification and so can be tailored for increased specificity. Bioaccumulation of metals can take place at many locations in the cell such as the cell wall or the cell surface and periplasmically, extracellularly, or intracellularly (cytoplasmic). A few disadvantages of biosorption are (1) its sensitivity to operating conditions such as pH and ionic strength and the presence of organic or inorganic ligands; (2) its lack of specificity in metal binding; (3) its requirement for large amounts of biomass if the biosorption capacity is low; (4) reusability of the biomass after desorption is possible only if weak chemicals are used for desorption; and (5) the biomass needs to be replaced after about 5 to 10 sorption-desorption cycles.

Potamogeton lucens is an excellent biosorbent for heavy metal ions. Sorption occurs mainly by ion exchange reactions with cationic weak exchanger groups present on the plant surface. Biosorption of heavy metals using biomass has also been found to be very effective in treating mine waste as long as the heavy metal species are free in solution and do not form soluble or precipitated species with sequestering compounds. The process involves diffusion, adsorption, chelation, complexation, coordination, or microprecipitation. It has been estimated that biosorptive processes could reduce capital and operating costs by 20 and 36%, respectively, and total treatment costs by 28% as compared with a conventional ion exchange process. Studies done with the aquatic macrophyte *Potamogeton lucens*, which had a carboxyl functional group, indicated that copper adsorption by the biomass is not affected by equimolar concentrations of metals such as sodium, calcium, or iron, or by a nonionic surfactant like pine oil. Anionic surfactants such as sodium oleate compete with the surface groups of the biomass for the free copper ions in solution (Schneider et al., 1999).

The root bark of the Indian sarsaparilla (*Hemidesmus indicus*) was used as a biosorbent for the successful removal of Pb, Cr, and Zn from aqueous solutions (Sekhar et al., 2003). *Spirulina* sp. (a cyanobacteria blue-green algae) was found to be an effective biosorbent and bioaccumulant of heavy metal ions such as Cr^{3+}, Cd^{2+}, and Cu^{2+}. Bioaccumulation follows the biosorption process, where metal ions that are bound to the cell wall because of ion exchange get transported into the interior of the cell (active uptake). The cell membrane is able to identify the metal species and to distinguish metal ions that are micronutrients from those that are toxins (Chojnacka et al., 2004).

Cadmium is widely used in rechargeable nickel cadmium batteries, pigments, stabilizers, coatings, alloys, and electronic components; hence wastewater from such industries may contain this metal as a pollutant. Wastewaters of dye and pigment production; film and photography processing; galvanometry and metal cleaning, plating, and electroplating; leather production; and mining operations will contain chromium (VI). A dead biomass of *Aeromonas caviae* was reported to biosorb hexavalent chromium isolated from raw water wells. Nonliving cells of *Bacillus licheniformis* and *B. laterosporus* were able to biosorb Cd(II) and Cr (Zouboulis et al., 2003). *Gloeothece magna*, a nontoxic freshwater cyanobacterium, adsorbed Cd(II) and Mn(II). Live microorganisms *Aspergillus niger* and *Pseudomonas aeruginosa* bioaccumulated 30 to 40% Cr, while several yeast, fungi, and bacteria exhibited the bioaccumulation feature for Cu ion (Malik, 2004).

Zinc is used in paints, dyes, tires, and alloys and to prevent corrosion. Untreated and acid-treated (in HNO_3 for 24 h) cassava waste biomass (*Manihot sculenta Cranz*) was able to biosorb Zn and Cd metal ions from the waste stream. Acid treatment inhibited desorption of the metal (Horsfall and Abia, 2003).

Wastewater from the electroplating, electronics, and metal cleaning industries contains high concentrations of nickel (II) ions. Batch biosorption capacities for the free biomass of *Chlorella sorokiniana* and the loofa sponge–immobilized biomass of *C. sorokiniana* were found to be 48.08 and 60.38 mg nickel (II)/g, respectively (Akhtar et al., 2004). Organisms that bioaccumulated Ni include the cyanobacteria such as *Anabaena cylindrical* and *A. flos aquae; the yeast Candida* spp.; the fungus *Aspergillus niger*, and the bacteria *Pseudomonas* spp. Ni (Malik, 2004).

Similar to those of other fungi, the cell walls of white rot fungi consist mostly of polysaccharides, peptides, and pigments that have a good capacity for binding heavy metals; hence a broad range of metals, including Cd, Cr, Cu, Ni, Pb, Hg, alkyl–Hg, and rare earth elements U and Th are known to be biosorbed. *Phanerochaete chrysosporium* mycelia have a biosorption capacity of about 60 to 110 mg of metals ions (Baldrian, 2003). *Trametes versicolor* exhibited biosorption capacity in the order Pb > Ni > Cr > Cd > Cu.

The use of active, growing cells for bioremediation of metal-contaminated effluent has several advantages including (1) the ability to self-replenish; (2) continuous metabolic uptake of metals after physical adsorption; (3) the potential for optimization through development of resistant species and cell-surface modification; (4) irreversibility since metals diffuse into the cells and get bound to intracellular proteins or chelatins before being incorporated into vacuoles and other intracellular sites; (5) avoidance of a separate biomass production process such as cultivation, harvesting, drying, processing, and storage; and (6) the possibility of developing a single stage-process. Limitations to bio-uptake by living cells are (1) the sensitivity of the system to operating conditions like pH and metal/salt concentration, and (2) the requirement for external metabolic energy (Malik, 2004).

Bioprocesses and Reactors

Generally reactors used for metal biosorption include rotating biological contactors, fixed bed, trickle filters, fluidized beds, air lift, and biofilm reactors. The living or dead biomass has been immobilized by encapsulation, cross-linking, or supports made from agar, cellulose, or alginates. A membrane bioreactor with *Alcaligenes eutrophus* supported on a tubular membrane made of polysulfone has been successfully tested for treatment of metal-contaminated waste effluents from several industries. Zinc effluent from a plating plant was reduced from 20 to below 1.00 ppm; Zn from a zinc factory was reduced from 10 to less than 0.05 ppm; Mg from the same effluent was reduced from 28 to 20 ppm; Cu from a nonferrous industry effluent was reduced from 8 to below 0.05 ppm; and Ni from a synthetic effluent was reduced from 10 to below 0.05 ppm (Diels et al., 1995).

The photofilm processing industry generates effluents in the form of used film fixer solutions that contain significant amounts of silver (greater than 3000 mg/L). The effluent also contains thiosulfate, a silver complexing agent used to remove unreacted or unexposed silver from photofilms, which interferes with the silver removal process. A chemoautotrophic bacterium *Thiobacillus thioparus* was able to oxidize thiosulfate to sulfate and sulfur. This treated water was contacted with a fungal culture *Cladosporium cladosporioides* in a continuous upflow biosorbent column packed with beads of the immobilized fungus for silver recovery (Pethkar and Paknikar, 2003). *Pseudomonas maltophila, Staphylococcus aureus,* and a coryneform organism were reported to accumulate more than 300 mg silver per gram.

A synthetic effluent containing several metals was treated by passing it through a column packed with vermicompost. The adsorption capacities of vermicompost for Cd(II), Cu(II), Pb(II), and Zn(II) ions were 33.01, 32.63, 92.94, and 28.43 mg/g, respectively. The ability of vermicompost to bind metals was attributed to the presence of negatively charged functional groups (Matos and Arruda, 2003).

A laboratory-scale up-flow algal column reactor packed with alginate–algal beads removed Cu and Ni completely from a synthetic solution. A rotating biological contactor achieved good Cu and Zn removal efficiencies (Malik, 2004). Moving-bed sand filters were used effectively with a mixed bacterial population to remove Ni from wastewater. A reactor with two strains, *Alcaligenes eutrophus* CH34 and *A. eutrophus* AE1308, removed metals such as Cd, Zn, Cu, Pb, Y, Co, Ni, Pd, and Ge via bioprecipitation. Similarly a metal accumulating strain and *Ralstonia eutropha* JMP134 together have been employed for bioaugmentation of Cd removal (Malik, 2004).

When an effluent containing copper ions and nitrates was treated in a bioelectrochemical reactor in the presence of denitrifying bacteria, the reactor could remove copper by electrochemical action and it could simultaneously perform bacterial denitrification with the help of the hydrogen generated by the electrolysis of water at the anode and the nutrients added externally (Watanabe et al., 2001).

Toxic Metals

Mercury

A *Pseudomonas putida* strain removed greater than 90% of mercury from a 40 mg/L solution in 24 h (Okino et al., 2000). An *Escherichia coli* variant containing simultaneously the merA and glutathione S-transferase genes was able to reduce mercury in the solution to Hg^0. Transgenic *Arabidopsis thaliana* plants containing a modified bacterial Hg 2+ reductase gene converted the toxic metal to Hg^0. Organomercurials are detoxified by organomercurial lyase; the resulting Hg^{2+} then is reduced to Hg^0 by mercuric reductase.

Arsenic

The large-scale treatment of timber with chromated copper arsenate and creosote oil by the wood preserving industry leads to a significant source of arsenic in the environment. *Acinetobacter, Edwardsiella, Enterobacter, Pseudomonas, Salmonella,* and *Serratia* species are resistant to arsenic. Several bacterial and fungi species are able to methylate arsenic compounds to volatile dimethyl- or trimethylarsine. Methanogenic bacteria perform methylation of inorganic arsenic under anaerobic conditions, coupling the methane biosynthetic pathway. The process consists of reduction of arsenate to arsenite followed by methylation to dimethylarsine. As(III) was oxidized to As(V) by heterotrophic bacteria (Illialetdinov and Abdrashitova, 1981) such as *Alcaligenes faecalis*. Aerobic chemolithoautotrophic microbes derive metabolic energy from the oxidation of As(III) (Santini et al., 2000). *Geospirillum anenophihs, G. barnseii,* and *Chysiogenes arsenatis* use As^{5+} as a terminal electron acceptor to support anaerobic growth, leading to the formation of soluble As^{3+}. This technique can be used for leaching arsenic from contaminated soil. Addition of an electron donor such as acetate can enhance arsenic reduction as well as promote the reduction of Fe^{3+} and Mn^{4+}, which bind As^{5+} to soil. *Methanobacterium* sp. in the presence of vitamin B12 as the methyl group donor is able to biomethylate AsO_3^- anaerobically to AsO_2^- followed by methylation to methylarsonic acid, dimethylarsenic acid, and finally to dimethylarsine (White et al., 1995).

Selinium

Wolinella succinogenes was able to reduce SeO_4^{2-} and SeO_3^{2-}. *Pseudomonas maltophila* O-2 was able to accumulate Se^0 both inside and outside the cells. *Thauera selenatis* is capable of anaerobically reducing SeO_4^{2-} to SeO_3^{2-} and further to Se^0 with concomitant reduction of NO_3^-. Reduction of elemental selenium to selenide (Se^{2-}) has been observed in cultures of *Thiobacillus ferroxidans*. Reduction of SeO_3^{2-} to Se^0 has been observed with fungi such as *Fusarium* sp., *Mortierella* sp., *Saccharomyces cerevisiae, Candida albicans,* and *Aspergillus funiculosus*, with both extracellular and intracellular deposition of Se^0 (White et al., 1995). *Aeromonas* sp., *Bacillus* sp., and *Pseudomonas* sp. microorganisms produce dimethylselenide derivatives of SeO_4^{2-} and SeO_3^{2-}. *Alternaria alternata* fungus methylated inorganic selenium.

Tellurium

Fungus *Penicillium* sp. is reported to produce dimethyltelluride and dimethylditelluride from tellurium (White et al., 1995). TeO_3^{2-} reduction to Te^0 by *Pseudomonas maltophila O-2, Rhodobacter sphaeroides* (deposited at intracellular cytoplasmic membrane), and fungus such as *Schizosaccharomyces* has been reported in literature (White et al., 1995).

Acid Mine Water

Mine water is generally very acidic (pH < 3.0) with high concentrations of metals such as Cu, Fe, Zn, A1, Pb, As, and Cd, and a high concentration of dissolved sulfates (greater than 3,000 ppm). The process to reduce the concentration of metals and sulfates that is generally adopted is addition of slaked lime to neutralize and precipitate large amounts of gypsum sludge contaminated with heavy metals. This process is expensive and labor intensive.

Bioremediation technologies include passive treatment systems (using wetlands or compost reactors under aerobic or anaerobic conditions) and active treatment methods using sulfate-reducing bacteria. In the former method, precipitated metals are retained (in the organic matrix) rather than recovered, and their long-term fate is unsure, while in the latter metals are precipitated and recovered as metal sulfides. Bioremediation using anaerobic sulfate-reducing bacteria (*Desulfibrio* sp.) has two advantages. First, sulfate can be reduced to sulfide, which reacts with dissolved metals like copper, iron, and zinc in the contaminated waters to form insoluble precipitates. Such processes have even been developed on a commercial scale to operate near mines. Second, system acidity is decreased by the reduction of sulfate to sulfide and by the carbon metabolism of the bacteria (Garcia et al., 2001). The bacteria require an anaerobic environment. Various organic waste materials, such as straw and hay, sawdust, peat, spent mushroom compost, and whey, have been used as electron donors for the sulfate reducers in the treatment of acid mine drainage. Hydrogen has also been used as the electron source to treat mine waste to reduce sulfate and precipitate Cu and Zn in a fixed bed bioreactor (Foucher et al., 2001).

Plants

Heavy metals have different patterns of behavior and mobility within a tree. For example, (1) lead, chromium, and copper tend to be immobilized and held in the roots; (2) Cd, Ni, and Zn are more easily translocated to the aerial tissues; and (3) Cd moves up into the harvestable parts of a tree (Pulford and Watson, 2003).

Thlaspi (Brassicaceae) can accumulate 3% Zn, 0.5% Pb, and 0.1% Cd in their shoots; *Alyssum* (Brassicaceae) accumulate Ni; and *Thlaspi caerulescens* is known to accumulate Zn. *Salix* were found to adsorb 30% heavy metals. *Salvinia minima* and *Spirodela punctata* removed 70 to 90% of lead and zinc, and water hyacinth removed arsenic, cadmium, lead, and mercury. *Asellus aquaticus*, a freshwater isopod, was able to bioaccumulate Pd, Pt, and Rh. *Microspora* (a macro alga) was found to adsorb lead. *Lemna minor* (an aquatic plant) adsorbed lead and nickel (Axtell et al., 2003). A marine algae *Chlorella* spp. NKG16014 biosorbed Cd.

Conclusions

Although biosorption with dead biomass appears to be very attractive, biosorption with live cells has other advantages: Metal biosorption can be combined with degradation of other organic contaminants present in the waste, and organisms can be genetically modified to improve their performance as well as survive harsh environments.

Research activities with respect to pollution control should be directed toward the following areas: (1) hastening the mobilization of metals, (2) designing metal-tolerant strains that can also adapt to performing biodegradation of organic pollutants, (3) breeding natural or engineered strains, that is, design biomass with specific metal-binding properties through the expression of metal-chelating proteins and peptides, (4) using live bacteria, (5) designing biosurfactants to assist in the solubilization and desorption of metals from polluted soils or sediments, (6) better understanding the cell microenvironment, (7) studying anaerobic respiration for the in situ treatment of organic and metal contaminants in the subsurface, and (8) improving process development techniques that combine biotreatment, separation, and recovery.

References

Akhtar, N., J. Iqbal, and M. Iqbal. 2004. Removal and recovery of nickel(II) from aqueous solution by loofa sponge-immobilized biomass of *Chlorella sorokiniana*: characterization studies. *J. Hazardous Mat.* B108:85–94.

Axtell, N. R., S. P. K. Sternberg, and K. Claussen. 2003. Lead and nickel removal using *Microspora* and *Lemna minor*. *Bioresour. Technol.* 89:41–48.

Baldrian, P. 2003. Interactions of heavy metals with white-rot fungi. *Enzyme Microbial Technol.* 32:78–91.

Chojnacka, K., A. Chojnacki, and H. Górecka. 2004. Trace element removal by *Spirulina* sp. from copper smelter and refinery effluents. *Hydrometallurgy* 73:147–153.

Diels, L., S. Van Roy, K. Somers, I. Willems, W. Doyen, M. Mergeay, D. Springael, and R. Leysen. 1995. The use of bacteria immobilized in tubular membrane reactors for heavy metal recovery and degradation of chlorinated aromatics. *J. Membrane Sci.* 100:249–258.

Eger, P. 1994. Wetland treatment for trace metal removal from mine drainage: the importance of aerobic and anaerobic processes. *Water Sci. Technol.* 29(4):249–256.

Foucher, S., F. Battaglia-Brunet, I. Ignatiadis, and D. Morin. 2001. Treatment by sulfate-reducing bacteria of Chessy acid-mine drainage and metals recovery. *Chem. Eng. Sci.* 56:1639–1645.

Fredrickson, J. K., and Y. A. Gorby. 1996. Environmental processes mediated by iron-reducing bacteria. *Curr. Opin. Biotechnol.* 7:287–294.

Garcia, C., D. A. Moreno, A. Ballester, M. L. Blazquez, and F. Gonzalez. 2001. Bioremediation of an industrial acid mine water by metal-tolerant sulphate-reducing bacteria. *Minerals Eng.* 14(9):997–1008.

Horsfall, M. Jr., and A. A. Abia. 2003. Sorption of cadmium(II) and zinc(II) ions from aqueous solutions by cassava waste biomass (*Manihot sculenta Cranz*). *Water Res.* 37:4913–4923.

Illialetdinov, A. N., and S. A. Abdrashitova. 1981. Autotrophic arsenic oxidation by a *Pseudomonas arsenitoxidans* culture. *Mikrobiologiia* 50:197–204.

Malik, A. 2004. Metal bioremediation through growing cells. *Environ. Int.* 30:261–278.

Matos, G. D., and M. A. Z. Arruda. 2003. Vermicompost as natural adsorbent for removing metal ions from laboratory effluents. *Process Biochem.* 39:81–88.

Meunier, N., J.-F. Blais, and R. D. Tyagi. 2004. Removal of heavy metals from acid soil leachate using cocoa shells in a batch counter-current sorption process. *Hydrometallurgy* 73:225–235.

Okino, S., K. Iwasaki, O. Yagi, and H. Tanaka. 2000. Development of a biological mercury removal-recovery system. *Biotechnol. Lett.* 22:783–788.

Pethkar, A. V., and K. M. Paknikar. 2003. Thiosulfate biodegradation/silver biosorption process for the treatment of photofilm processing wastewater. *Process Biochem.* 38:855–860.

Pulford, I. D., and C. Watson. 2003. Phytoremediation of heavy metal-contaminated land by trees — a review. *Environ. Int.* 29:529–540.

Xantini, J. M., L. I. Sly, R. D. Schnagl, and J. M. Macy. 2000. A new chemolithoautotrophic arsenite-oxidizing bacterium isolated from a gold mine: phylogenetic, physiological, and preliminary biochemical studies. *Appl. Environ. Microbiol.* 66:92–97.

Schneider, I. A. H., R. W. Smith, and J. Rubio. 1999. Effect of mining chemicals on biosorption of Cu (ii) by the non-living biomass of the macrophyte potamogeton lucens. *Minerals Eng.* 12(3):255–260.

Sekhar, C. K., C. T. Kamala, N. S. Chary, and Y. Anjaneyulu. 2003. Removal of heavy metals using a plant biomass with reference to environmental control. *Int. J. Miner. Process* 68:37–45.

Valls, M., and V. de Lorenzo. 2002. Exploiting the genetic and biochemical capacities of bacteria for the remediation of heavy metal pollution. *FEMS Microbiol. Rev.* 26:327–338.

Watanabe, T., H. Motoyama, and M. Kuroda. 2001. Denitrification and neutralization treatment by direct feeding of an acidic wastewater containing copper ion and high-strength nitrate to a bio-electrochemical reactor process. *Water Res.* 35(17):4102–4110.

White, C., S. C. Wilkinson, and G. M. Gadd. 1995. The role of microorganisms in biosorption of toxic metals and radionuclides. *Int. Biodeterior. Biodegrad.* 35(1–3):17–40.

Zouboulis, A. I., M. X. Loukidou, and K. A. Matis. 2003. Biosorption of toxic metals from aqueous solutions by bacteria strains isolated from metal-polluted soils. *Process Biochem.* 45:1–8.

Bibliography

Babel, S., and T. A. Kurniawan. 2003. Low-cost adsorbents for heavy metals uptake from contaminated water: a review. *J. Hazardous Mat.* B97:219–243.

Chang, I. S., P. K. Shin, and B. H. Kim. 2000. Biological treatment of acid mine drainage under sulphate-reducing conditions with solid waste materials as substrate. *Water Res.* 34(4):1269–1277.

Cinanni, V., I. A. Gough, and A. J. Sciuto. 1996. A water treatment and recovery plant for highly acidic heavy metal laden effluents. *Desalination* 106:145–150.

Dinçer, A. R. 2004. Use of activated sludge in biological treatment of boron containing wastewater by fed-batch operation. *Process Biochem.* 39:721–728.

Granato, M., M. M. M. Gonqalves, R. C. Villas Boas, and G. L. Sant'Anna Jr. 1996. Biological treatment of a synthetic gold milling effluent. *Environ. Pollution* 91(3):343–350.

Leighton, I. R., and C. F. Forster. 1997. The adsorption of heavy metals in an acidogenic thermophilic anaerobic reactor. *Water Res.* 31(12):2969–2972.

Lovley, D. R., and J. D. Coatest. 1997. Bioremediation of metal contamination. *Curr. Opin. Biotechnol.* 97:5265–5269.

Mhamed, Z. A. 2001. Removal of cadmium and manganese by a non-toxic strain of the freshwater cyanobacterium. *Goeothece. Mgna. Water Res.*, 35(18):4405–4409.

Moldovan, M., S. Rauch, M. Gómez, M. A. Palacios, and G. M. Morrison. 2001. Bioaccumulation of palladium, platinum and rhodium from urban particulates and sediments by the freshwater isopod *Asellus aquaticus. Water. Res.* 35(17):4175–4183.

Tsukamoto, T. K., and G. C. Miller. 1999. Methanol as a carbon source for microbiological treatment of acid mine drainage. *Water Res.* 33(6):1365–1370.

Yabe, M. J. S., and E. de Oliveira. 2003. Heavy metals removal in industrial effluents by sequential adsorbent treatment. *Adv. Environ. Res.* 7:263–272.

CHAPTER 14

Semiconductor Waste Treatment

The semiconductor industry had a phenomenal growth in the past 25 years. It is a $150 billion dollar industry, and because of its tremendous growth, it is also facing several environmental issues. Semiconductor manufacturing can be grouped broadly into three categories: (1) Silicon crystal wafer growth and preparation, (2) semiconductor or wafer fabrication, and (3) final assembly and packaging. The semiconductor fabrication processes are always performed in a clean room and include the following steps: oxidation, lithography, etching, doping (through processes such as vapor phase deposition and ion implantation), and layering (through processes such as metallization). Figures 14-1 to 14-3 provide a flowsheet of the entire process.

Silicon in the form of ingots is grown from seed crystals. Ingots are shaped into wafers through a series of cutting and grinding steps. The ends of the silicon ingots are removed, and individual wafers are cut from the ingot. The wafer is then polished using an aluminum oxide–glycerine solution. Further polishing is done using a slurry of silicon dioxide particles suspended in sodium hydroxide. Contaminants from the wafer are cleaned by either using a spray or immersing the wafers in acids, bases, or organic solvents.

To create the desired electronic components like transistors and resistors, impurities or dopants are introduced into the wafer to change the conductivity of the silicon. Deposition of thin films onto the silicon wafer substrate involves chemical vapor deposition, sputtering (electric deposition of a metal onto the substrate under conditions of high vacuum), and oxidation. The raw materials for deposition are in the form of gases, solid metal, and inorganic compounds. Diffusion of doping agents into the wafer layer is performed under high temperature conditions or through ion implantation, which involves bombarding the silicon wafer under high vacuum and temperature with a plasma of ionized doping agents. Photolithography is a process in which a pattern or mask is superimposed upon a photochemically coated wafer, and the etching or pattern from the mask is replicated on the underlying material. Both wet and dry etching methods

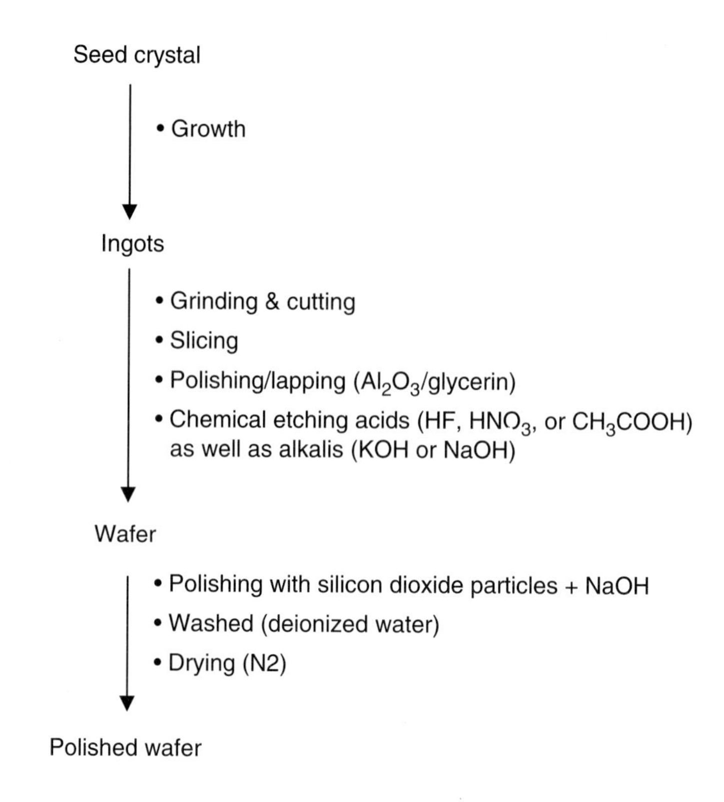

FIGURE 14-1. Silicon crystal growth and wafer preparation.

are employed; the former involves a sequence of various chemicals (typically acidic), and the latter involves wafers being processed in a chamber through which gases are pumped.

Chips or dies are mounted onto the surface of a ceramic substrate as part of a circuit, connected directly onto a printed wiring board, or incorporated into a protective package. Backside preparation involves coating with gold. Finally the wafer is separated into individual chips by sawing. The electroplating process and the final rinse is typically the primary source of process wastewater in the semiconductor assembly and packaging process.

Waste

Water usage in integrated circuit manufacture is among the highest in any industrial sector. The process requires large quantities of deionized water. Because of the purity required, process water is not recycled, and hence wastewater discharge is a major issue. Current use of ultrapure water (UPW) is 5 to 7 L/cm^2 of silicon, and in a wet bench, it is 53 L/wafer (300 mm).

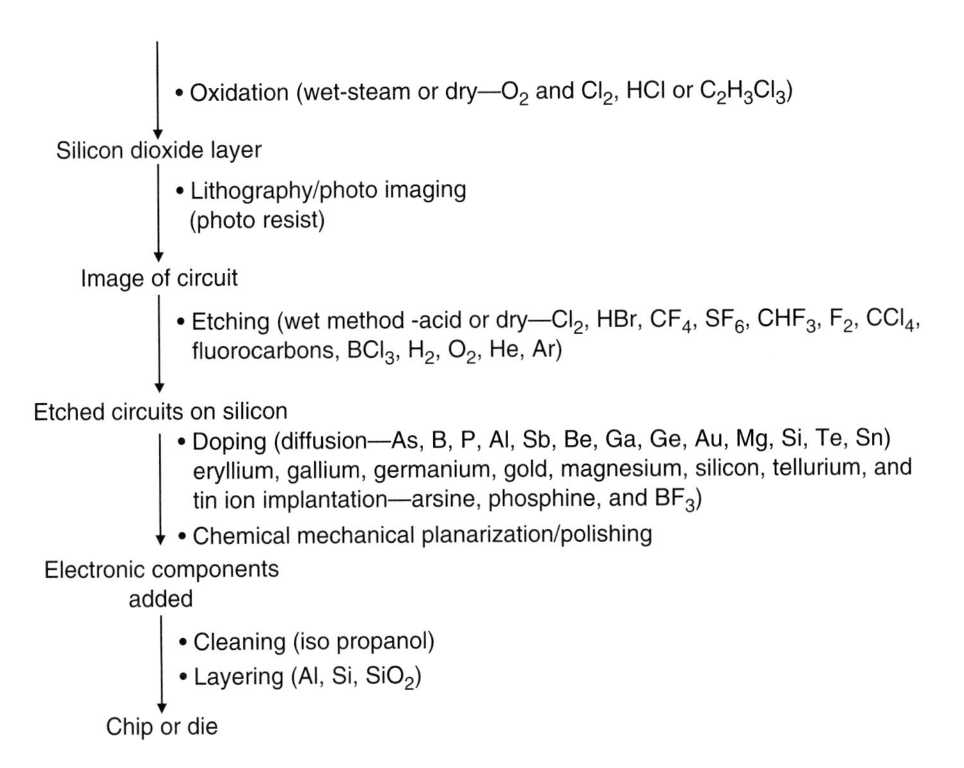

FIGURE 14-2. Semiconductor fabrication.

This works out to 20.45 million tons of water for producing 2.7 billion square centimeters of wafer. The semiconductor fabricators that use chemical mechanical planarization/polishing (CMP) consume 4.2 to 12 gallons of water per minute, which works out to more than 4.25 million gallons annually. Thus, at an average cost of $0.016 per gallon of UPW and the same amount for subsequent average waste disposal, operating a single polisher requires an expenditure of $136,000 per year in water-related costs alone.

From Figs. 14-1 to 14-3 one can see that unreacted highly toxic metals, liquids, and gases could be leaving the semiconductor manufacturing plant as waste. Hydrofluoric acid is the major inorganic acid present in the gaseous effluent stream, and calcium fluoride is also generated at 0.0018 kg per square centimeter of wafer (2000 data). Fumes from lead soldering, tin plating, and other vaporized metals used in the chemical vapor deposition also escape with the effluent gases. Disposal of these hazardous effluents such as waste solvents, dissolved organic compounds, acids, alkalis, photoresistant chemicals, dissolved metals (including arsenic, copper, chromium, and selenium), waste etchants, waste aqueous developing materials, and catalyst solutions pose a major problem. Chlorofluorocarbons (CFCs), halons, carbon

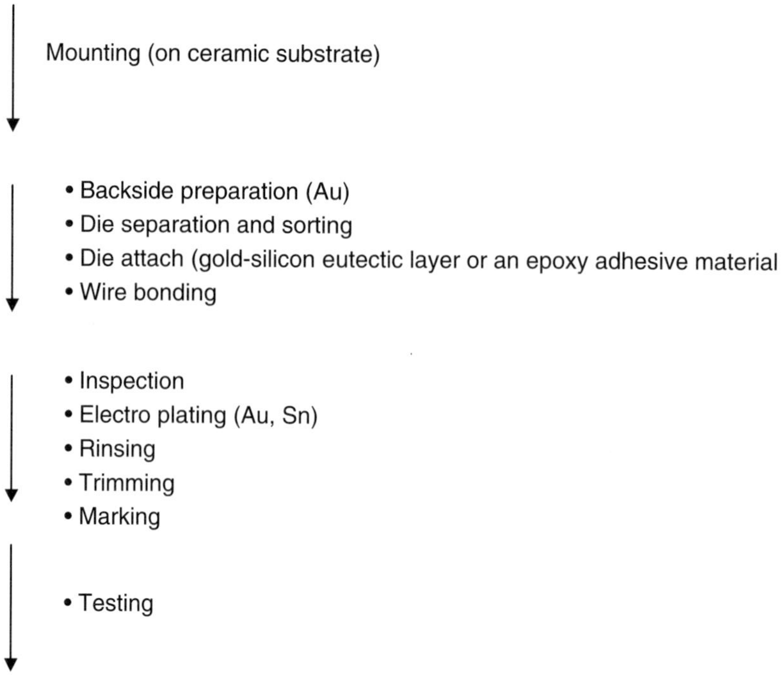

Mounting (on ceramic substrate)

- Backside preparation (Au)
- Die separation and sorting
- Die attach (gold-silicon eutectic layer or an epoxy adhesive material
- Wire bonding

- Inspection
- Electro plating (Au, Sn)
- Rinsing
- Trimming
- Marking

- Testing

FIGURE 14-3. Semiconductor assembly and packaging.

tetrachloride, and polychlorinated biphenyls have been banned or voluntarily phased out from the manufacturing process. Lead, cadmium, and mercury compounds used in packaging substrates, and perfluoro octyl sulfonates (PFOS), a component in some photoresists and antireflective coatings, have been grouped under the high-risk category (chemicals or materials have been targeted by a government authority for significant use restriction or potential ban). Perfluorocarbons (PFCs) and hydrofluorocarbons (HFCs), both of which have high global warming potential but shorter atmospheric lifetimes than the CFCs, have been grouped under the medium-risk chemicals (significant regulation of these compounds).

The manufacture of compound semiconductors such as gallium arsenide, indium phosphide, and indium antimonide require the use of a number of very hazardous gases, which include arsine, phosphine, trimethyl indium, trimethyl gallium, trimethyl aluminum, silane, and others. Disposal of unconsumed process gases and the products of the deposition process pose several problems. The worst long-term environmental concern among these is arsine, which will always produce an arsenic-tainted waste stream. In addition, the presence of phosphorous and hydrogen during pumping could also lead to pyrophoric conditions.

Current use of ultrapure water (UPW) is 5 to 7 L/cm^2 of silicon, and the goal is to reduce this level to 4 to 6 L /cm^2 by 2005. UPW use in a wet bench is 53 L/wafer (300 mm), which should be reduced to 43 L/wafer by 2005. The chemical use target is to reduce the quantity (in liters per square centimeter per mask layer) by 5% per year via more efficient use, recycle, and reuse systems. Reuse of wastewater (for cooling towers, for instance) should increase from current average levels of 65 to 70% in 2005, 80% in 2010, and 90% in 2013. Energy use for all fabrication tools is 0.5 to 0.7 kWh/cm^2, which should be brought to 0.4 to 0.5 kWh/cm^2 in 2005 and 0.3 to 0.4 kWh/cm^2 in 2008. By 2010, PFC emissions must be reduced by 10% from the 1995 baseline, as agreed to by the World Semiconductor Council. Through process optimization and alternative chemistries, recycling, and/or abatement, the industry must continue to diminish the emissions of byproducts with high global warming potential. The estimated cost to the United Kingdom economy could be as much as $761 million a year for complying with the "Waste Electrical and Electronic Equipment Directive" (European Commission 2002/95/EC and 2002/96/EC). A further $334 million a year might be needed by the industry to meet "Restriction of Use of Certain Hazardous Substances." Possible use of supercritical CO_2 for cleaning instead of water is being investigated to reduce water usage. Sulfur trioxide is being tried instead of wet chemicals as a cleaning agent for removing residual photoresist and organic polymers. This attempt could reduce the handling of large quantities of hazardous chemicals.

Physical and Chemical Treatment Methods

Several physical and chemical methods that are being practiced for treating semiconductor waste effluent include coagulation and precipitation, ion exchange, adsorption with activated carbon, membrane filtration, and chemical oxidation. Heavy metals can be precipitated as insoluble hydroxides at high pH or sometimes as sulfides. But the disposal of this highly concentrated toxic sludge poses another problem. If the sludge is not considered hazardous, then a gravity settling system can be both economical and safe. To treat a CMP waste that contains copper, a complete system that involves removal of activated carbon oxidant, filtration of slurry particles, and ion exchange to extract copper from the effluent is necessary for its removal. Strongly complexed copper is hard to precipitate or remove, and large-scale ion exchange process is expensive. Arsenic is one of the pollutants found in the wastewater. The general method used to remove this metal is by flocculation, and other methods that have been practiced include adsorbents, such as activated carbon, amorphous aluminum hydroxide, or activated alumina. The difficulty with the removal of metal anions is the fact that they do not precipitate out as hydroxides by simple pH adjustment (Reker et al., 2003).

TABLE 14-1
CMP Process Effluent Contaminants[a]

Interconnect material	Cu^{2+}, complexed Cu^{2+}, Cu_2O, CuO, $Cu(OH)_2$, WO_3, Al_2O_3, $Al(OH)_3$, Fe^{2+}/Fe^{3+}
Barrier or liner material	Tantalum, titanium oxides, oxynitrides
Abrasives	SiO_2, Al_2O_3, MnO_2, CeO_2
Oxidizers	Hydroxylamine, $KMnO_4$, KIO_4, H_2O_2, NO_3^-
Strong acids and weak buffering acids	HF, HNO_3, H_3BO_3, NH_4^+, citric acid
Strong bases	NH_3, OH
Organic materials dispersants/ surfactants	Poly(acrylic acid), quaternary ammonium salts, alkyl sulfates
Corrosion inhibitors	Benzotriazole, alkyl amines
Metal complexing agents	EDTA, ethanol amines
Acids	Poly(acrylic), oxalic, citric, acetic, peroxy acetic

[a]Golden et al., 2000.

Silica and fluoride in the wastewater could be made to react with lime to form insoluble silicates and calcium fluoride salts. Coagulation and settling of these solid insoluble particles in settling tanks could be initiated by the addition of polyacrylamide. Membrane filtration for recovery of metal has several problems, which include difficulty in retaining small-sized metal particles, abrasion of the membrane, and lack of resistance to pH fluctuations.

Chemical mechanical polishing is carried out to reduce wafer topological imperfections and to improve the depth of focus of lithography processes through better planarity. CMP process effluent contains many contaminants, some of which are shown in Table 14-1.

CMP wastewater treatment involves neutralization of ion and particle surface charge by oppositely charged inorganic and organic materials. When excess coagulant is added, the particles and some ions are trapped within a gel-like matrix and agglomerate. This process is known as "sweep coagulation." Typical inorganic coagulants used for this purpose are aluminum sulfate and ferric chloride, both of which form insoluble hydrated hydroxide gels at pH 5 to 8. Addition of organic flocculants such as polyacrylamide further destabilizes the coagulated agglomerate for gravity settling or active filtration. A new technique that is being researched is called electrocoagulation and electrodecantation, which uses electric fields to agglomerate charged silica particles instead of adding polymers. Commonly used techniques to separate the floc from the clarified water include gravity settling, cross-flow filtration, and single-pass low-pressure filtration. Removal of copper from the wastewater to a 50 ppb level was achieved using polymeric metal removal agents (a polymer containing sulfide functionality) even in the presence of ammonia and other competing materials. Copper removal has also been achieved by pH adjustment followed by ion exchange. The drawbacks of this approach include the large amounts of acid and base needed in the pH adjustment steps, the need for frequent ion bed regeneration, and the bed damage due to the presence of suspended solids.

Adsorption of metals from liquid streams using treated sawdust is found to be very effective. Hg (II) is effectively removed using polymerized sawdust or peanut hulls treated with bicarbonate. Divalent Cu, Pb, Hg, Fe, Zn, and Ni and trivalent Fe are removed using untreated sawdust as well as sawdust treated with a reactive monochlorotriazine type dye. The treated sawdust showed better adsorption efficiency than the untreated sawdust (Shukla and Sakhardande, 1991). A column packed with a resin of sawdust, onion skin, and polymerized corncob could remove 86% of Pb, 79% of Ca, 77% of Ni, 75% of Zn, 71% Mg, 65% Mn, and 60% Cu divalent ions. Sawdust modified with iron hexamine gel efficiently removed very toxic metals like Hg, Cr, and Cd. Heavy metal cations are capable of forming complexes with O^-, $N,^-$, S^-, and P^- containing functional groups. The cell walls of sawdust consist of cellulose, lignin, and many hydroxyl groups, which are present as part of tannins or other phenolic compounds. It is speculated that ion exchange or hydrogen bonding may be the principal mechanisms for the binding of these metals to sawdust. Polacrylamide-treated sawdust was very effective in removing Cd and Hg(II), while rubber wood sawdust could effectively adsorb Co(II), Cr(II), and Cr(VI). Treatment of exposed sawdust with nitric acid completely removes the metal ions (Yu et al., 2001). The binding capacity of various ion exchange resins for Cu (II) varies between 0.01 and 0.1 g per gram of the resin.

Dimethyl sulfoxide (DMSO) is a widely used organic solvent in the semiconductor industry; hence finds it way into the effluent and requires costly treatment. Fenton treatment was also investigated using H_2O_2: Fe^{2+} at the ratio of 1,000:1,000 mg/L for wastewater containing 800 mg DMSO/L. Such a treatment achieved a total organic carbon (TOC) removal efficiency of 26%, and the biological oxygen demand/chemical oxygen demand (BOD:COD) ratio of the wastewater was increased from 0.035 to 0.87 when the reaction was carried out at pH 3 and the coagulation at pH 7. An increase in BOD:COD ratio makes this process an attractive pretreatment step before biological treatment (Park et al., 2001). Sulfuric acid is used for wafer cleaning, and its disposal involves neutralization; the quantity of waste therefore exceeds the quantity of the used sulfuric acid. Generally sulfuric acid makes up about 17% of the entire quantity of waste acid in semiconductor industrial waste. Atmospheric and vacuum distillation and recovery of sulfuric acid has been attempted successfully.

Biochemical Methods

Isopropyl alcohol and acetone are common solvents in the cleaning steps, and large quantities of their vapors are released into the atmosphere. A trickle bed air biofilter packed with about 7.8 L of coal (voidage=0.44) achieved a 90% removal efficiency for this vapor mixture with influent carbon loadings of the alcohol and acetone below 80 and 53 $g/m^3/h$, respectively, at a temperature of 30°C, relative humidity of 90%, and an empty-bed residence time of 25 s. The biofilter was seeded with activated sludge from a wastewater treatment plant. The nutrient to the trickle biofilter feed contained inorganic salts (Mg, Na, K, Mn, and ammonium sulfates, chlorides, and phosphates) and $NaHCO_3$ as a buffer. The carbon mass ratio of the influent air stream to nitrogen, phosphorus, sulfur, and iron of the nutrient solution was equal to 100:10:1:1:0.5, respectively (Chang and Lu, 2003).

Complex effluents having a COD of 80,000 mg/L and isopropyl alcohol (ipa) of 35,000 mg/L cannot be treated effectively with one technique alone but can be successfully treated using a process that combines physical, chemical, and biological methods (Fig. 14-4). The initial treatment consisted of air stripping the effluent using a packed column at a temperature of 70°C to recover 95% of the ipa at 9% purity. Fenton oxidation of this stripped stream was carried out after diluting it with other effluents. The use of 5 g/L of $FeSO_4$ and 45 g/L of H_2O_2 for the oxidation achieved a 95% reduction in COD and a 99% reduction in the color of the effluent. Using sludge from a

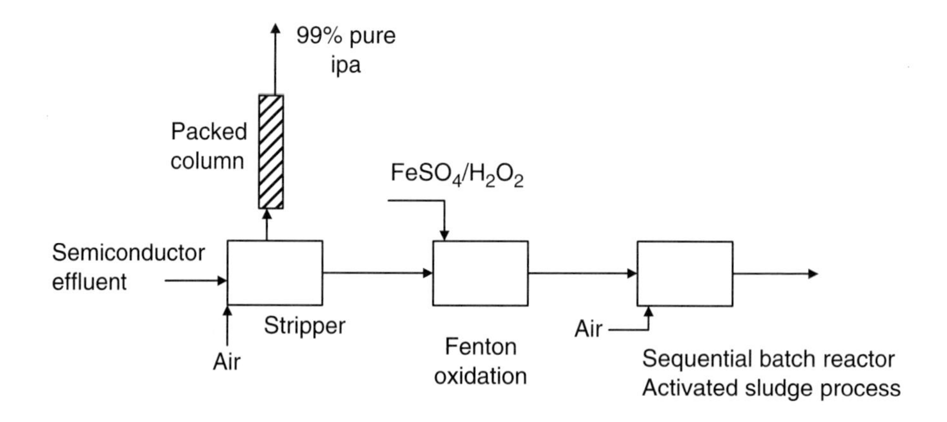

FIGURE 14-4. Combined physical, chemical, and biological treatment.

municipal wastewater treatment plant, an aerobic sequencing batch reactor with a 12-h cycle, and mixed liquor suspended solids (MLSS) of 3,000 mg/L was able to achieve an 85% reduction in COD. The combined treatment was capable of lowering the wastewater COD concentration from 80,000 mg/L to below 100 mg/L and completely eliminated the wastewater color (Lin and Jiang, 2003). Activated sludge entrapped in polyethylene glycol prepolymer pellets was applied to the continuous treatment of organic wastewater discharged from a semiconductor plant that had a BOD of 150 to 200 mg/L at a loading rate of 5.21 kg BOD/m^3/day achieving BOD removal efficiencies of 95 to 97% (Hashimoto and Sumino, 1998).

Biological breakdown of DMSO produces dimethylsulfide (DMS), which ultimately produces 2 mol of formaldehyde and 1 mol of sulfide. Formaldehyde is converted to CO_2 or used for cell synthesis, and sulfide is oxidized to sulfate. Enzyme systems such as methionine sulfoxide reductase, methionine sulfoxide-peptide-reductase, biotin sulfoxide reductase, anaerobic DMSO reductase, anaerobic trimethylamine reductase, and aerobic DMSO reductase are reported to mediate DMSO reduction to DMS (Griebler and Slezak, 2001). Microorganisms that use DMSO as a terminal electron acceptor are anaerobically grown *Escherichia coli* HB101, anaerobic rumen bacterium *Wolinella succinogenes, Rhodopseudomonas capsulata,* and *Escherichia coli*. Wastewater containing 800 mg/L of DMSO was treated successfully in an activated sludge process to achieve TOC, soluble COD (SCOD), and soluble BOD (SBOD) removal efficiencies of 90, 87, and 63%, respectively, at a hydraulic retention time (HRT) of 24 h at a loading rate of 0.8 kg DMSO/m^3/day. Most of the sulfur in DMSO was oxidized to sulfate (Park et al., 2001).

Biosorption

Metal recovery can be achieved with the use of plant, algal, or microbial biomass; this adsorption process is termed "biosorption." Pretreatment enhances the metal-binding ability. Dead microorganisms or their derivatives (bacteria, fungi, yeast, algae, and higher plants) can complex metal ions through the action of ligands or functional groups located on the outer surface of the cells. Biosorptive processes can reduce capital costs by 20%, operational costs by 36%, and total treatment costs by 28% when compared with conventional approaches (Volesky, 2001). *Mucor rouxii,* a soil fungus, can biosorb copper and silver found in CMP effluent. Biosorption of metals is also discussed in Chapter 13, Treatment of Waste from Metal Processing and Electrochemical Industries.

Aspergillus oryzae and *Rhizopus oryzae* are able to biosorb copper (II) very effectively from wastewater (Huang and Huang, 1996). Acid-washed *A. oryzae* mycelia exhibited maximum biosorption capacity when compared with the other adsorbents. Acid washing can be used as a pretreatment step and also as a regeneration step in the heavy metal removal process. A column reactor packed with 2- to 3-mm diameter pellets of *A. oryzae* was also effective in removing Cu (II). Sodium alginate–immobilized Soil 5Y cells and immobilized *Pseudomonas aeruginosa* PU21could biosorb 0.14 and 0.15 g Cu per gram of the biomass, respectively, at pH 5 (Ogden et al., 2001). There were two distinct adsorption phases—an initial rapid uptake followed by a gradual uptake; the former was probably due to the adsorption of copper ions onto the cell walls. The immobilized Soil 5Y–biosorbed Cu (II) could be desorbed by treating it with HCl (achieving 90% recovery). Other organisms that could adsorb Cu^{2+} are *Bacillus* bacteria, reaching adsorption equilibrium in 10 min at pH 7.2; planktonic *Thiobacillus ferrooxidans* cells, reaching adsorption equilibrium in 15 min, and immobilized *Zoogloea ramigera* cells, which produce an extracellular polysaccharide layer and reach their maximum copper adsorption capacity in 2 h. Brown seaweed *Sargassum* sp. (*Chromophyta*) harvested from the sea (northeastern coast of Brazil) could biosorb copper ions with a high biosorption capacity (1.48 mmol/g at pH 4.0). Other biosorbents reported in the literature were *Rhizopus arrhizus* (0.25 mmol/g), *Pseudomonas aeruginosa* (0.29), *Phanerochaete chrysosporium* (0.42), pretreated *Ecklonia radiata* (1.11), and *Ulothrix zonata* (2.77) (Antunes et al., 2003).)

NaOH-pretreated *Mucor rouxii* biomass showed a high adsorption capacity for the removal of lead, cadmium, nickel, and zinc from aqueous solution. Recovery of these biosorbed metal ions was achieved with nitric acid treatment. Caustic regeneration of eluted biomass rehabilitated the metal ion biosorption capacity even after five cycles of reuse (Yan and Viraraghavan, 2003). Live biomass had a higher biosorption capacity than dead biomass (i.e., 35.69, 11.09, 8.46, and 7.75 mg/g at pH 5.0 for Pb^{2+}, Ni^{2+}, Cd^{2+}, and Zn^{2+}, respectively, as against 25.22, 16.62, 8.36, and 6.34 mg/g, respectively, with dead biomass). Biosorption depended on an intermediate pH; a value of 6.0 was found to be the maximum. At low pH (~2 to 4), the binding sites get protonated due to a high concentration of protons and the negative charge intensity on the sites is decreased, resulting in the reduction or inhibition of the binding of metal ions (which are positively charged). Yeast extract, peptone, and glucose medium, or yeast and malt broth medium had no effect, whereas dextrose and peptone medium decreased the biosorption capacity of the fungus. Biosorption capacity of Pb remained almost constant even in the presence of other ions. The biosorption capacity of Ni, Cd, and Zn decreased in the presence of other ions, indicating the operation of a competitive adsorption mechanism. Heavy metals such as Ni, Zn, Cd, Ag, and Pb were biosorbed by a *Rhizopus arrhizus* biomass under pH-controlled conditions. The maximum sorption capacity for Pb was observed at a pH 7.0 (200 mg/g) (Fourest et al., 1994). Dead *R. nigricans* obtained as a waste by-product from the pharmaceutical fermentation industry has been found to adsorb Pb over a range of metal ion concentrations, adsorption time, pH, and co-ions (Li et al., 1998). The uptake process obeys both the Langmuir and Freundlich isotherms.

Phanerochaete chrysogenum, a waste byproduct from antibiotic production, has been surface modified with surfactants and investigated for the removal of arsenic from the waste effluent (Loukidou et al., 2003). At pH 3, the removal capacities were 37.85 mg As/g for cationic surfactant hexadecyl-trimethylammonium bromide–modified biomass, 56.07 mg As/g for polyelectrolyte Magnafloc-modified biomass, and 33.31 mg As/g dodecylamine-modified biomass. The adsorption capacity of activated chitosan for arsenic is higher than any other adsorbent, such as fly ash, bauxite, or alumina (197.6 mg/g at pH 3.0, 30 mg/g at pH 2.5, 12.6 mg/g at pH 3.5, and 12.3 mg/g at pH 2.6, respectively). The metal-loaded biomass following biosorption (dodecylamine-modified biomass) was separated by dispersed-air flotation, leading to 75% arsenic anion removal (Loukidou et al., 2001). Physical and chemical pretreatment processes enhance the biosorption capacity of cations. Physical methods include heat treatment, autoclaving, freeze-drying, or boiling, whereas chemical methods include contact with acids, alkalis, or organic chemicals.

Conclusions

The semiconductor industry uses large quantities of water and a wide range of heavy metals, acids, alkalis, and toxic and hazardous organic and inorganic chemicals. Hence this industry is facing serious environmental problems. Recovery and reuse of water, acids, and other chemicals could solve many of its waste disposal problems, but the need for high purity water and chemicals makes the industry hesitant to reuse the recovered chemicals. Biofilters or biotrickling filters appear to be good technologies for treating vapors and gaseous effluents from the semiconductor plant. Coagulation followed by settling and filtration of the liquid effluent is effective and cheap for removing the hazardous material from the effluent, but the disposal of the toxic sludge generated is a serious problem that has not been solved. Bioremediation for liquid effluent appears to be very limited except for the use of biosorption for extracting metals. Phytoremediation also appears to be a good technique for treating contaminated soil and solid wastes.

References

Antunes, W. M., A. S. Luna, and C. A. Henriques. 2003. An evaluation of copper biosorption by a brown seaweed under optimized conditions. *Electronic J. Biotech.* 6(3).

Chang, K., and C. Lu. 2003. Biofiltration of isopropyl alcohol and acetone mixtures by a trickle bed air biofilter. *Process Biochem.* 39:415–423.

Fourest, E., C. Canal, and J.-C. Roux. 1994. Improvement of heavy metal biosorption by mycelial dead biomasses (*Rhizopus arrhizus, Mucor miehei* and *Penicillium chrysogenum*: pH control and cationic activation. *FEMS Microbiol. Rev.* 14(4):325–332.

Golden, J. H., R. Small, L. Pagan, and C. Shang. 2000. Evaluating and treating CMP wastewater, *Semiconductor Int.* Jan. 10.

Griebler, C., and D. Slezak. 2001. Microbial activity in aquatic environments measured by dimethyl sulfoxide reduction and intercomparison with commonly used methods. *Appl. Environ. Microbiol.* 67(1):100–109.

Hashimoto, N., and T. Sumino. 1998. Wastewater treatment using activated sludge entrapped in polyethylene glycol prepolymer. *J. Ferment. Bioeng.* 86(4):424–426.

Huang, C., and C. P. Huang. 1996. Application of *Aspergillus oryze* and *Rhizopus oryzae* for Cu(II) removal. *Water Res.* 30(9):1985–1990.

Li, Z., Z. Li, Y. Yaoting, and C. Changzhi. 1998. Removal of lead from aqueous solution by non-living *Rhizopus nigricans. Water Res.* 32(5):1437–1444.

Lin, S. H., and C. D. Jiang. 2003. Fenton oxidation and sequencing batch reactor (SBR) treatments of high-strength semiconductor wastewater. *Desalination.* 154(2):107–116.

Loukidou, M. X., K. A. Matis, and A. I. Zouboulis. 2001. Removal of arsenic from contaminated dilute aqueous solutions using biosorptive flotation. *Chem. Ing. Tech.* (73)6:596.

Loukidou, M. X., K. A. Matis, and A. I. Zouboulis, and M. Liakopoulou-Kyriakidou. 2003. Removal of As(V) from wastewaters by chemically modified fungal biomass. *Water Res.* 37(18):4544–4552.

Ogden, K. L., A. J. Muscat, and L. C. Stanley. 2001. Investigating the use of biosorption to treat copper CMP water. MicroMagazine.com. <http://www.MicroMagazine.com>.

Park, S. E., T.-I. Yoon, J.-H. Bae, H.-J. Seo, and H.-J. Park. 2001. Biological treatment of wastewater containing dimethyl sulphoxide from the semi-conductor industry. *Process Biochem.* 36:579–589.

Reker, M., M. Lenart, and S. Harnsberger. 2003. Treatment and water recycling of copper cmp slurry waste streams to achieve compliance for copper and suspended solids. *Semiconductor Fabtech.* 8:144–148.

Shukla, S. R., and V. D. Sakhardande. 1991. Dyestuffs for improved metal adsorption from effluents. *Dyes and Pigments* 17(1):11–17.

Volesky, B. 2001. Detoxification of metal-bearing effluents: biosorption for the next century. *Hydrometallurgy* 59:203–216.

Yan, G., and T. Viraraghavan, 2003. Heavy-metal removal from aqueous solution by fungus *Mucor rouxii. Water Res.* 37:4486–4496.

Yu, B., Y. Zhang, A. Shukla, S. S. Shukla, and K. L. Dorris. 2001. The removal of heavy metals from aqueous solutions by sawdust adsorption—removal of lead and comparison of its adsorption with copper. *J. Hazardous Mat.* 84(1):83–94.

CHAPTER 15

Waste from Nuclear Plants

Introduction

The nuclear industry provides products that play a vital role in society. This is a unique industry that provides products both for the protection and destruction of society. They provide stable nuclides used in medicine (imaging and diagnostic) and nuclear explosives used by the military. It is one of the major energy sources for the production of electricity to meet the world's needs.

There are three types of nuclear wastes, based on their radionuclide characteristics:

- Uranium-contaminated waste
- Plutonium-contaminated waste
- Other radionuclide-contaminated waste

Of these types of wastes, uranium- and plutonium-contaminated wastes are potentially hazardous to human and animal health. Other nuclide wastes are low-level waste, having lower radioactivity. Although there are natural sources of radioactivity, the release of anthropogenic radionuclides into the environment is significant and a subject of intense public concern. Plutonium (Pu) contamination of soils, sediments, and/or water is an important consideration because this transuranic element can influence populations inhabiting the contaminated environment. A long half-life ($t_{1/2} = 2.41 \times 10^4$ years for ^{239}Pu) and potential health effects of Pu have resulted in extensive field and laboratory studies to resolve its environmental behavior (Garland et al., 1981).

Waste Management

Radioactive waste management involves the treatment, storage, and disposal of liquid, airborne, and solid effluents from the nuclear industry's

Bacterial oxidation of pyrite: (*Thiobacillus ferroxidans*)

$$2FeS_2 + H_2O + 7\frac{1}{2}O_2 \longrightarrow Fe_2(SO_4)_3 + H_2SO_4$$

Chemical oxidation and solubilization of the uranium by ferric sulfate:

$$UO_2 + Fe_2(SO_4)_3 \longrightarrow UO_2SO_4 + 2FeSO_4$$

Chemical oxidation of the pyrite by ferric sulfate:

$$FeS_2 + 7\,Fe_2(SO_4)_3 + 8H_2O \longrightarrow 15\,FeSO_4 + 8H_2SO_4$$

Bacterial reoxidation of ferrous sulfate: (*Thiobacillus ferroxidans*)

$$4FeSO_4 + 2H_2SO_4 + O_2 \longrightarrow 2Fe_2(SO_4)_3 + 2H_2O$$

FIGURE 15-1. Solubilization of a radionuclide—uranium to uranyl sulfate.

operations. Four methods are employed involving chemical transformations, namely:

- Limit generation
- Delay and decay
- Concentrate and contain
- Dilute and disperse

Limiting the generation of waste is the first and most important consideration in managing radioactive wastes. Delay and decay is frequently an important strategy because much of the radioactivity in nuclear reactors and accelerators is very short lived. Concentrating and containing is the objective of treatment activities for longer-lived radioactivity. The waste is contained in corrosion resistant containers and transported to disposal sites. Leaching of heavy metals and radionuclides from these sites is a problem of growing concern. Microorganisms corrode even the high-grade metal containers and solubilize the metal ions. Ferric sulfate formed in situ by the biological oxidation of pyrite (by *Thiobacillus ferroxidans*) converts uranium present in these sites to soluble uranyl sulfate (Fig. 15-1). For wastes having low radioactivity, dilution and dispersion are adopted.

Bioremediation

Chemical approaches are available for metal and radionuclide remediation but are often expensive to apply and lack the specificity required to treat target nuclides against a background of competing metal ions. In addition, such approaches are not applicable to cost-effective remediation of large-scale subsurface contamination in situ. Biological approaches, on the other hand, offer the potential for the highly selective removal of toxic metals and radionuclides coupled with considerable operational flexibility; they can be used both in situ and ex situ in a range of bioreactor configurations. A good degree of mineralization is achieved during biodegradation of radioactive waste.

Reactions mediated by microorganisms include solubilization or volatilization of metals ions (radionuclide ions) from organic and inorganic complexes, compounds, and minerals by production of acids or chelating agents (Francis, 1994), as well as removal from aqueous solution by a number of mechanisms that include biosorption, accumulation, and chemical precipitation. Chemical transformations such as oxidation and reduction can also be catalyzed by a range of microorganisms; these reactions can alter a number of important properties, such as speciation and water solubility, that influence biotic effects and environmental mobility of these ions (Gadd, 1993; Lovley, 1995). The different reactions or transformations that microorganisms bring about on metal ions or radionuclide ions are:

- Biosorption and accumulation
- Translocation
- Reduction and precipitation
- Solubilization

Immobilization—Biosorption and Accumulation

Biosorption is microbial uptake of radionuclide species, both soluble and insoluble, by physicochemical mechanisms, such as adsorption. Biosorption can also provide nucleation sites and stimulate the formation of extremely stable minerals. The constituent biomolecules of microbial cell walls have great affinity for radionuclides and are of greatest significance in biosorption. Once inside the cells, metals and radionuclides may be bound, precipitated, localized, or translocated. Microorganisms can form aggregates with other colloidal materials (clay minerals) and thus help in the transport of radionuclides. Many microbial exopolymers act as polyanions under natural conditions, and negatively charged groups can interact with cationic metal and radionuclide species, thereby achieving the biosorption on the cell walls (Geesey and Jang, 1990). The carboxyl groups on the peptidogylcan are the main binding site for cations in gram-positive cell walls, with phosphate groups contributing significantly in gram-negative species (Beveridge and Doyle, 1989). Chitin is an important structural component of fungal cell walls, and this polymer is an effective biosorbent for radionuclides. Actinide accumulation by fungal biomass is one such example (Tobin et al., 1994).

Fungi, including yeasts, have received attention in connection with metal biosorption, particularly because waste biomass arises as a byproduct from several industrial fermentations, while algae have been viewed as a renewable source of metal-sorbing biomass. Both freely suspended and immobilized biomass from bacterial, cyanobacterial, algal, and fungal species have received attention. One drawback of this method of remediation is the treatment (disposal) of the radionuclide accumulated biomass. A chemical or physical treatment of the radioactivity in the biomass becomes unavoidable.

Macskie and Dean (1989) have developed a biofilter to remove and recover heavy metals from synthetic aqueous solutions. The active agent in the metal uptake is a phosphatase overproduced at the cell surface by bacteria (growing on the inner rim of a tube), a *Citrobacter* sp., originally isolated from a contaminated soil sample. The process of metal uptake relies on in situ cumulative deposition of insoluble metal phosphatase tightly bound to the cell surface. Soluble metals are converted to insoluble metal phosphates by a biocatalytic process that readily operates at low metal concentrations unmanageable by classical precipitation, thus overcoming the chemical constraints of the solubility product of the metal phosphate in the bulk solution. The wastewater containing the heavy metal pollutant is passed through the pipe. All the heavy metal ions get bound to the phosphatase on the cell surface. Since high loads of phosphate are produced in a localized environment, metals can be precipitated at very low metal concentrations. After the metals have been concentrated, they can be safely disposed of as metal byproducts to be reused elsewhere.

Transport

The uptake and transport of radionuclides by microorganisms is dependent on the pH and monovalent cation (K^+) concentration. Many times the entry of radionuclides into the microbial cell occurs via active transport systems for K^+ or NH_4^+. In a sense radionuclides are competitive inhibitors of the K^+ channel. For example, Cs^+ accumulation is particularly dependent on external pH and monovalent cation concentration, especially K^+ (Avery et al., 1992; Perkins and Gadd, 1995). Cyanobacteria and algae are also capable of Cs^+ accumulation (Avery et al., 1992; Garnham et al., 1993). In eukaryotic microorganisms, such as microalgae and fungi, vacuoles appear to be a preferential intracellular location for Cs^+ (Avery, 1995). Metals or radionuclides may also precipitate within cells as sulfides, oxides, and phosphates. Microorganisms are also known to produce specific biomolecules (peptides) to bind to radionuclides. The fruiting bodies of fungi are also known to have high concentrations of radionuclides. ^{137}Cs accumulation by macrofungi (mushrooms) following the Chernobyl accident in 1986 is well documented. Grazing of these fruiting bodies by animals may lead to radionuclide (cesium) transfer along the food chain (Dighton and Terry, 1996).

Reduction and Precipitation

Reduction is one of the most important chemical transformations catalyzed by microorganisms, affecting the solubility of radionuclides. Under anaerobic conditions, the oxidized form of the metal becomes the TEA (terminal electron acceptor). For example, a strain of *Shewanella putrefaciens* reduced U(VI) to U(IV) (Lovely et al., 1991), giving rise to a black precipitate of U(IV) carbonate because U(IV) compounds are less soluble than U(VI) compounds. *Geobacter metallireducans* also reduces U(VI) to U(IV) species. These transformations play a significant role in the environment because they immobilize uranium.

Because many radionuclides of concern are both redox active and less soluble when reduced, bioreduction offers much promise for controlling the solubility and mobility of target radionuclides in contaminated sediments.

The first demonstration of dissimilatory U(VI) reduction was by the Fe(III)-reducing bacteria *G. metallireducens* and *S. oneidensis* (Lovely et al., 1991), which conserved energy for anaerobic growth via reduction of U(VI). It should be noted, however, that the ability to reduce U(VI) enzymatically is not restricted to Fe(III)-reducing bacteria. Other organisms, including a *Clostridium* sp., *Desulfovibro desulfuricans,* and *D .vulgaria*, also reduce U(VI).

Although ^{238}U remains the priority pollutant in most medium- and low-level radioactive wastes, other actinides, including ^{230}Th, ^{237}Np, ^{241}Pu, and ^{241}Am, can also be present. Fe(III)-reducing bacteria have the metabolic potential to reduce Pu(V) and Np(V) enzymatically. This is significant in that the tetravalent actinides are amenable to bioremediation because of their high ligand complexing abilities (Lloyd and Macaskie, 2000) and are also immobilized in sediments containing active biomass (Peretrukhin et al., 1996).

The most obvious applications of microbially mediated precipitation of toxic metals and radionuclides are those involving sulfide precipitation, phosphatase-mediated precipitation, and chemical reduction. Organisms capable of sulfide production (*Thiobacillus ferrooxidans*) are receiving considerable attention in bioremediation, both in reactor and in situ treatment systems. A promising application of biological metal reduction is uranium precipitation from nuclear effluents.

Solubilization

Microorganisms and plants are known to produce chelating agents that complex with metals and radionuclides. These complexes are usually soluble in water. Once in solution, they may either get converted to their corresponding hydroxides or they may be absorbed by plants. Leaching may also be brought about by autotrophic bacteria under aerobic conditions. Such processes are catalyzed mainly by *thiobacilli*, such as *Thiobacillus ferrooxidans*. In fact, this organism is used on a commercial scale for the extraction of uranium from ore (Francis, 1990). Heterotrophic bacteria produce a large number of diverse chelating agents, such as dicarboxylic acids, glucuronic acids, protocatechuic acid, and salicylic acid, to complex with metals or radionuclides. Uranyl complexes with oxalic acid, citric acid, and succinic acids have been reported. Alongside these chelating agents, microorganisms are known to excrete "siderophores" under iron-limiting conditions. Solubilization of Pu(IV) with siderophores has been reported (Birch and Bachofen, 1990) and is an important means of remediation of Pu(IV).

Phytoremediation

Phytoremediation is a technology that should be considered for remediation of contaminated sites because of its cost effectiveness, aesthetic advantages, and long-term applicability. This technology can be applied for metal pollutants that are amenable to phytostabilization, phytoextraction, phytotransformation, rhizosphere bioremediation, or phytoextraction (Schnoor, 1997).

Lee et al. (2002) observed that plutonium uptake and accumulation by the Indian mustard plant (*Brassica juncea*) was higher than that by the sunflower plant (*Helianthus annuus*). They also observed that Pu uptake was dependent on the chelating agent (nitrate, citrate, etc.) present in the soil.

Composting

Composting is generally achieved by converting solid wastes into stable humus-like materials via biodegradation of putrescible organic matter (Huang et al., 2000). The composting process consists of microbiological treatment in which aerobic microorganisms use organic matter as a substrate. The main products of the composting process are fully mineralized materials, such as CO_2, H_2O, NH_4^+, stabilized organic matter heavily populated with competitive microbial biomass, and ash. Compost has the potential of improving soil structure, increasing cation exchange capacity, and enhancing plant growth. Ipek et al. (2002) showed that beta-radioactivity was greatly decreased by aerobic composting.

Bioremediation holds the key to radioactive waste management. Chemical approaches, though effective, are not economical and cannot be applied to larger field areas. A *combination of phytoremediation alongside bioremediation* would certainly contain the hazardous radioactive wastes, thereby providing the much needed safety cover for the communities living near these contaminated sites.

References

Avery, S. V. 1995. Caesium accumulation by microorganisms, uptake mechanisms, cation competition, compartmentalization and toxicity. *J. Ind. Microbiol.*, 14:76–84.

Avery, S. V., G. A. Codd, and G. M. Gadd. 1992. Interactions of cyanobacteria and microalgae with caesium. In: *Impact of heavy metals on the environment*, J. P. Vernet (ed.), pp. 133–182, Amsterdam: Elsevier.

Beveridge, T. J., and R. J. Doyle. 1989. *Metal ions and bacteria*, New York: Wiley.

Birch, L., and R. Bachofen, 1990. Complexing agents from microorganisms. *Experienta*. 46:827–834.

Dighton, J., and G. Terry, 1996. Uptake and immobilization of caesium in UK grassland and forest soils by fungi following the Chernobyl accident. In: *Fungi and environmental change*, J. C. Frankland, N. Magan, and G. M. Gadd (eds.), pp. 184–200. Cambridge: Cambridge University Press.

Francis, A. J. 1990. Microbial dissolution ad stabilization of toxic metals and radio nuclides in mixed wastes. *Experientia*. 46:840–851.

Francis, A. J., 1994. Microbial transformations of radioactive wastes and environmental restoration through bioremediation. *J. Alloys Compounds*. 213:226–231.

Gadd, G. M. 1993. Microbial formation and transformation of organometallic and organometalloid compounds. *FEMS Microbiol. Rev.* 11:297–316.

Garland, T. R., D. A. Cataldo, and R. E. Wildung, 1981. Absorption, transport, and chemical fate of plutonium in soybean plants. *J. Agric. Food. Chem.* 29(5):915–920.

Garnham, G. W., G. A. Codd, and G. M. Gadd. 1993. Uptake of cobalt and cesium by microalgal- and cyanobacterial-clay mixtures. *Microb. Ecol.* 25:71–82.

Geesey, G., and L. Jang. 1990.Extracellular polymers for metal binding. In: *Microbial Mineral Recovery*, 223–247. New York: McGraw-Hill.

Huang, J. S., C. H. Wang, and C. G. Jih. 2000. Empirical model and kinetic behavior of thermophilic composting of vegetable waste. *J. Environ. Eng.* 126:1019–1025.

Ipek, U., E. Obek, L. Akca, E. I. Arslan, H. Hasar, M. Dogne, and O. Baykara. 2002. Determination of degradation of radioactivity and its kinetics in aerobic composting. *Bioresource Technol.* 84, 283–286.

Lee, J. H., L. R. Hossner, M. Attrep, Jr., and K. S. Kung. 2002. Comparative uptake of plutonium from soils by Brassica juncea and Helianthus annuus. *Environ. Pollution* 120:173–182.

Lloyd, J. R., and L. E. Macaskie. 2000. In: *Environmental microbe-metal interactions*, 277–327. Washington, DC: ASM Press.

Lovely, D. R., E. J. P. Phillips, Y. A. Gorby, and E. Land. 1991. Microbial reduction of uranium. *Nature*. 350:413–416.

Lovley, D. R. 1995. Bioremediation of organic and metal contaminants with dissimilatory metal reduction. *J. Ind. Microbiol.* 14:85-93.

Peretrukhin, V. F., N. N. Khizhniak, N. N. Lyalikova, and K. E. German. 1996. Biosorptioin of technetium-99 and some other radionuclides by bottom sediments, which were taken from lake White Kosino of Moscow region. *Radiochem*. 38:440–443.

Perkins, J., and G. M. Gadd. 1995. The influence of pH and external K^+ concentration on caesium toxicity and accumulation in *Escherichia coli* and *Bacillus subtilis*. *J. Ind. Microbiol.* 14:218–225.

Schnoor, J. L. 1998. Phytoremediation ground-water remediation technologies analysis center, Technology Evaluation Report, TE 98-01, Pittsburgh, PA.

Tobin, J., C. White, and G. M. Gadd, 1994. Metal accumulation by fungi: applications in environmental biotechnology. *J. Ind. Microbiol.* 13:126–130.

Bibliography

Mackie, L. E., and A. C. R. Dean. 1989. *Adv. Biotechnol. Proc.* 12:159–201.

CHAPTER 16

Cyanide Waste

Cyanide is used in the production of organic chemicals such as nitrile, nylon, acrylic plastics, and synthetic rubber. It is also used in the electroplating, metal processing, steel hardening, and photographic industries. The wastes from such industries not only contains cyanide but also significant amounts of heavy metals such as copper, nickel, zinc, silver, and iron. Since cyanide ions are highly reactive, metal complexes of variable stability and toxicity are readily formed. Ore processing in gold and silver mining operations uses dilute solutions of sodium cyanide (100 to 500 ppm), which is inexpensive ($1.75/kg, 2003 price) and highly soluble in water, and under mildly oxidizing conditions, dissolves the gold contained in the ore. Each year 2 to 3 million tons of cyanide are industrially produced. Food processing industries that handle crops such as cassava and bitter almonds also generate considerable quantities of cyanide waste because of the presence of the cyanogenic glucosides that are present in the plant material.

Physical Processes

In nature, cyanide is oxidized to more stable products, which are relatively nontoxic when compared with the free cyanide. Cyanide treatment involves either a destruction-based process or a physical process of cyanide recovery. Cyanide and its related compounds such as ammonia, cyanate, nitrate, and thiocyanate can be destroyed by one of several processes. They include INCO SO_2/air (which uses SO_2 and air in the presence of a soluble copper catalyst to oxidize cyanide to the less toxic cyanate), copper-catalyzed hydrogen peroxide (which uses hydrogen peroxide as the oxidizing agent instead of SO_2 and air), Caro's acid, alkaline breakpoint chlorination (a two-step process in which the first step involves conversion to cyanogen chloride followed by hydrolysis of the cyanogen chloride to cyanate), and activated carbon adsorption followed by recovery of cyanide by desorption (Akcil, 2003). Chemical and physical processes to degrade cyanide and its related compounds are expensive, complex to operate, and add toxic chemicals to the environment. Chlorination is not effective when cyanide species are complexed with metals such as nickel and silver because of their slow reaction rates. The process also produces sludge, which requires licensed disposal.

The total chemical cost for chlorine, hydrogen peroxide, SO_2/air, and biological processes are $15.8, 6.5, 1.2, and 0.6 per kilogram of cyanide destroyed (Mosher and Figueroa, 1996). The selection of the technique will depend on the chemical characterization of the untreated solution or slurry, as well as its quantity and environmental setting; the capital, equipment, and reagents available; the operating and maintenance costs; licensing fees; and review of the applicable regulations.

Bioprocess

Biological treatment involves the acclimation and enhancement of indigenous microorganisms to fix or biotransform the toxic cyanide to less toxic derivatives. Biotreatment is less expensive and simple to operate. Thiocyanate is used in several industrial processes, including photofinishing, herbicide and insecticide production, dyeing, acrylic fiber production, thiourea manufacture, metal separation and electroplating, and in soil sterilization and corrosion inhibition; hence it is found in wastewaters. Thiobacilli, pseudomonads, and *Arthrobacter* spp. are capable of degrading thiocyanate. Cyanate is an intermediate product in the first stage of thiocyanate hydrolysis and is further hydrolyzed to ammonia and bicarbonate (Hung and Pavlostathi, 1997).

Although methanogens are inhibited by cyanide, a 90% cyanide removal and simultaneous reduction of chemical oxygen demand (COD) and methane production were achieved when effluent was exposed to sludge adapted to cyanide (taken from an upflow anaerobic sludge blanket reactor). Cyanide inhibition on methanogenic activity was more pronounced for acetoclastic than for hydrogenotrophic methanogens (Gijzen et al., 2000). Two *Pseudomonas* sp., CM5 and CMN2 without acclimation, were able to degrade cyanide in a solution of whey from a concentration of 80 and 160 ppm to less than 1 ppm in batch mode. During metabolism, the microorganisms used cyanide as a nitrogen and carbon source, converting it to ammonia and carbonate (Akcil et al., 2003).

Burkholderia cepacia strain C-3 isolated from soil with a carbon source was able to biodegrade cyanide at a pH of 10. Cu^{2+} or Fe^{2+} at a concentration of 1 m*M* inhibited both the growth of the bacteria and cyanide degradation. The highest growth was observed in the presence of Mg^{2+}. Phenol inhibited the reaction, while ethanol and methanol had no effect. Fructose, glucose, and mannose were the preferred carbon sources for cyanide biodegradation (Adjei and Ohta, 2000).

Mechanism of Action

The cyanide oxygenase from the bacterium *P. fluorescens* NCIMB 11764 converted free cyanide to carbon dioxide and ammonia. *P. putida* followed a two-step enzymatic pathway for cyanide degradation: cyanide hydratase transformed cyanide to formamide, and amidase degraded it further to formate and ammonia. *Alcaligenes xylosooxidans* sub sp., *A. denitrificans,* and *P. fluorescens* converted cyanide to ammonia and formate in a single step using cyanide dihydratase without producing formamide. *Stemphylium loti, Fusarium lateritium,* and *Gloeocerocospora sorghi* do not possess the amidase enzyme necessary to convert formamide to ammonia, which leads to the accumulation of formamide. *F. lateritium* and *G. sorghi* only detoxify cyanide to formamide by the action of cyanide hydratase, and none of these fungi utilized cyanide as a source of carbon or nitrogen (Kwon et al., 2002). *Fusarium oxysporum, Gliocladium virens, Trichoderma koningii,* and *F. solani* IHEM 8026 grow on KCN as a sole nitrogen source. *Trichoderma* strains have the cyanide-degrading enzymes, cyanide hydratase and rhodanese. Cyanide hydratase activity in *G. sorgh, S. loti, Colletotrichum graminocola, F. moniliforme, F. lateritium, F. solani,* and *Helminthosporium maydis* was different in uninduced and induced mycelium (Ezzi and Lynch, 2002). The enzyme cyanide hydratase present in fungi *F. solani* isolated from cyanide-contaminated soil specifically converted HCN to formamide but not the CN ion (Dumestre et al. 1997).

A microbial consortium composed primarily of *Pseudomonas* and *Bacillus* sp., degraded thiocyanate. *P. stutzeri* utilized potassium thiocyanate as a nitrogen and sulfur source and succinate as a carbon and energy source. *Thiobacillus thioparus* was able to assimilate 500 mg/L of potassium thiocyanate within 60 h, but thiocyanate degradation was inhibited by the presence of thiosulfate. Thiobacilli and pseudomonads utilized thiocyanate as the nitrogen and sulfur source and tolerated thiocyanate at concentrations of up to 5.8 g/L (Hung and Pavlostathi, 1997). *Escherichia coli, Flavobacterium* sp., and *P. fluorescens* had the enzyme cyanase that was responsible for catalyzing the hydrolysis of cyanate to ammonia and bicarbonate.

Metal-Cyanide Effluent

Several water management and treatment alternatives are possible in mining operations, including land application, biological treatment, breakpoint chlorination, hydrogen peroxide, and the SO_2/air process. The treatment methods must take care of cyanide, metals, thiocyanate, ammonia, and nitrate as well as high levels of total dissolved solids and sulfate. Except for breakpoint chlorination, chemical oxidation processes involving hydrogen peroxide and sulfur dioxide do not remove thiocyanate, ammonia, and

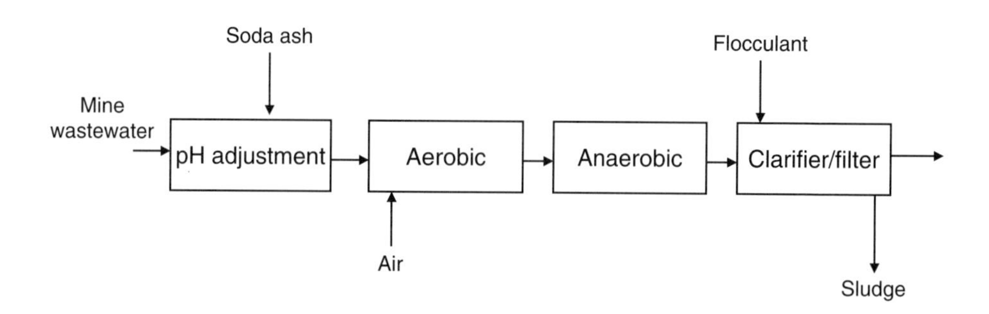

FIGURE 16-1. General treatment procedure.

nitrate. But the former is very expensive and produces high residual total dissolved solid and chloride concentrations. Biological treatment techniques are of recent origin, and a combined biological and chemical treatment approach has been found to have several advantages (see Fig. 16-1). Biological methods are environmentally friendly and cost less to operate, but capital costs are higher.

A typical biological and a chemical treatment procedure may involve a three-step approach.

- A combined activated sludge treatment step for the conversion of thiocyanate to ammonia and its oxidation to nitrate
- A denitrification treatment step leading to nitrogen gas
- A high density sludge ferric sulfate treatment step to precipitate arsenic and other metals in their sulfate forms

The types of reactors available for aerobic operation include rotating biological contactors, packed beds, biological filters, sequencing batch reactors, facultative lagoons, and activated sludge systems (Akcil, 2003). The aerobic treatment requires about 1 to 5 mg/L phosphate as the nutrient, 2 ng/L of dissolved oxygen, and a process temperature above 10°C. Commercial processes that have been operating are in-plant cyanide destruction; in situ cyanide destruction of spent heap leach piles; metal and sulfate removal using active (in-plant) sulfate reduction; and passive processes such as wetlands and ecological engineering for metals polishing.

To recover precious metals from ore in a heap leach operation, a dilute cyanide leaching solution is sprayed on the ore, which is heaped on an impermeable pad. As the solution percolates through the heap, precious metals are complexed, dissolved, and then recovered. A heap that has been treated several times by the cyanide solution has very little precious metal left and hence is discarded. The heap is considered closed if the leachate from a heap contains less than 0.2 mg/L of cyanide; this is accomplished by rinsing the heap to remove residual cyanide and then destroying the cyanide. Mosher and Figueroa (1996) report costs for cyanide detoxification and closure of a

1.2 million ton heap leach as $0.29, $0.21, $0.22, and $0.12 per ton of heap treated by chlorine, hydrogen peroxide, SO_2/air, and biological processes, respectively. The total estimated cost is the sum of the costs for capital, engineering, operation, maintenance, reagents, and licensing fees.

In biological treatment of cyanide effluent from metal ore mines, the bacteria convert free and metal-complexed cyanides to bicarbonate and ammonia, and the freed metals are either adsorbed within the biofilm or precipitated from solution. Free cyanide is the most readily degradable and iron cyanide the least, and the degradability of Zn, Ni, and Cu metal cyanide complexes fall in between. Iron cyanides have been shown to degrade the least. Ammonia to nitrate conversion follows a two-step aerobic process with nitrite as the intermediate. The nitrate is reduced to nitrogen gas under anoxic conditions. *Pseudomonas* sp. help in complete oxidation of cyanide, thiocyanate, and ammonia. Other gram-negative bacteria that play a crucial role in the process are *Achromobacter, Flavobacterium, Nocardia, Bdellovibrio, Mycobacterium,* and two nitrifiers, *Nitrosomonas* and *Nitrobacter.* Cyanide and thiocyanide serve as energy and food sources for the destruction stage bacteria and can be toxic to the nitrifying bacteria. Ammonia and bicarbonate serve as food and energy sources for the nitrifying bacteria.

Copper and zinc cyanide complexes are found in wastewaters originating from the electroplating and mining industries. *Citrobacter* sp. MCM B-181, *Pseudomonas* sp. MCM B-182, *Pseudomonas* sp. MCM B-183, and *Pseudomonas* sp. MCM B-184 were capable of degrading free as well as metal-cyanide complexes by utilizing metal cyanides as a nitrogen source and glucose or sugarcane molasses as a carbon source; ammonia and carbon dioxide were formed as degradation products. The degradation was not followed by metal biosorption onto the bacterial cells but by the precipitation from solution of copper and zinc as their respective hydroxides. A rotating biological contactor achieved 99.9% removal of 0.5 m*M* metal cyanide and a COD removal efficiency of 85% in 15 h using a consortium containing all four microorganisms (Patil and Paknikar, 2000). The biodegradation process was affected by the presence of metals such as Cr and Fe. A synthetic solution made up of sodium cyanide in water, ferrous sulfate, copper sulfate, zinc sulfate, and potassium thiocyanante mixed with sludge from a municipal activated sludge plant was successfully treated, achieving a greater than 90% removal efficiency in a biofilter packed with a support media. The total metal content was approximately 36% of the dry biomass, with first Zn being preferentially adsorbed, followed by Cu. Free cyanide, thiocyanate, and copper, iron, and zinc metallocyanides from a synthetic gold milling effluent mixed with sewage was treated in a stirred aerated bioreactor to achieve more than 95% removal of free cyanide, thiocyanate, copper, and zinc (Granato et al., 1996). Iron removal was low, about 68%.

Biodegradation of ferrous (II) cyanide ions was achieved using *Pseudomonas fluorescens* immobilized on calcium–alginate gel in a packed bed column reactor. *Cryptococcus humicolus* MCN2 yeast strain could degrade tetracyanonickelate (II) in batches. Seventy percent of the degradation of cyanide occurred in the lag phase of cell growth. Ammonia was produced because most of the cyanide had disappeared, and its production rate coincided with the formamide degradation rate. Ammonia was assimilated directly into the cell biomass. CO_2 generation was also proportional to the cell growth (Kwon et al., 2002).

Conclusions

Cyanide removal from waste currently relies on chemical treatment technologies, but recently biological treatment processes have been used successfully in large-scale operations. Proper closure and disposal of the spent heap leach ore that could contain adsorbed cyanide is another major problem that needs to be addressed. A combination of chemical and biological treatment technology could be highly effective and economically viable. In addition to cyanide, related compounds found in the mining effluents such as cyanate, thiocyanate, ammonia, and nitrate also must be treated and disposed of safely.

References

Adjei, M. D., and Y. Ohta. 2000. Factors affecting the biodegradation of cyanide by *Burkholderia cepacia* Strain C-3. *J. Biosci. Bioeng.* 89(3):274–277.

Akcil, A., A .G. Karahan, H. Ciftci, and O. Sagdic. 2003. Biological treatment of cyanide by natural isolated bacteria (*Pseudomonas* sp.). *Minerals Eng.* 16:643–649.

Akcil, A. 2003. Destruction of cyanide in gold mill effluents: biological versus chemical treatments. *Biotech. Adv.* 21:501–511.

Dumestre, A., N. Bousserrhine, and J. Berthelin. 1997. Biodegradation of free cyanide by the fungi *Fusaium solani*: relation to pH and cyanide speciation in solution. *Earth Planetary Sci.* 325:133–138.

Ezzi, M. I., and J. M. Lynch. 2002. Cyanide catabolizing enzymes in *Trichoderma* spp. *Enzyme Microbiol Technol.* 31:1042–1047.

Gijzen, H. J., E. Bernal, and H. Ferrer. 2000. Cyanide toxicity and cyanide degradation in anaerobic wastewater treatment. *Water Res.* 34(9):2447–2454.

Granato, M., M. M. M. Goncalves, R. C. Villas Boas, and G. L. Sant'Anna. 1996. Biological treatment of a synthetic gold milling effluent. *Environ. Pollution* 91(3):343–350.

Hung, C. H., and S. G. Pavlostathis. 1997. Aerobic biodegradation of thiocyanate, *Water Res. World* 31(11):2761–2770.

Kwon, H. K., S. H. Woo, and J. M. Park. 2002. Degradation of tetracyanonickelate (II) by *Cryptococcus humicolus* MCN2. *FEMS Microbiol. Let.* 214:211–216.

Mosher, J. B., and L. Figueroa. 1996. Biological oxidation of cyanide a viable treatment option for the minerals processing industry. *Minerals Eng.* 9(5):573–581.

Patil, Y. B., and K. M. Paknikar. 2000. Development of a process for biodetoxification of metal cyanides from waste waters. *Process Biochem.* 35:1139–1151.

Bibliography

Evangelho, M. R., M. M. M. Goncalves, G. L. Sant'Anna, Jr., and R. C. Villas Boas. 2001. A trickling filter application for the treatment of a gold milling effluent. *Int. J. Mineral Process* 62:279–292.

CHAPTER 17

Treatment of Waste from Food and Dairy Industries

Introduction

Wastewaters produced by the food industry are characterized by their organic content; most are composed of easily biodegradable compounds such as carbohydrates, proteins, and in some cases, lipids. Organic suspended solids are often present in these effluents (e.g., the organic content generated by fish meal processing). Other food processing industries (e.g., olive oil processing) use a number of chemicals during processing and they all become part of the effluent. Some of these chemicals are phytotoxic. It is estimated that at least 10% of the total wastes produced by industrial and commercial activity are from the food and dairy industry. Food wastes can cause "oxygen sag," where a few organisms survive. For many food processing plants, a large fraction of the solid waste produced at the plant comes in the early stages of processing when the desired food constituents are separated from the undesired ones. Undesirable constituents include tramp material (soil and extraneous plant material); spoiled food stocks; and fruit and vegetable trimmings, peel, pits, seeds, and pulp. In some food processing plants, caustic peeling is used to remove skins from soft fruit and vegetables such as tomatoes. High-moisture solid waste materials can also be generated by water cleanup and reuse operations in which the dissolved or suspended solids are concentrated and separated from wastewater streams. Apart from these, many materials commonly generated in the food industry are cardboard, plastics, and metal cans. These are best recovered for reuse or recycled to minimize the waste.

Dairy Industry

Dairy processing (cheese, casein, butter production) effluents predominantly contain milk and milk products that have been lost in the processing.

TABLE 17-1
Characterization of the Effluents from Dairy Factories[a]

Origin	*COD*	*BOD*	*Fats*	N_t	P_t	*pH*	*TSS*	*VSS*
Dairy factory	4,000	2,600	400	55	35	8–11	675	635
Whey	61,250	—	—	2,500	533	4.6	5,077	4,900
Cheese factory	4,430	3,000	754	18	14	7.32	1,100	—
Yogurt and buttermilk	1,500	1,000	—	63	7.2	—	191	—

[a]COD—chemical oxygen demand (mg O_2/L); BOD—biological oxygen demand (mg/L); N_t—total nitrogen (mg/L); P_t—total phosphorous (mg/L); TSS—total suspended solids (mg/L); VSS—volatile suspended solids (mg/L).

Milk lost into the effluent stream can amount to 0.5 to 2.5% of the incoming milk, but can be as high as 3 to 4%. Although all compounds are biodegradable, some of them, such as lactose, are readily consumed in biological treatment, whereas protein and especially fat are not easily degraded. In order to understand the environmental impact of these effluents, it is useful to briefly consider the nature of milk. Apart from water, which makes up about 87.5% of its weight, raw milk typically contains 13% total solids, 3.9% fat, 3.4% protein, 4.8% lactose, and 0.8% minerals. The quality control process of the raw milk prior to its use causes the generation of a particularly complex effluent that contains raw milk and a mixture of different chemicals. The liquid waste in a dairy originates from the manufacturing process, utilities, and service sections. The various sources of waste generation from dairy are spilled milk, spoiled milk, skimmed milk, whey, wash water from milk cans, equipment, bottles, and floor washing. Whey is a high-strength waste product of cheese manufacture, and it is the most difficult to degrade. It contains milk proteins, water soluble vitamins, and mineral salts. Table 17-1 shows a summary of different wastewaters from dairy factories.

Both aerobic and anaerobic processes are employed for the treatment of these wastes. Aerobic treatment is characterized by relatively high energy consumption, and biomass production is not preferred. Anaerobic processes, on the other hand, prove most suitable for the treatment of dairy wastes. Milk fat is quite difficult to degrade biologically because of the potential toxic effects exerted by the fatty acids that result from the breakdown of fat molecules. This necessarily calls for a suitable bioreactor design to avoid undesirable fat accumulation. The treatment of cheese whey waste waters by anaerobic degradation is constrained by the drop in pH that inhibits further conversion of acids to methane. This can be taken care of by buffering the solution in a hybrid reactor. It is clear that buffering action is needed initially for maintaining the pH, but at a later stage, a mature microbial population improves the stability (Ghaly, 1996). Apart from the hybrid reactor, other alternate reactor types have also been tried for the treatment of dairy-based wastewaters. In the study carried out by Guitonas et al. (1994), a fixed bed reactor with cells immobilized on rice straw was used for the treatment of milk-based synthetic organic waste. The advantage of this system was a lower adaptation time with change in the organic loading rate.

Meat Processing Industry

The meat processing industry is large, common to many countries, and generates large volumes of wastewater that require considerable treatment before release into the environment. The effluent contains high volumes of carbohydrates, proteins, fats, and other organic materials, in addition to a high concentration of phosphate, acetic acid, butyric acid, and chloride. The concentration of pollutants in various wastewater streams from slaughterhouses or rendering plants is summarized in Table 17-2.

Screening, settling, and dissolved air flotation are still widely used for the removal of suspended solids and fats, oils, and greases. Anaerobic systems are well suited to the treatment of slaughterhouse wastewater. They achieve a high degree of BOD removal at a significantly lower cost than comparable aerobic systems and generate a smaller quantity of highly stabilized, more easily dewatered sludge. Furthermore, the methane-rich gas that is generated can be captured for use as a fuel. However, anaerobic treatment suffers from the disadvantage of odor generation from the ponds, thus making the development of alternate designs very essential. The high-rate anaerobic treatment systems such as the upflow anerobic sludge blanket (UASB) and fixed bed reactors are less popular for slaughterhouse wastewaters because of the presence of large amounts of fat, oil, and suspended matter in the influent. The anaerobic contact reactor appears to be more suitable compared with UASB because the latter is constrained by the lack of formation of granules and there is also loss of sludge due to high fat concentrations (Rajeswari et al., 2000). Hence, a pretreatment step for removal of fats and suspended solids becomes essential if an UASB is to be used. However, for a low COD load, the more efficient UASB appears to result in a high COD reduction.

TABLE 17-2
Analysis of Wastewater from Slaughterhouses

Parameter[a]	*Wastewater*
BOD, mg/L	1,600–3,000
COD, mg/L	4,200–8,500
Oil and grease, mg/L	100–200
Total suspended solids, mg/L	1,300–3,400
Total Kjeldahl nitrogen mg/L	114–148
NH_4-nitrogen mg/L	65–87
Total phosphorous mg/L	20–30
Volatile fatty acids, mg/L	175–400

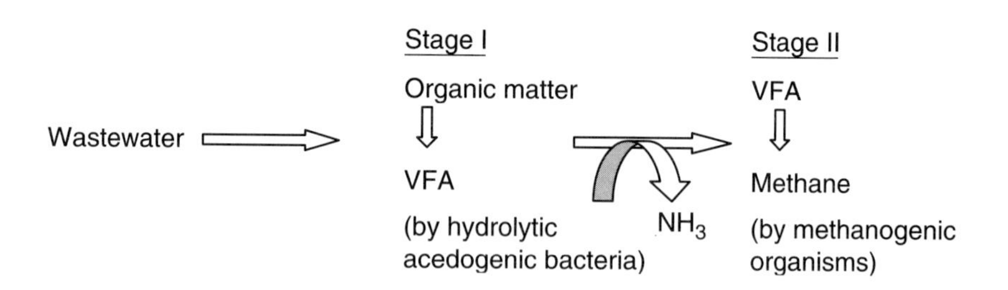

FIGURE 17-1. Two-phase system for wastewaters with high concentrations of organic solids.

Two-phase reactor systems (Fig. 17-1) are best suited for degradation of food wastes. In stage one, hydrolytic and acidogenic bacteria (anaerobic) degrade organic suspended solids to volatile fatty acids (VFAs). These VFAs are then further degraded to methane by the methanogenic (anaerobic) organisms. A two-stage system for treating high-strength wastewater from an abattoir has been tried by Rivera et al. (1997). The system consists of an anaerobic digester followed by an artificially constructed wetland that utilizes the root zone of hydrophytes planted in a gravel substrate. The treatment efficiency was high, with COD and BOD reductions of 87.4 and 88.5%, respectively.

General Treatment Methods

The four Rs of waste management (recover, reduce, reuse, and recycle) are best suited for the food industry. All the same, the waste generated from this industry (as discussed earlier) is best treated by bioremediation methods. Two-phase reactor systems are best suited for degradation. Apart from the well-known methods of bioremediation, newer methodologies have been adopted for improving the efficiency of transformation, reducing sludge formation, and aiding in the formation of sludge that can be used for farming purposes. Composting is one such option of disposal. However, odor and leaching of soluble constituents are limiting factors. Composted material is valued as a soil amendment or potting soil, but widespread use and marketability are constrained by shipping cost. Composition of the composting feedstock needs to be controlled to obtain the appropriate physical mix to allow the natural composting aerobic bioprocesses to proceed. Examples in the literature show that the full range of food processing wastes can be composted, including fruit and vegetable wastes such as peelings and skin; whole fish and fish offal; meat processing wastes such as paunch contents, blood, fats, intestines, and manure; and grain processing wastes such as chaff, hulls, pods, stems, and weeds (Schaub and Leonard, 1996).

Residues from extraction of oils such as cotton, olive, and palm contain tannins and phenolics that are toxic to plants and animals. Apart from the general methods, these wastes can also be detoxified by growing mushrooms (e.g., *Pleurotus* and *Lentinula* species). While actively growing, these mushrooms produce enzymes that can degrade lignins, phenolics, and tannins. Producing a crop of mushrooms while disposing of an otherwise hazardous waste has become a popular "research model" in recent years. *Pleurotus* cultivation may even aid removal of pollutants from contaminated waste sites.

Food waste can be treated by a two-stage anaerobic process, followed by an aerobic treatment to completely mineralize the pollutants. Also, recent developments such as composting, phytoremediation, and mushroom culturing have substantial potential in cleanup of these wastes.

References

Ghaly, A. E. 1996. A comparative study of anaerobic digestion of whey and dairy manure in a two-stage reactor. *Bioresource Tech.* 58:61–72.

Guitonas, A., G. Pashalidis, and A. Zouboulis. 1994. Treatment of strong wastewater by fixed bed anaerobic reactors with organic support. *Water Sci. Tech.* 29(9): 257–263.

Rajeswari, K. V., M. Balakrishnan, A. Kansal, K. Lata, and V. V. N. Kishore. 2000. State of the art of anaerobic digestion technology for industrial wastewater treatment. *Renewable and Sustainable Energy Reviews* 4:135–156.

Rivera, F., A. Warren, C. R. Curds, R. Colin, E. Robles, A. Gutierrez, E. Galleges, and A. Calderon. 1997. The application of the root zone method for the treatment and reuse of high-strength abattoir waste in Mexico. *Water Sci. Tech.* 35(5): 271–227.

Schaub, S. M. and J. J. Leanard. 1996. Composting: an alternative waste management option for food processing industries. *Trends Food Sci. Tech.* 7:263–268.

Bibliography

APHA – AWA – WPCF. 1985. *Standard methods for examination of water and waste water,* 16th ed. Washington, DC: APHA-AWA-WPCF 1985.

Guerrero, L., F. Omil, R. Mendez, and J. M. Lema. 1999. Anaerobic hydrolosis and acidogenesis of wastewaters from food industries with high content of organic solids and protein. *Water Res.* 33(15):3218–3290

Johns, M. R. 1995. Anaerobic digestion of organic solids from slaughterhouse wastewater. *Bioresource Technol.* 54:203–216.

F. Omil, J. M. Garrido, B. Arrojo, and R, Mendez. 2003. Anaerobic filter reactor performance for the treatment of complex dairy wastewater at industrial scale. *Water. Res.* 37:4099–4108.

CHAPTER 18

Sugar and Distillery Waste

Sugarcane is one of the most common raw materials used in sugar and ethanol production. More than 30 billion liters of spent wash are generated annually by 254 cane molasses–based distilleries in India alone (0.2 to 1.8 m^3 of wastewater per ton of sugar produced). The effluent has a pH of 4 to 7, a COD of 1,800 to 3,200 mg/L, and a BOD of 720 to 1,500 mg/L; its total solids are 3,500 mg/L, total nitrogen 1,700 mg/L, and total phosphorus 100 mg/L. Several other countries in the world, such as Thailand, Malaysia, Taiwan, and Brazil, also produce sugar from sugarcane. The wastewater contains not only a high concentration of organic matter but also a large amount of dark brown pigment called melanoidin.

Alcohol Distillery Effluent

The Americas account for 66% of the world's ethanol production, followed by Asia-Pacific, which produces about 18%. The total production of alcohol in India during the year 1994–1995 was 1165 million liters. The residue of the distillation process is the spent wash, which is a strong organic effluent. The other wastes from the process include yeast sludge (which is usually mixed with spent wash), floor washes, waste cooling water, and waste from the operations of yeast recovery or byproducts recovery processes. About 12 to 16 L of waste liquid effluent is generated for 1 L of alcohol. The distillery wastewater, known as spent wash, is characterized by its color, high temperature, low pH, and high ash content; it contains a high percentage of dissolved organic and inorganic matter (7 to 10%), of which 50% may be reducing sugars and 10 to 11% may be proteins.

The metals present in spent wash in milligrams per liter are Fe, 348, Mn, 12.7, Zn,4.61, Cu, 3.65, Cr, 0.64, Cd, 0.48, and Co, 0.08, with the electric conductivity in the range of 15-23 dsm^{-1}. Indian spent wash contains very large amounts of potassium, calcium, chloride, sulfate, and BOD (around 50,000 mg/L) compared with spent wash in other countries. Organic compounds

TABLE 18-1
Typical Indian Distillery Effluent, pH 4 to 5.5

Compound	*Concentration, mg/L*
COD	100,000–150,000
BOD	35,000–50,000
Total solids	80,000–120,000
Total suspended solids	8,000–22,000
Total volatile suspended solids	6,000–22,000
Total dissolved solids	90,000–95,000
Chloride	900–3,400
Total phosphorous	30–40
Sulfate as SO_4	1,100–18,000
Nitrogen oxide	60–90
Potassium	52–62

extracted from spent wash using alkaline reagents are humic in nature, similar to those found in the soil excepting that fulvic acid predominates over humic acid. The characteristics of a typical Indian distillery effluent are given in Table 18-1. Normally 200% oxygen must be fed into the effluent to meet the oxygen demand, or, put another way, the total oxygen input required is 93.30 kg/m^3. In practice, the best of the best conventional aeration systems gives 1 kg to a maximum of 1.2 kg of O_2. The total energy required for this process would be 93.30 KWh/m^3.

Treatment of Distillery Effluent

Physicochemical treatment, including sedimentation with the addition of coagulant and other additives such as alum, lime, ferric chloride, and activated charcoal, has been found to be unsatisfactory. Despite the installation of huge anaerobic lagoons, aeration tanks, and solar drawing pits, the problems of pollution have not been solved yet. The concentration of spent wash and its use as an animal feed additive is a common practice among countries producing alcohol from beet molasses in Europe and North America. Many distilleries allow their effluents to be used for soil treatment in the form of direct irrigation water, spent wash cake, and spent wash–press mud compost. The methods that are commonly employed are given below. Distilleries practice these methods individually or in combination.

- Anaerobic, methanogenic digestion of slops, followed by aerobic digestion
- Evaporation of slops, followed by aerobic composting using a cellulosic carrier material
- Evaporation of slops, followed by incineration of the concentrate, with or without generation of steam, along with gas cleaning
- Evaporation of slops, so the concentrate can be used as an additive for cattle feed
- Disposal of slops into the deep sea after some treatment

Molasses contains appreciable amounts of calcium salts, which cause deposition and scaling of heat exchangers. Since the conventional aerobic processes for primary treatment of distillery waste are not cost effective and require large land areas, the main emphasis has been on anaerobic processes, since they have the dual advantages of pollution control and fuel production. A general estimate suggests that the cost of an anaerobic biological digester is recovered within 2 to 3 years of installation as a result of substantial savings of coal and other fuels. It is estimated that these distilleries have the potential to generate a total of 560×10^6 m^3 per annum of biogas if all of them would opt for anaerobic digestion. Assuming the calorific value of biogas as 5,300 kcal/m^3, this amounts to 830 Gigawatt hour/annum and translates to 158 MW of power. Anaerobic digestion also reduces by a considerable amount the sludge that is produced when compared to that produced by the aerobic process. The anaerobic processes have a few disadvantages. The process is slow because the rates of reaction and synthesis are low, long startup periods are required, and further treatment becomes inevitable since the reduction in COD achieved is only on the order of 85%. Generally industries have resorted to a subsequent aerobic digestion or biocomposting. The effluent also has a caramel color that is found to contaminate the groundwater. A number of process packages on biomethanation of distillery-spent wash have been developed by international consultants; their salient features are listed in Table 18-2.

Indian Scene

India has more than 200 distilleries, less than half of which have some technology to address the issue of contaminated wastewater. In India primary spent wash is generally put through an anaerobic digestion step to utilize its high COD load to produce methane. The secondary spent wash produced by the anaerobically digested primary molasses spent wash (DMSW) effluent is darker in color and needs huge volumes of water to dilute it; currently its use as irrigation water is causing gradual soil darkening (see Fig. 18-1). Its disposal into natural bodies of water may result in their eutrophication. The color leads to a reduction of sunlight penetrating the rivers, lakes, or lagoons, which in turn decreases both photosynthetic activity and dissolved

TABLE 18-2
Technologies Available for Effluent Treatment

Process	*Residence time, days*	*Organic loading, kg/COD/m³/day*	*Biomass produced, m³/kg COD destroyed*	*Methane content of biogas, %*
		For upflow anaerobic sludge blanket (UASB) reactors		
Sulzer	5–6	14–20	0.5	75
Biotin	3–5	10–25	—	65–80
Biothane (Esmil)	5–6	10–25	—	65–70
Euro–Consult	5–6	10–25	0.35	80
Biomagaz	5–6	—	0.35	69
Anupuls (Degremont)	2.5–5	10	0.35	65
BIMA/BVT	5–6	5–10	—	70
		For immobilized beds		
Bacardi	—	12–8	0.13	58–60
SGN	—	20–25	—	55–65
Anoxal	—	14–16	0.32	70
—	—	—	—	70
		For fluidized beds		
Degremont	0.2	25	0.37	>65
Dorr-Oliver	—	10	0.37	65

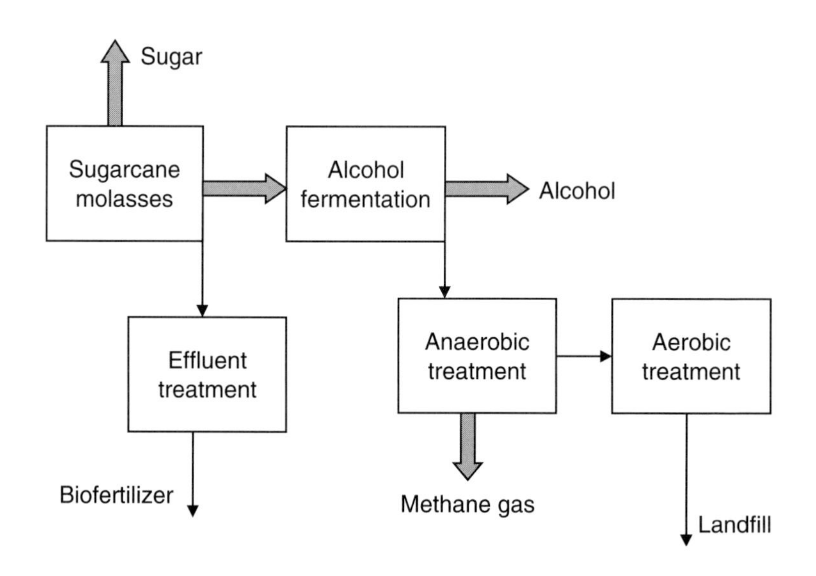

FIGURE 18-1. Typical process flow for sugarcane and distillery waste.

TABLE 18-3
Technologies Followed by the Indian Distillery Industries

Scale	*Reactor type*	*Volume*	*COD loading rate, kg/m^3/d*	*CH_4 yield*	*COD reduction, %*
Pilot	Fixed film	100 L	25	0.4 m^3/kg COD reduced	60–70
Pilot	Fixed film	100 m^3/day		0.4–0.45 m^3/kg COD reduced	60–70
Pilot	Hybrid	0.15–1.6 mld	48–50	3.10 m^3/(m^3 day)	70
Full	USAB	3.75 mld		13,100 m^3/(m^3 day)	65–70
Full	USAB	7 mld		24,000 m^3/(m^3 day)	90–92
Full	CSTR	9.5 mld		0.5 m^3/kg COD reduced	70

oxygen concentrations, causing harm to aquatic life. Disposal on land is also hazardous, causing a reduction in soil alkalinity and manganese availability, inhibition of seed germination, and the ruin of vegetation. The decolorization of molasses spent wash by physical or chemical methods and subsequently directly applied as a fertilizer has also been attempted and found to be unsuitable. Anaerobic treatment of distillery wastewater has been tried in pilot and full-scale operations. Some of these are hybrid, fixed film, and continuous stirred reactors. Performance of the few reactors and their scale is summarized in Table 18-3. Since it is highly acidic and hot, the effluent invariably needs pretreatment for pH and temperature. Also lime scrubbing of the biogas is needed for H_2S removal before it can be used for power generation.

International Status

Effluent from baker's yeast grown on sugar beet molasses has been treated anaerobically (USAB with internal and external sludge recirculation facilities); this is then followed by an aerobic digester, generating biogas at the rate of 0.65 m^3/per kg of COD, achieving a 60 to 70% reduction in COD. Goodwin et al. (2001) have reported anaerobic biotreatment of malt whisky distillery pot ale using a UASB system. The sludge developed in the reactor was flocculent and did not form compact granules. Garcia et al. (1998) have studied the anaerobic digestion of wine distillery wastewater in a downflow fluidized bed. The system achieved 85% total organic carbon (TOC) removal.

Advantages are low energy and the settling of solids to the reactor bottom, where they can easily be drawn out. A downflow fluidized bed reactor utilizes floatable particles as carriers. Bed volume increases because of gas production. Kida et al. (1999) achieved efficient removal of organic matter and NH_4^+ from pot ale by a combination of methane fermentation and a biological denitrification-nitrification process. The effluent was treated in an upflow anaerobic filter achieving an approximately 80% reduction in total organics. Benitez et al. (2003) have used a combination of ozonation and aerobic biodegradation of wine vinasses in discontinuous and continuous mode of operation. The oxidation of organic substrates of wine distillery waste using ozone led to a 31 to 85% reduction in COD for a hydraulic retention time of 24 to 72 h.

Microorganisms

Several microorganisms have been reported as suitable for treating and decolorizing the wastewater generated from a molasses plant. They include *Aspergillus fumigatus* (Ohmomo et al, 1987) , *Coriolus versicolor* (Aoshima et al., 1985), *Phanerochaete chrysospotium* (Fahy et al, 1997), and the filamentous fungi (Sirianuntapiboon et al., 1988). Anaerobic digestion studies carried out by Sirianuntapiboon et al. (2004) with an acetogenic bacteria strain No.BP103 showed a 76% decolorization yield when cultivated at 30°C for 5 days in a molasses pigment medium. In addition, this strain could decolorize 32 and 73% of molasses pigments in stillage and anaerobically treated molasses wastewater, respectively, when supplemented with nutrients.

When alcohol distillery wastewater (cane molasses vinasse) was treated in a UASB reactor under thermophilic conditions (55°C) at an influent concentration of 10 g COD/L, the BOD removal was good (80%), but the COD removal was low (39 to 67%). The poor COD elimination was attributed to the low degradability of the waste itself. Phenolic compounds present in vinasse, which are produced through oxidation and cause a dark brown color, are refractory as well as toxic for methanogens. The researchers observed more of *Methanosarcina*-like coccoids and very little of *Methanothrix*-like bamboo-shaped rods, which is more sensitive to toxic compounds. The temperature optimum for the former is 50 to 58°C and for the latter 60 to 65°C; hence operating at the elevated temperature would favor methane generation. *Methanothrix* in granular sludge is most essential for the establishment of a high performance UASB process. High concentrations of bivalent cations, such as Mg^{2+} and Ca^{2+}, induced development of single cells of the *Methanosarcina* species, which are more easily washed out from the UASB reactor than large clumps or packets (Harada et al., 1996). Nagano and Kobayashi (1990) and Romero et al. (1990) also reported that during anaerobic treatment of alcohol effluents COD removal is low while BOD removal is high.

Effluent from a malt whisky manufacturing plant has been treated anaerobically in different reactor configurations including UASB, upflow anaerobic filter, and batch stirred reactor. Overall COD and BOD removal efficiencies of greater than 98% were achieved for effluent from a malt whisky manufacturer in a UASB reactor followed by a batch aerobic reactor (Uzal et al., 2003). An aerobic jet loop reactor with hydraulic retention times that varied from 2.1 to 4.4 days was able to achieve about 98% degradation of the effluent from a winery. *Pseudomonas, Saccharomyces cerevisiae,* and yeast-like fungi, such as *Trichosporon capitatum* and *Geotrichum peniculatum,* were present in the activated sludge.

The white-rot fungi namely, *Coriolus versicolor* and *Phanerochaete chrysosporium* could achieve 54 and 38% decolorization efficiencies, respectively, and 60 and 49% reductions in COD, respectively, in 10 days. The major shortcomings of the process were the need to add extra carbon and the effluent needed to be diluted (Kumar et al., 1998). Chandra and Singh (1999) carried out chemical decolorization of anaerobically treated distillery effluent using chemical and biological methods. Maximum decolorization and COD reduction of 98 and 88%, respectively, were achieved by treatment with hydrogen peroxide and calcium oxide.

References

Aoshima, I., Y. Tozawa, S. Ohmomo, and K. Ueda. 1985. Production of decolorizing activity for molasses pigment by *Coriolus versicolor* Ps4a. *Agri. Biol. Chem.* 49:2041–2045.

Benitez, J. F., F. J. Real, J. Garcia, and J. M. Sanchez. 2003. Kinetics of the ozonation and aerobic biodegradation of wine vinasses in discontinuous and continuous processes. *J. Hazard. Mater.* 101(2):203–218.

Chandra, R., and H. Singh, 1999.Chemical decolourisation of anaerobically treated distillery effluent. *Indian J. Environ. Protection* 19(11):833–837.

Fahy, V., F. J. FitzGibbon, G. McMullan, D. Singh, and R. Marchant. 1997. Decolorisation of molasses spent wash by *Phanerochaete chrysospotiu. Biotechnol. Letters* 19:97–99.

Garcia, C. D., P. Buffiere, R. Moletta, and S. Elmaleh. 1998. Anaerobic digestion of wine distillery waste water in down-flow fluidised bed. *Water Res.* 32(12): 3593–3597.

Goodwin, J. A. S., J. M. Finlayson, and E. W. Low. 2001. A further study of the anaerobic biotreatment of malt whisky distillery pot ale using UASB system. *Bioresource Tech.* 78:155–159.

Harada, H., S. Uemura, A. Chen, and J. Jayadevan, 1996a. Anaerobic treatment of a recalcitrant distillery wastewater by a thermophilic UASB reactor. *Bioresource Tech.* 55.215–221.

Kida, K., S. Morimura, Y. Mochinaga, and M. Tokuda. 1999. Efficient removal of organic matter and NH_4^+ from pot ale by a combination of methane fermentation and biological denitrification and nitrification process. *Process Biochem.* 34:567–569.

Kumar, V., L. Wati, P. Nigam, I. M. Banat, B. S. Yadav, D. Singh, and R. Marchan, 1998. Decolorization and biodegradation of anaerobically digested sugarcane molasses spent wash effluent from biomethanation plants by white-rot fungi, *Process Biochem.* 33(1): 83–88.

Namasivayam, C., A. Kanagarathinam, and K. Ranganathan. 1994. Treatment of distillery wastewater using 'waste' coirpith, impregnated with 'waste' Fe^{3+}/Cr^{3+} hydroxide. *Chem. Environ. Res.* 3(1,2): 43–52.

Ohmomo, S., Y. Kaneko, S. Sirianuntapiboon, P. Somchai, P. Atthasampunna, and I. Nakamura. 1987. Decolorization of molasses waste-water by a thermophilic strain, *Aspergillus fumigatus* G-2-6. *Agri. Biol. Chem.* 51: 3339–3346

Romero, L. I., D. Sales, and E. Martinez de la Ossa. 1990. Comparison of three practical processes for purifying wine distillery wastewater. *Proc. Biochem. Int.* 25(3):93–96.
Sirianuntapiboon, S., P. Somchai, S. Ohmomo, and P. Atthasampunna. 1988. Screening of filamentous fungi having the ability to decolorize molasses pigments. *Agri. Biol. Chem.* 52:387–392.
Sirianuntapiboon, S., P. Phothilangka, and S. Ohmomo. 2004. Decolorization of molasses wastewater by a strain No.BP103 of acetogenic bacteria. *Bioresource Tech.* 92, 31–39.
Uzal, N., C. F. Gokcay, and G. N. Demirer. 2003. Sequential (anaerobic/aerobic) biological treatment of malt whisky wastewater. *Process Biochem.* 39:279–286.

Bibliography

Akunna, J. C., and M. Clark. 2000. Performance of a granular bed anaerobic baffled reactor (GRABBR) treating whisky distillery wastewater. *Bioresour. Tech.* 74:257–2561.
Bardiya, M. C., R. Hashia, and S. Chandna. 1995. Performance of hybrid reactor for anaerobic digestion of distillery effluent. *J. Indian Assoc. Environ. Manage.* 22(3):237–239.
Ciftci, T., and I. Ozturk. 1995. Nine years of full-scale anaerobic–aerobic treatment experiences with fermentation industry effluents. *Water Sci. Tech.* 32(12):131–136.
Goodwin, J. A. S., and J. B. Stuart. 1994. Anaerobic digestion of malt whisky distillery pot ale using upflow anaerobic sludge blanket reactor. *Bioresource Tech.* 49:75–81.
Jalgaonkar, A. D. 1995. Power generation based on distillery spentwash. *Wealth from waste.* S. Khanna and K. Mohan, eds., 245–252. New Delhi: Tata Energy Research Institute.
Lata, K., A. Kansal, M. Balakrishnan, K. V. Rajeshwari, and V. V. N. Kishore. 2002. Assessment of biomethanation potential of selected industrial organic effluents in India. *Resources, Conservation and Recycling* 35:147–161.
Pathe, P. P., T. Nandy, and S. N. Kaul. UASB reactor for the treatment of sugar effluents. 1995. *Indian. J. Environ. Protection* 15(3):174–180.
Petruccioli, M., J. Cardoso Duarte, A. Eusebio, and F. Federici. 2002. Aerobic treatment of winery wastewater using a jet-loop activated sludge reactor. *Process Biochem.* 37:821–829.
Radwan K. H., and T. K. Ramanujam. 1995. Treatment of sugar cane wastewater using modified biological contactor. *Indian J. Environ. Health* 37(2):77–83.
Ramendra, A. M. 1992. Anaerobic and aerobic fermentation—a proven biotechnology for distillery effluent treatment, *Indian J. Environ. Protection.* 2(11):835–838.
Reddy, U., and B. Shivalingaiah. 1997. Studies on the treatment of sugar industry wastewater. *Proceeding of the International Conference on Industrial Pollution and Control Technologies,* November 17–19, Hyderabad, India. Pollution Control Board of India: N. Delhi, pp. 177–180.
Tokuda, M., Y. Fujiwara, and K. Kida. 1999. Pilot plant test removal of organic matter, N and P from whisky pot ale. *Process Biochem.* 35:265–267.
Vaidyanathan, R., T. Meenambal, and S. Jayanthi. 1996. Evaluation of biokinetic coefficients for the rational design of anaerobic digester to treat sugar mill wastewater. *J. Indian. Water Works Assoc.* 28(1):21–24.

CHAPTER 19

Paper and Pulp

The pulp and paper making industry is very water intensive (about 60 m^3 water per ton of paper produced), and in terms of freshwater use ranks third after the primary metals and chemical industries. The major raw material used by the pulp and paper industry is wood, which is composed of cellulose fibers. The wood is broken down to separate the cellulose from the noncellulose material; the cellulose is then dissolved chemically to form a pulp. The pulp slurry is then vacuum dried on a machine to produce a paper sheet. Dyes, coating materials, and preservatives are also added at some point in the process. Lignin is a complex aromatic polymer that is an integral cell wall constituent that gives strength and rigidity to the tissues and allows vascular plants to resist microbial attack. The presence of residual lignin affects some properties of the manufactured pulp and paper products. Therefore, lignin is selectively removed during pulping without significant degradation of the cellulose fibers. The paper industry has several sectors such as packaging board, newsprint, boxes, printing and writings, and tissues. The world production of paper and board is about 320 million tons per year (1996 data). North America produces more than half, Western Europe about 20%, and Japan about 12%. The consumption of water varies depending on the type of paper being produced. Manufacture of tissue, printing and writing paper, newsprint, and packaging material requires about 60, 35, 30, and 18 m^3 of water per ton (Thompson et al., 2001).

Wood pulp is prepared either by mechanical or chemical means. The mechanical pulping process involves passing a debarked block of wood through a rotating grindstone where the fibers are stripped off and suspended in water. In chemical pulping, large amounts of chemicals are added to break down the wood in the presence of heat and pressure. In the kraft pulping process, sodium hydroxide and sodium sulfide are added to dissolve the nonfibrous material. The effluent generated (black liquor) contains high dissolved solids and alkali-lignin and polysaccharide degradation by-products, and has a high pH. A chemical mechanical pulping process is a combination of the two where the wood is first partially softened by chemicals and then

TABLE 19-1
Different Operations and Various Chemicals Used in Paper and Pulp Manufacture

Operation	*Chemicals used*
Chemical pulping	Acids, alkalies, lime, sulfurous acid, sodium hydroxide, sodium sulfide
Bleaching	Chlorine, bleaching agents, sulfates, solvents (chloroform)
Paper making	Pigments
Sizing and starching	Waxes, glue, synthetic resins, hydrocarbons
Coloring and dyeing	Inks, paints, solvents, rubbers, dyes
Cleaning and degreasing	Tetrachloroethylene, trichloroethylene, methylene chloride, carbon tetrachloride, trichloroethane

mechanical methods are used. Thermomechanical pulping involves carrying out the process at 100°C. The yield of this process is about 93% based on dry wood, implying that 30 to 70 kg/ton is lost in water, leading to a chemical oxygen demand (COD) of 1,000 to 5,600 mg/L. The papermaking process produces an effluent that contains about 50% cellulose. This contaminated water is referred to as "whitewater." It consists of 40% lignin, 40% carbohydrates, and the rest extractives.

White paper is produced by bleaching, which involves addition of a bleaching agent such as chlorine, hydrogen peroxide, or sodium peroxide. After filtration, coloring materials are added. Coatings and preservatives are added during or after the paper making process. Recycled paper is also a source of cellulose fiber for corrugated paper and newsprint. The recycled fiber needs to be de-inked using flotation, which is followed by washing and screening. Typical paper industry operation and the type of chemicals used in each operation are listed in Table 19-1.

The quality of untreated effluent from pulp and paper manufacture varies, depending on the paper type as shown in Table 19-2. The COD of the effluent is as high as 11,000 mg/L. Dissolved small organic molecules in the effluent give a high biological oxygen demand (BOD), while more complex lignin molecules do not increase BOD but create a high COD and dark color. The wastewater sludge from the de-inking operation contains heavy metals. The pulping process generates a considerable amount of wastewater (about 200 m^3 per tonne of pulp produced). There are two main differences in the quality of the wastewaters from pulping and papermaking operations (Billings and DeHaas, 1971), namely, (1) pulp wastewater contains dissolved wood-derived substances that are extracted from the wood during the pulping and bleaching processes and (2) pulping effluent will have some discoloration due to the dissolved lignin, more so when chemical pulping

TABLE 19-2
Characteristics of Effluents from the Manufacture of Pulps (Billings and DeHaas, 1971)

	Suspended solids, kg/tonne	*BOD, kg/tonne*
Bleached ground wood, textile, re-inked	20–360	11–220
Bine papers	22–45	7–18
Book and publication papers	22–45	9–22
Tissue	13–45	4–13
Coarse paper (boxboard, insulating, corrugated)	22–45	9–110
Newsprint	9–26	4–9

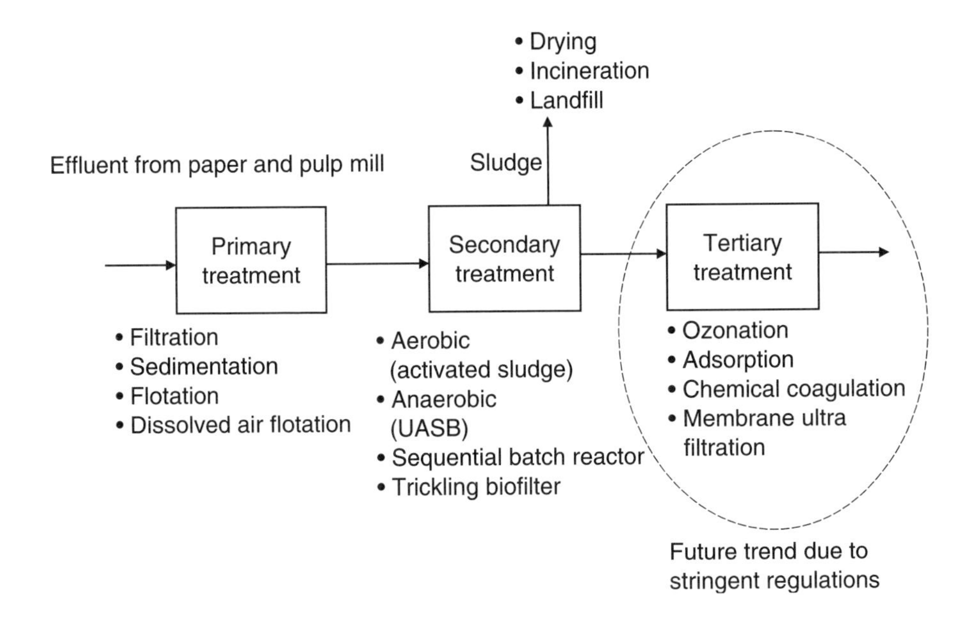

FIGURE 19-1. Various treatment procedures for paper and pulp industry effluent.

methods are employed. When colored paper is manufactured, the effluent will have some discoloration because of the dyes used in the manufacturing process. Figure 19-1 gives a broad overview of the various treatment procedures for the effluent that is generated by the paper and pulp manufacturing industry.

Bioprocesses

The anaerobic degradation of complex organic compounds is affected by the presence of sulfate. The first steps in the anaerobic degradation process are the hydrolysis and fermentation of biopolymers like carbohydrates and proteins to intermediates such as propionate, butyrate, formate, and H_2+CO_2 by fermenting bacteria. In the absence of sulfate, propionate and butyrate are degraded by acetogens to acetate, formate, and hydrogen, which are then converted by methanogens (see Table 19-3). In the presence of sulfate, the sulfate reducers will compete with these bacteria for propionate, butyrate, acetate, hydrogen, and formate by coupling the oxidation of these compounds to sulfate reduction. COD can be degraded via sulfate reduction if the COD to sulfate (g/g) ratio is below 0.66 (mol COD/mol sulfate < 0.5). *Methanosaeta* spp. were the dominant acetate degraders, and *Methanobacterium* spp. the dominant hydrogen- and formate-consuming methanogens. *Desulfobulbus* spp. and *Syntrophobacter* spp. were necessary for propionate degradation.

TABLE 19-3
Anaerobic Degradation Reactions (Elferink et al.,1998)

Syntrophic acetogenic reactions
$\text{propionate}^- + H_2O \rightarrow \text{acetate}^- + HCO_3^- + H^+ + 3H_2$
$\text{butyrate}^- + 2H_2O \rightarrow 2\ \text{acetate}^- + H^+ + 2H_2$
$\text{propionate}^- + 2HCO3^- \rightarrow \text{acetate}^- + 3\ \text{formate}^- + H^+$
$\text{butyrate}^- + 2HCO_3^- \rightarrow 2\ \text{acetate}^- + 2\ \text{formate}^- + H^+$
Methanogenic reactions
$\text{acetate}^- + H_2O \rightarrow CH_4 + HCO_3^-$
$H_2 + 0.25HCO_3^- + 0.25H^+ \rightarrow 0.25CH_4 + 0.75H_2O$
$\text{formate}^- + 0.25H_2O + 0.25H^+ \rightarrow 0.25CH_4 + 0.75HCO_3$
Sulfidogenic reactions
$\text{propionate}^- + 0.75SO_4^{2-} \rightarrow \text{acetate}^- + HCO_3^- + 0.75HS^- + 0.25H^+$
$\text{butyrate}^- + 0.5SO_4^{2-} \rightarrow 2\ \text{acetate}^- + 0.5HS^- + 0.5H^+$
$\text{butyrate}^- + 1.5SO_4^{2-} \rightarrow \text{acetate}^- + 2HCO_3^- + 1.5HS^- + 0.5H^+$
$\text{acetate}^- + SO_4^{2-} \rightarrow 2HCO_3^- + HS^-$
$H_2 + 0.25SO_4^{2-} + 0.25H^+ \rightarrow 0.25HS^- + H_2O$
$\text{formate}^- + 0.25SO_4^{2-} + 0.25H^+ \rightarrow 0.25HS^- + HCO_3$

Butyrate was probably degraded by syntrophic butyrate degraders such as *Syntrophospora* and *Syntrophomonas* (Elferink et al., 1998).

Very little BOD or COD removal was observed when kraft or sulfite chlorine bleaching effluents were treated anaerobically, since they are inhibitory to methanogenic bacteria. This inhibition is attributed to organohalogens such as chlorophenols and halomethanes present in the bleaching effluents. Also methane productivity is poor unless the wastewater is not highly diluted. The alkaline steps in the bleaching process extract wood resin compounds into the effluent, which are also inhibitory to methanogens. Presently there is pressure to replace chlorine with chlorine-dioxide bleaching (elemental chlorine-free) or totally replace all chlorinated bleaching agents (totally chlorine-free) with ozone and hydrogen peroxide (Vidal et al., 1997). But effluents from the chlorine and elemental chlorine-free bleaching process have similar methanogenic toxicities.

The lag phase for methane production in a batch process depends on how much of the kraft bleaching plant effluent is treated. The higher the fraction of effluent, the longer the lag phase, and once methane production starts, the rate is similar despite the differences in the length of the lag phase. This shows that the toxicity of effluent to aceticlastic methanogens is bacteriostatic rather than bacteriocidal. Aceticlastic methanogens do not easily acclimate to the toxic compounds in the effluent. We can, therefore, conclude that ethane production obtained after long lag phases in batch assays is due to a slow degradation of the toxic compounds that finally eliminates the inhibition. When the toxic compounds had been degraded to a level below the toxic threshold, methane production started (Yu and Welander, 1996).

Bioconversion of various waste paper materials by *Trichoderma viride* cellulase depends on the composition of the enzyme system, as well as the structure of cellulose. The crystalline section is difficult to hydrolyze, and the amorphous section is more susceptible to cellulase attack. Biodegradability of various paper materials tested with *Trichoderma viride* cellulase (0.2 mg/mL) incubated at 50°C for 2 h indicated that cardboard exhibited the highest efficiency followed by office paper and then foolscap (van Wyk and Mohulatsi, 2003). A decrease in hydrolytic efficiency was observed for all wastepaper materials because of the accumulation of sugar produced during biodegradation.

The solid industrial waste from the primary, secondary, and tertiary treatment stages from a pulp and paper factory contained 20% solids at pH 6. When the sludge was treated with bicarbonate, a 60% reduction in COD (initial COD = 1,216 mg/g) and 50% in extractable organic halogen (EOX) (initial EOX = 1,546 mg/kg) were observed under anoxic conditions over a period of 5 months (Ratnieks and Gaylardeb, 1997). Aerobic biological treatment by lagoons and activated sludge systems have been widely used to treat pulp and paper mill effluents to achieve BOD removal efficiencies between 65 and 99% and COD reductions between 25 and 65%. It is well established that fatty acids and resin acids are biodegraded aerobically, while chlorinated organic compounds are poorly degraded by these methods. Activated sludge treatment with addition of N and P to maintain a ratio of 100:5:0.3 was effective in treating *Pinus radiata* bleached kraft mill wastewater, achieving 90% BOD removal and 60% COD removal efficiencies in a reactor for hydraulic retention time (HRT) between 6 and 16 h (Diez et al., 2002). Generally the activated sludge processes were carried out at 35°C. Thermophilic conditions increased the effluent turbidity and decreased the sludge settleability, leading to sludge loss. Experimental studies carried out by Vogelaar et al. (2002) in an activated sludge tubular reactor operated at 30 and 55°C indicated that total COD removal was 58 and 48% and colloidal COD removal was 86 and 70%, respectively. Sludge production was the same at both temperatures.

Several thousand species of white rot fungi, most of them Basidiomycotina (which attack either hardwood or softwood) and a few Ascomycotina (which attack only hardwood) degrade lignin. Brown rot fungi extensively degrade cellulose and hemicelluloses in wood and to a very limited extent lignin. The soft rot fungi Ascomycotina or Deuteromycotina degrade both hardwood and softwood. White rot wood fungi including *Phanerochaete chrysosporium* and *Trametes versicolor* are capable of degrading wood and its constituents, such as cellulose and lignin, using the cellulose fraction as a source of carbon. Lignin peroxidases and laccase have the ability to depolymerize high molecular weight lignins and simultaneously polymerize the low molecular weight products formed. Kraft black liquor treated by the white rot fungi *Trametes elegans* showed a reduction in lignin weight (the average molecular weight decreased from 9,032 to 7,698) and an increase in polydispersity (from 1.6 to 3.2) after 15 days of incubation, indicating the lignin present in the effluent is degraded without the addition of extra nutrient (Lara et al., 2003).

Composting proceeds through three phases: (1) the mesophilic phase, (2) the thermophilic phase (lasts from a few days to several months), and (3) the cooling and maturation phase (lasts for several months). During composting, organic matter is transformed by microorganisms into CO_2, biomass, heat, and a humuslike end product (Tuomela et al., 2000). Aerobic conditions are prevalent in the thin topmost crust where the cellulose is oxidized to carbon dioxide; the anaerobic activity starts very close to the surface (Durrant, 1996). Two thermophilic actinomycetes degrade 0.7 to 2.5% of lignin in 42 days at 50°C, and a thermophilic fungus *Thermomyces lanuginosus* degrades 4.2% of lignin in a compost environment.

Table 19-4 shows the results from anaerobic treatment of black liquor and bleach effluent wastes that came from a pulp and paper mill and were inoculated with sludge from a batch anaerobic reactor. There have been many reports of adsorbable organic halogen (AOX) reduction using a variety of microorganisms; sample results are: a combination of aerobic and anaerobic treatments can achieve 65% removal; treatment of bleach pulp effluents by *Phaenerochaete chrysosporium* achieves 40–60% reduction;

TABLE 19-4
Anaerobic Treatment of Waste from a Pulp and Paper Mill (Ali and Sreekrishnan, 2000)

	Black liquor	*Bleach effluent*
Initial conditions		
COD, mg/L	24,500	2,500
Total solids	31.5	3.1 g/L
After inoculation, pH 7 at 37°C		
COD reduction, %	43	31
Methane production increase, %	33	27
After addition of 1% w:v glucose		
COD reductions, %	71	66
Adsorbable organic halide reduction, %	73	73

Trametes versicolor, 52–59%; *Ceriporiopsis subvermispora*, 32%; *Saccharomyces cerevisiae*, 64%; and a mixture of thermophilic aerobic and anaerobic microbes, 36–56%.

Bleach plant effluents from the pulp and paper industry are highly colored and also partly toxic as a result of the presence of chloroorganics. Several microorganisms have been used for removing color from bleach effluent. A few promising ones are immobilized *P. chrysosporium* (50% color removal in 3 to 6 h) and *T. versicolor* (71% color removal in 16 h), and alginate-immobilized *C. versicolor* (60% color removal in 30 h) and *Rhizomucor pusillus* (90% color adsorbed in 3 h). It is reported that chlorinated phenols and adsorbable chlororganics were first adsorbed onto the fungal biomass followed by breakdown, leading to color removal (Christov and van Driessel, 2003).

Bioreactors

A moving bed biofilm reactor (MBBR) is a completely mixed, continuously operated system in which the biomass is grown on small carrier elements and circulated in the liquid. In an anaerobic or anoxic reactor, mechanical agitation is used to circulate these elements; in an aerobic reactor, aeration is used. No sludge is recycled here. Generally the carrier elements have a high biofilm growth area, are cylindrical in shape, and are made of polyethylene. Thermophilic aerobic treatment of thermomechanical pulp white water in such a reactor led to about a 65% reduction in soluble COD at a temperature of 55°C (Jahren et al., 2002). The biodegradation was carried out with the addition of nitrogen- and phosphorous-containing nutrients. The removal

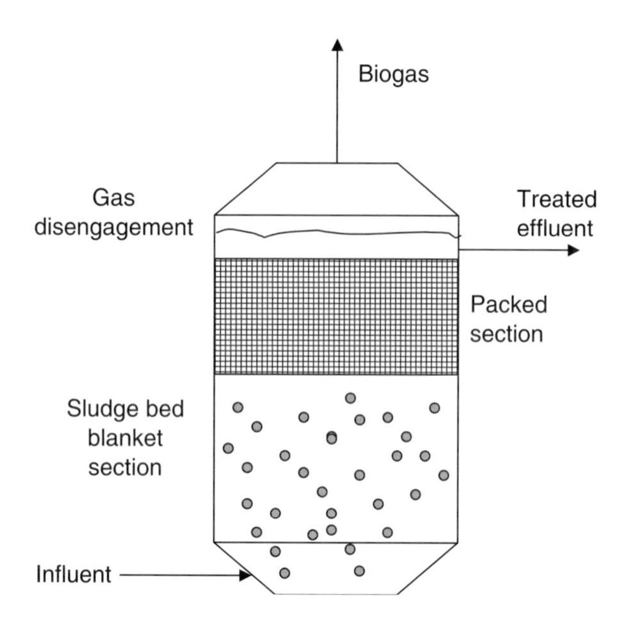

FIGURE 19-2. Hybrid upflow sludge bed filter (USBF) anaerobic reactor.

rate at thermophilic conditions (55°C) was 1.1 to 1.8 kg when compared with 0.18 to 0.2 soluble COD (SCOD)/kg volatile suspended solids (VSS)· day at mesophilic conditions (20 to 40°C), indicating the advantages of carrying out the reaction at thermophilic conditions. No difference in sludge yield is observed at these temperature conditions.

Fiberboard manufacturing produces considerable amounts of wastewater, ranging from 3 to 15 m^3 wastewater per ton of board produced for medium density fiberboard and wet process fiberboard, respectively, with almost 40 g COD/L, a pH of 3.0, and the presence of phenolic and tannin compounds. Such effluent has been treated in an upflow anaerobic sludge blanket (UASB) reactor at high organic loading rates (OLR), achieving COD removal efficiencies of 90% and phenolics removal of 90%. A hybrid upflow sludge bed filter (USBF) anaerobic reactor consists of a UASB with a packed section located above the top of the sludge bed (Fig. 19-2). PVC corrugated rings were used as packing material in the anaerobic filter zone. Wastewater from a fiberboard manufacturing plant was treated in such a reactor, achieving COD removal efficiencies of about 90% OLR, 6.5 to 8.5 kg COD/m^3 day. Eighty percent of the COD fraction was converted into methane (Fernandez et al., 2001). The total suspended solids (TSS) removal efficiency was 54%; COD and color removal efficiencies of 10% were achieved by adding 10 mg/L of neutral polyelectrolyte in the pretreatment section.

Black alkaline liquor from kraft pulp mills is generally treated in activated sludge plants and aerated lagoons. Anaerobic treatment of this effluent was successfully carried out achieving a COD removal efficiency on the order of 86% and a methane production efficiency of 36.9 mmol/mL at a hydraulic loading rate (HLR) of 0.60 per day (Buzzini and Pires, 2002). Yeast extract, ammonium chloride, and monobasic anhydrous sodium phosphate were added to supply the recommended amount of nitrogen and phosphorus, and ethanol was added as an additional carbon source. The conventional activated sludge process for the treatment of wastewater is not suitable for high organic loading and shows bulking problems during the summer months as a result of low dissolved oxygen content. Anaerobic upflow filters operated at mesophilic (35°C) and thermophilic (55°C) temperatures lead to a 75 and 90% decrease in the SCOD of a simulated paper mill wastewater at an HRT of 25 h at an OLR of 2 g/L. The methane content of the biogas produced by the mesophilic digester was higher (85%) than that produced by the thermophilic digester (81%) (Ahn and Forster, 2002).

Effluents from a hardwood-based bleached kraft mill (BOD, COD, and adsorbable organic halogen of 200 to 300, 500 to 600, and 5 to 10 mg/L, respectively) that were treated in a sequencing batch reactor reached COD removal efficiencies of 75% at 35°C, and the efficiencies decreased with increase in temperature. AOX removal decreased with increasing temperatures (from 70 to 60%). The SBR was operated in four cycles of fill, aeration, settle, and withdrawal with a total cycle time of 8 h (Tripathi and Allen, 1999). BOD, COD, and AOX removal efficiencies of 99, 85, and 75%, respectively, were obtained when the same effluent was treated in an activated sludge reactor operated with 38 h of hydraulic retention time and 10 to 15 days of solids retention time.

An anaerobic baffled reactor (ABR) consists of a series of vertical baffles that force the wastewater to flow under and over them; therefore, the wastewater comes into contact with a large active biological mass. This reactor system has several advantages: it is simple in design, requires no gas separation system, and the over- and underflow of liquid reduces bacterial washout and enables it to retain active biological solids without the use of any fixed media. Diluted black liquor mixed with digested sewage sludge in the ratio of 3:1 (v:v) having a resultant soluble COD of 2 g/L was treated in an ABR to achieve a COD reduction of 60% and a methane yield of 0.147 m^3/kg COD removed at an HRT of 2 days (Grover et al., 1999).

Calcium carbonate is used in the coating process, which leads to high concentrations of calcium ions in the wastewater of papermaking industries. Accumulation of calcium scale in a UASB reactor can significantly decrease granule activity. Pretreatment for removal of calcium hardness includes a softening process that uses lime soda or fluidized sand-coated calcium carbonate. Another novel approach was to couple a CO_2 stripper to a UASB; the former helped in the removal of Ca ions in the form of calcium carbonate, preventing it from accumulating in the reactor and causing clogging. When a UASB and a CO_2 stripper were combined to treat a synthetic effluent containing 5,000 mg/L calcium hardness and 3,000 mg/L COD, 60% calcium and 90% COD removals were achieved. In the absence of the CO_2 stripper, the COD removal efficiency dropped down to 70% (Kim et al., 2003).

Treatment of bleach plant effluent using *Coriolus versicolor,* a white rot fungus and *Rhizomucor pusillus* strain RM7, a mucoralean fungus in an aerobic rotating biological contactor showed 58 and 68% decolorization, respectively. Addition of glucose stimulated color removal by *C. versicolor,* but not with *R. pusillus.* In addition, *C. versicolor* removed 55% of AOX and 70% COD compared with 40 and 59%, respectively, by *R. pusillus.* It appeared that definite differences exist between the decoloring mechanisms of the white rot fungus and the mucoralean fungus; the former is based on adsorption and biodegradation while the latter is only adsorption (Driessel and Christov, 2001). In the Mycor process (called FPL/NCSU Mycor method), *P. chrysosporium* is immobilized on the surface of rotating disks to achieve 80% color removal and convert 70% of organic-bound chlorine to chloride at an HRT of 2 days (Christov and van Driessel, 2003).

Conclusions

The paper and pulp process uses plenty of water, and the waste generated from this industry contains solvents, chlorinated compounds, resins, and most importantly lignin, which is highly resistant to degradation. Chlorinated compounds are also toxic to many microorganisms. Physicochemical treatment methods are expensive. Conventional biological methods such as activated sludge and aerated lagoons help in reducing the COD load and toxicity, but these methods cannot effectively remove the color from bleach plant effluents; in addition they consume energy for aeration. White rot fungus appears to be efficient in color removal. Anaerobic degradation is affected by the presence of sulfate. Destruction of adsorbable organic halogen is another aspect that needs to be addressed. Carrying out biodegradation at thermophilic conditions would be advantageous since the waste stream from the paper mill is generally around 50°C, but finding efficient microorganisms that can perform well and overcome the other problems mentioned earlier is still a research challenge.

References

Ahn, J. H., and C. F. Forster. 2002. A comparison of mesophilic and hermophilic anaerobic upflow filters treating paper and pulp liquors. *Process Biochem.* 38:257–262.

Ali, M., and T. R. Sreekrishnan. 2000. Anaerobic treatment of agricultural residue based pulp and paper mill effluents for AOX and COD reduction. *Process Biochem.* 36:25–29.

Billings, R. M., and G. G. DeHaas. 1971. Pollution control in the pulp and paper industry. In:, *Industrial pollution control handbook,* ed. H.F. Lund, 18.1–18.28. New York: McGraw-Hill.

Buzzini, A. P., and E. C. Pires. 2002. Cellulose pulp mill effluent treatment in an upflow anaerobic sludge blanket reactor. *Process Biochem.* 38:707–713.

Christov, L., and B. van Driessel. 2003. Wastewater bioremediation in the pulp and paper industry. *Indian J. Biotech.* 2:444–450.

Durrant, L. R. 1996. Biodegradation of lignocellulosic materials by soil fungi isolated under anaerobic conditions. *Int. Biodeterior. Biodegrad.* 46:189–195.

Elferink, S. J. W. H. O., W. J. C. Vorstman, A. Sopjes, and A. J. M. Stams. 1998. Characterization of the sulfate-reducing and syntrophic population in granular sludge from a full-scale anaerobic reactor treating papermill wastewater. *FEMS Microbiol. Ecol.* 27:185–194.

Fernandez, J. M., F. Omil, R. Mendez, and J. M. Lema. 2001. Anaerobic treatment of fibreboard manufacturing wastewaters in a pilot scale hybrid USBF reactor. *Water Res.* 35(17):4150–4158.

Grover, R., S. S. Marwaha, and J. F. Kennedy. 1999. Studies on the use of an anaerobic baffled reactor for the continuous anaerobic digestion of pulp and paper mill black liquors. *Process Biochem.* 34:653–657.

Jahren, S. J., J. A. Rintala, and H. Odegaard. 2002. Aerobic moving bed biofilm reactor treating thermomechanical pulping whitewater under thermophilic conditions. *Water Res.* 36:1067–1075.

Kim, Y. H., S. H. Yeom, J. Y. Ryu, and B. K. Song. 2004. Development of a novel UASB/CO2-stripper system for the removal of calcium ion in paper wastewater. *Process Biochem.* 39:1393–1399.

Lara, M. A., A. J. Rodriguez-Malaver, O. J. Rojas, O. Holmquist, A. M. Gonzalez, J. Bullon, N. Penaloza, and E. Araujo. 2003. Black liquor lignin biodegradation by *Trametes elegans*. *Int. Biodeterior. Biodegrad.* 52:167–173.

Ratnieks, E., and C. C. Gaylardeb. 1997. Anaerobic degradation of paper mill sludge. *Int. Biodeterior. Biodegrad.* 39(4):287–293.

Tripathi, C. S., and D. G. Allen. 1999. Comparison of mesophilic and thermophilic aerobic biological treatment in sequencing batch reactors treating bleached kraft pulp mill effluent. *Water Res.* 33(3):836–846.

Tuomela, M., M. Vikman, A. Hatakka, and M. Itavaara. 2000. Biodegradation of lignin in a compost environment: a review. *Bioresour. Tech.* 72:169–183.

van Driessel, B., and L. Christov. 2001. Decolorization of bleach plant effluent by mucoralean and white-rot fungi in a rotating biological contactor reactor. *J. Biosci. Bioeng.* 92(3):271–276.

van Wyk , J. P. H., and M. Mohulatsi. 2003. Biodegradation of wastepaper by cellulase from *Trichoderma viride*. *Bioresour. Tech.* 86:21–23.

Vidal, G., M. Soto, J. Field, R. Mendez-Pampfn, and J. M. Lema. 1997. Anaerobic biodegradability and toxicity of wastewaters from chlorine and total chlorine-free bleaching of eucalyptus kraft pulps. *Water Res.* 31(10):2487–2494.

Yu, P., and T. Welander. 1996. Toxicity of kraft bleaching plant effluent to aceticlastic methanogens. *J. Fermentation Bioeng.* 82(3):286–290.

Bibliography

Babuna, F. G., O. Ince, D. Orhon, and A. Simsek. 1998. Assessment of inert COD in pulp and paper mill wastewater under anaerobic conditions. *Water Res.* 32(11):3490–3494.

Diez, M. C., G. Castillo, L. Aguilar, G. Vidal, and M. L. Mora. 2002. Operational factors and nutrient effects on activated sludge treatment of *Pinus radiata* kraft mill wastewater. *Bioresour. Tech.* 83:131–138.

Thompson, G., J. Swain, M. Kay, and C. F. Forster. 2001. Review paper—The treatment of pulp and paper mill effluent: a review. *Bioresour. Tech.* 77:275–286.

Vogelaar, J. C. T., E. Bouwhuis, A. Klapwijk, H. Spanjers, and J. B. van Lier. 2002. Mesophilic and thermophilic activated sludge post-treatment of paper mill process water. *Water Res.* 36:1869–1879.

CHAPTER 20

Paint Industries

The paint and coating application areas comprise (1) architectural coatings or house paints, which includes waterborne latex, exterior and interior solvent-borne paints, lacquers, and wood and furniture finishes; (2) industrial coatings including automotive, metal, machinery, and equipment finishes, paper coatings, electric insulating varnishes, and magnetic wire coatings; (3) special purpose coatings like industrial maintenance paints, marine coatings, traffic and metallic paints, automobile refinishing coatings, aerosol paints, and multicolor paints; and (4) miscellaneous products like paints used for graphics and artwork. A typical paint and coatings manufacturing operation involves formulation, milling, or grinding of pigments, mixing, filtering, filling, and equipment cleaning. The production process for a liquid paint starts with the dispersion of pigments, solvents, resins, and additives in a mill such as ball or bead mill, or a high-speed disperser. Diluents, resins, bactericides, fungicides, etc., are added to the dispersion mill effluent in a process known as letdown. When the formulation achieves the desired properties, mixing is stopped, the paint is filtered, and the final product is stored in cans for shipment. Paint manufacture requires several hundred raw materials, which include antifoams, defoamers, dispersants, surfactants, driers, antiskinning agents, extenders, fillers, pigments, flame or fire retardants, flatting agents, latex emulsions, oils, preservatives, bactericides, fungicides, resins, rheological and viscosity control agents, silicone additives, titanium dioxides, and colors.

Types of Pollutants

A variety of hazardous solid, liquid, and gaseous wastes is generated during the manufacturing operation. Solid waste is generated from used containers, spent filters, dried paints, pallets, and packaging materials. Equipment cleaning, spillage, and off-spec materials generate liquid waste. The various operations also lead to discharge of pollutants into the atmosphere. For example, (1) many raw materials used to manufacture paint are volatile organic compounds (VOCs) and evaporate readily in the atmosphere when

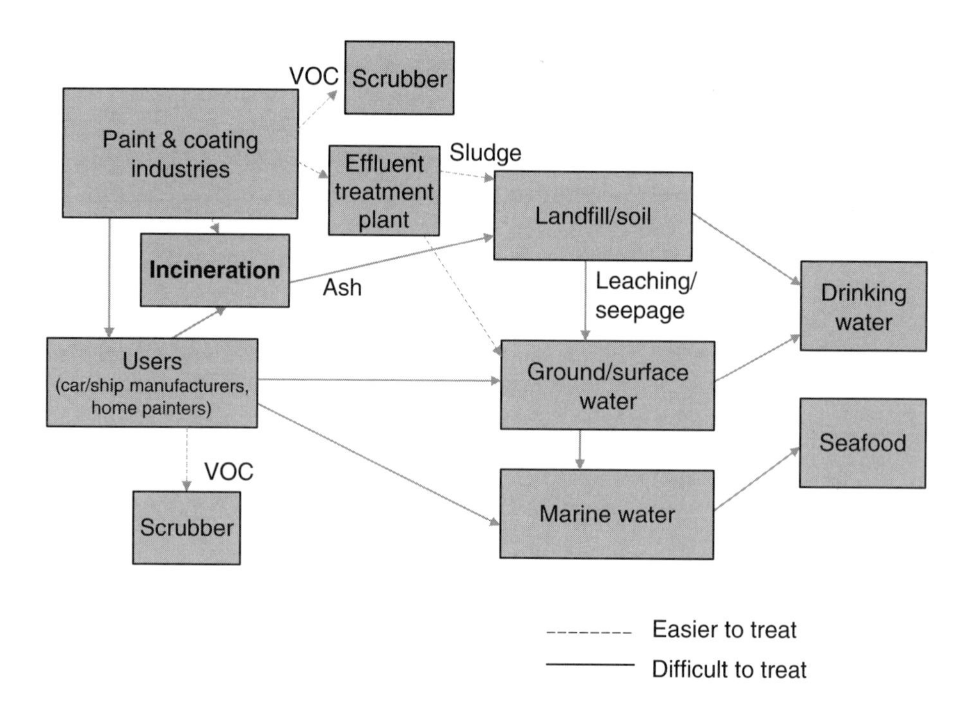

FIGURE 20-1. Movement of raw materials used in paint into the environment.

the ingredients are exposed to air, (2) pigment dust (particulate matter) is generated during the manufacturing process, and (3) solvents used for cleaning the equipment have high evaporation rates. The industrial and home users of these paints and coatings also generate various types of waste. If these wastes are not properly treated and detoxified, they can enter the environment as shown in Fig. 20-1.

Tributyltin is a herbicide used in paints as an antifouling agent to prevent marine organisms from growing, and it is known for its acute toxicity, imposex (the occurrence of induced male sex characteristics superimposed on normal female gastropods), and bioaccumulation; it causes increased shell thickness and decreases the reproductive capability of various water organisms. It is found in large harbors and dense shipping lanes, as well as in coastal areas with coral reefs and in seafood products. Its use was banned after the year 2003. Diuron and Irgarol herbicides are being used as alternatives for preventing algae growth on the surfaces of boats and ships. Diuron has been detected in the coastal waters of the Mediterranean Sea in concentrations of more than 2 mg/L (permissible concentration 430 ng/L), and Irgarol 1051 in concentrations of up to a few hundred nanograms per liter has been detected in European, Japanese, and Australian seas. In several locations and at several times, concentrations exceeded the maximum permissible limit of 24 ng/L.

The degradation products of Irgarol 1051 are known to exhibit toxicities similar to the original compound (Lamoree et al, 2002). Diuron (in the range of 3 mg/L) and Irgarol were also identified in the Japanese aquatic environment. Diuron around 1 to 40 μg/L has been detected in fresh- and groundwater in many western European countries and the United States (Okamura et al., 2003). Microgram levels of these two herbicides were also detected in the coastal waters of the United Kingdom. A reduction in *Fucus vesiculosus*, a perennial macro alga found in the Baltic Sea, has been observed in the inner parts of the archipelagos along the Swedish coast; this reduction is attributed to pollutants such as copper and Irgarol found in the anti-fouling paints (Karlsson and Eklund, 2004). Copper-based antifouling paint caused toxic effects on brine shrimp nauplii. The copper released from the paint entered their cells and caused decreased enzymatic activity.

Automobiles get three layers of paint: the primer, the base coat, and the clear coat. The primer is either solvent or powder based, the base coat is waterborne, and the clear coat is also solvent or powder based. Powder-based coatings generate the highest VOCs when compared with the other two, namely, 0.06 to 0.12 kg SOx, 0.06 kg NOx, and 0.04 kg of particulate matter per kilogram of each coating. Suspended solids (0.01 to 0.03 kg/kg of coating) and metals (about 0.004 kg/kg of coating) contribute primarily to the contamination of wastewater.

Paint solvents contribute about 45% of the VOCs in Seoul, South Korea's atmosphere (1997 data). Aromatics account for 95%, and the remainder are alkanes. Toluene was the most abundant compound, followed by *m*- and *p*-xylene, and then *o*-xylene. Benzene and styrene contributed less than 1% (Na et al., 2004). All paints, regardless of carrier, use the same basic chemical categories, namely, resins and crosslinkers (binder system), pigments, and modifying additives.

Polyurethanes are used in the manufacture of car paint. Once the paint is sprayed on the cars, thermal degradation of the polyurethane occurs, generating many new molecules of isocyanates, which are a result of secondary reactions such as chain breaking, isomerization, and dehydrogenation. Workers in car paint shops are thus exposed to additional amounts of reacted and unreacted isocyanates contained in the paint formulation.

The lacquers or paints used in the furniture industry contain isopropanol, butanol, butyl and ethyl acetates, toluene, ethylbenzene, xylenes, and aromatic hydrocarbon solvents. The residues contain unknown complex organic mixtures, and they are found to be more toxic than the constituent basic chemicals. Latex paints generally consist of organic and inorganic pigments and dyestuffs, extenders, cellulosic and noncellulosic thickeners, latexes, emulsifying agents, antifoaming agents, preservatives, solvents, and coalescing agents. The wastewater is alkaline and contains high BOD, COD, suspended solids, toxic compounds, and color. High-quality water with 85% of the COD removed was recovered for recycling purposes from an electrocoat painting bath by reverse osmosis. A typical paint stripping facility would generate effluent consisting of methylene chloride, phenol, and other organic compounds in concentrations of about 5,000, 1,800 and 2,200 mg/L, respectively (Arquiaga et al., 1995).

Microorganisms thrive in water-based paint by consuming oxygen. Once all the oxygen is consumed, anaerobic growth commences; during that process both bacteria and fungi produce cellulase, which breaks down long-chain cellulosic thickening agents, producing small oligomeric residual units. Fermentative bacteria break down the cellulose to glucose and the glucose to acid and carbon dioxide. Under anaerobic conditions, *Desulphovibrio desulphuricans* can use oxygen from sulfates, which generates hydrogen sulfide. Acid production by microorganisms causes a decrease in pH.

Biochemical Treatment

The general wastewater treatment facility for treating the effluent from a paint manufacturing plant consists of an equalization basin, a primary settling tank, a pH neutralization tank, an aeration tank, a secondary settling tank, and holding tanks. The BOD, COD, and TSS of the effluent were reduced from 588, 5,632 and 2,864 to 50, 100 and 100 mg/L, respectively, after the combined chemical and biochemical treatment. The concentrations of metals such as Pb, Cr, Cu, Mn, Ni, Zn, and Fe remained the same in the untreated and the treated effluent. The BOD, COD, and TSS of the effluent could be reduced to 28, 65, and 5 mg/L by coagulation-flocculation (combination alum and polyelectrolyte-anionic polyacrylamide) and cross-flow microfiltration using a cellulose acetate membrane with pore size 0.2 μm at a pressure drop of 0.3 bar. Sulfuric acid or calcium hydroxide was added to adjust the pH of the wastewater. The microfiltration treatment procedure also removed metals and bacterial contamination from the waste stream (Dey et al., 2004).

Paint-stripping wastewater contaminated with phenol was treated in reactors (e.g., activated sludge and rotating biological contactor) that predominantly contained *Pseudomonas* gram-negative bacteria. Gram-positive bacteria occurred less frequently and were solely represented by the genus *Bacillus*. Other genera such as *Acinetobacter, Moraxella, Paracoccus, Acetobacter, Flavobacterium, Klebsiella, Enterobacter,* and *Vibrio* were also found but in fewer numbers. The size of the microbial communities in the continuous flow rotating biological contactor was a maximum of 10^{10} bacteria/g, followed by batch and continuous flow activated-sludge reactors, and least in fill and draw rotating biological contactor (10^7 bacteria/g) reactors. This difference could be explained by two factors. (1) There is higher concentration of toxic paint stripping chemicals in the batch reactor than in the continuous flow reactors (53% in the former as against 20% in the latter). This happened because of the dilution in the continuous reactor. (2) Continuous-flow wastewater systems favor the attachment of bacteria to surfaces instead of being washed away in the batch reactor. Both activated sludge and rotating biological contactor reactors could effectively degrade paint stripping effluents mixed with domestic effluents (Arquiaga et al., 1995).

Typical volatile compounds from paint preparation include toluene (approximately 25%), methyl ethyl ketone (approximately 23%), *m*-and *p*-xylene (approximately 20%), and other organics such as ethylbenzene, *o*-xylene, 1-butanol, acetone, ethane, etc. Biofilters are well suited to treating VOCs found at paint spray booths, paint manufacturing plants, or filling stations. Unlike bacterial-based biofilters, fungal-based systems function even better in slightly dry conditions and at low pH. A preadapted compost-based media bed shows good resilience to operating conditions that could easily destroy systems based on bacteria alone. Generally the VOCs are converted to carbon dioxide and water. Although steam is costly, it is effective for media wetting, moisture, and temperature control. An industrial-scale biofilter with multiple layers of compost material supported on plastic spheres with a cross-flow air and water spray humidification system degraded 75% of the VOCs. In large units, maintaining wet conditions and uniform temperature are the two challenges. Capital cost is significantly less than for a thermal oxidizer, and operating costs are less than 10% of a comparably sized regenerative thermal oxidizer. Biofilters neither produce toxic or hazardous products, as in the case of incomplete combustion reactions, nor create NOx or SOx as thermal oxidizer technologies do. More details about biofilters are given in Chapter 30, Gaseous Pollutants and Volatile Organics.

A laboratory-scale biofilter packed with cubed polyurethane foam media populated by a mixed culture of fungi was able to degrade 98% of *n*-butyl acetate, methyl ethyl ketone, methyl propyl ketone, and toluene (solvent emissions from industrial painting operations) at a total VOC loading rate of 94.3 g/(m^3 h). The mixed culture of fungal species predominantly included *Cladosporium sphaerospermum, Penicillium brevicompactum, Exophiala jenselmei, Fusarium oxysporum, F. nygamai, Talaromyces flavus,* and *Fonsecaea pedrosi*. Weekend shutdowns did not affect the performance of the biofilter, and in less than 3 h, the VOC removal efficiency reached its original value (Moe and Qi, 2004). The longer the shutdown, the larger the decline in removal efficiency following restart and the longer the reacclimation time required by the biofilter to recover. While removal efficiencies of acetate and ketones recovered in very short time after restart, the removal efficiency for toluene took a few days to reach its original value.

A compost-based lab-scale hybrid bioreactor could achieve more than 80% removal efficiency of a paint VOC mixture consisting of toluene, xylene, methyl propyl ketone, butyl acetate, and ethyl 3-ethoxy-propionate with a total concentration of approximately 100 ppmv at a gas residence time of 46 s. Hydrophilic components of the gas stream were degraded completely, while minimum degradation of the hydrophobic components was observed in the bioreactor. Inoculation of a microbial solution cultivated with toluene vapor as the sole carbon source raised the degradation efficiency to 90%. The hybrid bioreactor consisted of a single column divided into two sections; the first was packed with a structured plastic media and was operated as a trickling filter, and the second section was packed with a compost-based material and operated as a biofilter.

A buffered solution containing phosphates was continuously recirculated and sprayed from the top of the column. Water and additional nitrogen sources were also added to the packing materials. Air was introduced from the bottom with the VOCs (Song et al., 2002).

Conclusions

Paint contains several hundred chemicals ranging from solvents to toxic chemicals and metals. These chemicals find their way into the environment through different routes. VOCs are generated during the manufacturing process as well as during usage; biofilters appear to be a promising technology for its degradation. Water-based paint industries would like to recycle their wastewater, but the major hurdle here is the microbial contamination of the recycled water and the presence of suspended matter.

References

Arquiaga, M. C., L. W. Canter, and J. M. Robertson. 1995. Microbiological characterization of the biological treatment of aircraft paint stripping wastewater. *Environ. Pollution*. 89(2):189–195.

Dey, B. K., M. A. Hashim, S. Hasan, and B. Sen Gupta. 2004. Microfiltration of water-based paint effluents. *Adv. Environ Res.* 8(3–4):455–466.

Karlsson, J., and B. Eklund. 2004. New biocide-free anti-fouling paints are toxic. *Marine Pollution Bull.* 49(5–6):456–464.

Lamoree, M. H., C. P. Swart, A. van der Horst, B. van Hattum. 2002. Determination of diuron and the antifouling paint biocide Irgarol 1051 in Dutch marinas and coastal waters. *J. Chrom. A* 970:183–190.

Moe, W. M., and B. Qi, Performance of a fungal biofilter treating gas-phase solvent mixtures during intermittent loading. 2004. *Water Res.* 38, 2259–2268.

Na, K., Y. P. Kim, I. Moon, and K.-C. Moon. 2004. Chemical composition of major VOC emission sources in the Seoul atmosphere. *Chemosphere* 55:585–594.

Okamura, H., I. Aoyama, Y. Ono, and T. Nishida. 2003. Antifouling herbicides in the coastal waters of western Japan. *Marine Pollution Bull.* 47: 59–67

Song, J. H., K. A. Kinney, J. T. Boswell, and P. C. John. 2002. Performance of a compost-based hybrid bioreactor for the treatment of paint spray booth emissions. http://www.environmental-expert.com, paper # 43055. Air & Waste Management Association National Conference, Baltimore, MD, June 24–26.

Bibliography

Papasavva, S., S. Kia, J. Clayal, and R. Gunther. 2001. Characterization of automotive paints: an environmental impact analysis. *Progr. Organic Coatings* 43:193–206.

Thom, R., T. Barton, and J. Boswell. 2001. Biofiltration of VOCs for paint, manufacturing and coatings applications, http://www.environmental-expert.com, paper # 1124. Paint Research Association 18th International Conference, Belgium, November 12–14.

Thomas, K. V., T. W. Fileman, J. W. Readman, and M. J. Waldock. 2001. Antifouling paint booster biocides in the UK coastal environment and potential risks of biological effects. *Marine Pollution Bull.* 42(8): 677–688.

CHAPTER 21

Pharmaceuticals

Drugs in the Environment

Pharmaceutical and antibiotic residues from human and animal medical care enter the water and soil from (1) the effluent treatment plants of manufacturing facilities, (2) the municipal sewage treatment plant, (3) hospital waste treatment plants, or (4) animal farms as shown in Fig. 21-1. Treating effluent from a pharmaceutical plant that manufactures drugs and antibiotics is relatively easier than treating waste from a hospital or municipal sewage plant; in the former case the substances that need to be degraded are well known. The waste from hospital or municipal sewage plants may contain low concentrations of many different pharmaceuticals and their metabolites, which makes the task very difficult.

Excess medication excreted by humans and animals, as well as unused or expired medicines, find their way into municipal sewage effluent treatment plants. Since the 1980s, pharmaceuticals like clofibrate, various analgesics, cytostatic drugs, antibiotics, and others have been reported to be present in the surface waters of many European countries. This has raised growing concern that some of these persistent products may find their way back into the drinking water. Genotoxic substances may represent a health hazard to humans and may have adverse effects on other organisms. Since antibiotics mainly interfere with bacterial metabolism, it can be assumed that bacterial communities in aquatic ecosystems feel the primary effects of antibiotic-containing effluents. One of these effects is the increase in resistance to certain antibiotics, which in turn gives rise to infections that are difficult to treat. Antibiotics are consumed by humans and are used in livestock and poultry production and fish farming. The increasing use of these drugs during the last five decades has caused genetic selection of more harmful bacteria [reported veterinary drug usage in the European Union (EU) was 1,600 tons in 1999]. When animal excreta, which contain unmetabolized drugs, are applied to agricultural fields as fertilizer or manure, they contaminate the soil, and possibly the groundwater, depending upon their mobility. Terrestrial and aquatic organisms are affected as a result of leaching from the fields. Solid waste from industrial effluent treatment plants are disposed as

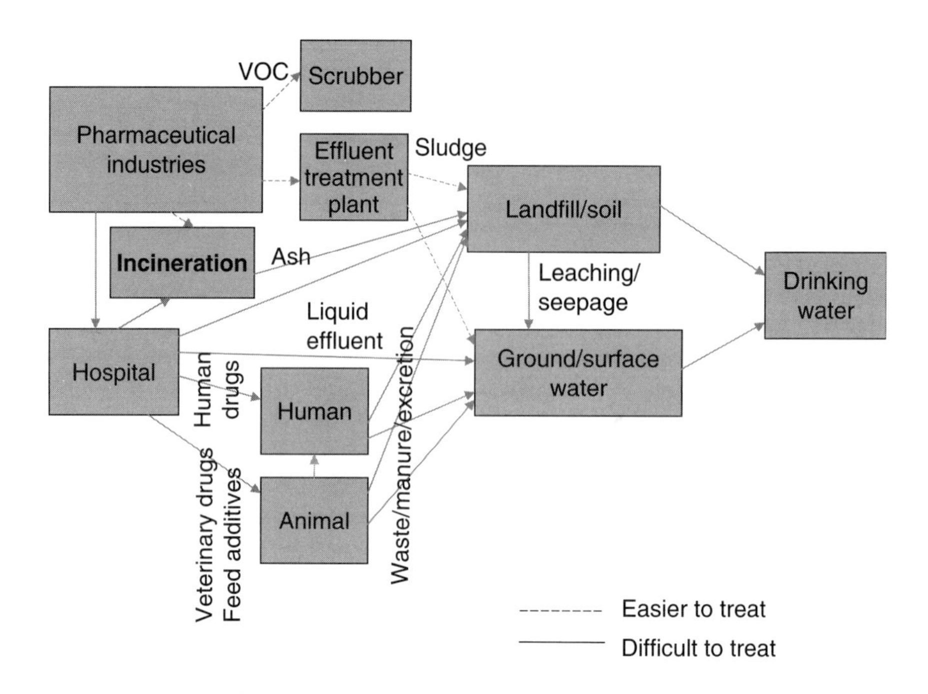

FIGURE 21-1. Movement of drugs and pharmaceutical products from source to environment.

landfill, which may lead to leaching of unmetabolized drugs into the groundwater. Of the drugs that are administered during fish farming, 70% of them are released into the environment, especially into the sediments near the fish farms. The modes of action of most pharmaceuticals in humans, animals, and fish are often poorly understood. The possible effects and side effects on nontarget receptor organisms and the synergistic effects produced as a result of mixing these drugs are also not known. The growth promoters, antibiotics, and other veterinary drugs given to poultry and cattle also end up in humans as a result of meat consumption. Natural and synthetic estrogens produce deleterious effects, such as feminization and hermaphroditism, in aquatic organisms. The persistence of a drug in a sediment or soil depends on its photostability, its binding and adsorption capability, its degradation rate, and its solubility in water. Strongly sorbing pharmaceuticals tend to accumulate in soil or sediment; in contrast, highly mobile pharmaceuticals tend to leach into groundwater and be transported by drainage and surface runoff.

The antibiotic tetracycline and its derivatives chlortetracycline and oxytetracycline are widely used in stockbreeding and aquaculture. The average concentration of oxytetracycline in German surface waters has been estimated at 0.01 μg/L (Backhaus and Grimme, 1999). In Germany,

0.165 mg/L of clofibric, a lipid-regulating agent, was found in river, ground, and drinking water, and of the 32 drugs that belong to the class of antiphlogistics, lipid regulators, psychiatric drugs, antiepileptic drugs, beta-blockers and sympathomimetics, 80% of them were found in the sewage treatment plant effluent with concentration levels on the order of 6 μg/L (Ternes, 1998). The sewage treatment plant, which treated household effluent, consisted of three tanks: preliminary clarification, final clarification, and aerator. The concentration of tetracycline and its derivative oxytetracycline in the river Lee near London has been estimated at 9.5 μg/L and tetracycline in British surface waters at about 1 μg/L. In the United Kingdom, drugs like diazepam, methaqualone, and penicilloyl antibiotics were found in potable water and groundwater. A nationwide study carried out by the U. S. Geological Society in 2002 found pharmaceuticals, hormones, and other organic wastewater contaminants in surface water (Smith, 2002). Apart from ceftriaxone and tilmicosin, several drugs, animal growth promoters and antibiotics, were found in nanogram levels in river sediment and river or drinking water in Italy. The concentrations found were several orders of magnitudes lower than the amount to produce any pharmacological effect, but possible effects of life-long exposures of these pharmaceutics on humans are not known (Zuccato et al., 2000).

Effect on Plants

Erythromycin, tetracycline, and ibuprofen affect the growth of the cyanobacterium *Synechocystis* sp. PCC6803 and the duckweed *Lemna minor* FBR006. Sulfadimethoxine alters the normal postgerminative development and growth of roots, hypocotyls, and leaves in *Panicum miliaceum, Pisum sativum,* and *Zea mays.* The bioaccumulation of this drug in these plants (root to stalk leaf bioaccumulation ratio is 2 to 20 μg/g) can affect other communities. *Azolla filiculoides* Lam. is a water fern that can take in 58 to 2,000 μg of this sulfa drug per gram for varying drug concentrations of 50 to 400 mg/L. A higher proportion of the drug was degraded in the presence of plants, between 50 and 56% at a concentration of 50 to 400 mg/L, while the degradation was 5 to 30%, in their absence (Forni et al., 2002). The drug affected the growth rate (as biomass yield per week) and nitrogen fixation.

Biodegradation of Pharmaceutical Products

The biodegradation of antibiotics and pharmaceuticals depends on the temperature, availability of organic and inorganic nutrients, concentration of the chemical, and presence of oxygen. The biodegradation rate of sulfonamides in activated sludge is identical for several of them. Nitrifying sludge degrades drugs such as chloramphenicol and oxytetracycline, but they are not mineralized. Estrogens and progestogens would be adsorbed onto sludge particles in the wastewater treatment plant and would not be biotransformed. Degradation studies carried out in artificial marine sediment under controlled laboratory conditions in a closed system indicated that oxytetracycline is highly persistent in the marine sediments (greater than 10 months), while sulfadimethoxine and ormethoprim are very short-lived (less than 21 to 62 days). Flumequine, sulfadiazine, and oxolinic acid are also not degraded and preserve their antibacterial activity for more than 180 days (Diaz-Cruz et al., 2003).

Aerobic batch biodegradation studies of veterinary and antimicrobial growth promoters such as metronidazole, oxytetracycline, olaquindox, and tylosin in the concentration range of 1 to 1,000 μg/L indicated that these drugs are moderately persistent in surface water except for olaquindox, which is more biodegradable than aniline. The half-life for aerobic biodegradation is 4 to 8 days for olaquindox, 10 to 40 days for tylosine, 14 to 104 days for metronidazole, and 31 to 40 days for oxytetracycline. Addition of 1 g/L of activated sludge from a wastewater treatment plant decreased the half-life by half. Under anaerobic conditions, the biodegradation rate decreased and the lag phase increased (Ingerslev et al., 2001).

A sewage treatment plant near Frankfurt, Germany, was able to eliminate through sorption on activated sludge and biodegradation more than 95% of the propranolol entering the effluent stream at a rate of about 520 g/day and 90% of the ibuprofen at a rate of about 250 g/day. Carbamazepine, clofibric acid, phenazone, and dimethylaminophenazone showed low biodegradation (Ternes, 1998). Antibiotics such as clofibric acid and diclofenac find their way in the aquatic environment, and the former has been found even in the North Sea. In an oxic biofilm reactor, clofibric acid and diclofenac were not degraded and ibuprofen was degraded to 30 to 36% of its initial concentration. When the biofilm reactor (BFR) is operated under anoxic conditions (anoxic means denitrification conditions in the absence of O_2 and presence of nitrate), ibuprofen degradation was low (only 21%) and some appreciable degradation was observed for diclofenac and clofibric acid (about 30%) (Zwiener and Frimmel, 2003). Hydroxyibuprofen was identified as the major metabolite of ibuprofen biodegradation in the oxic BFR. The BFR used for these studies used pumice stones as support material for microorganism growth. The biofilm was grown on the support material from activated sludge from a municipal sewage plant. Addition of acetone inhibited the degradation of ibuprofen.

The growth of *Pseudomonas putida* was inhibited by 50% at a concentration of 80 and 10 μg/L in the case of ciprofloxacin and ofloxacin, respectively (Kummerer et al., 2000). These antibiotics were not biodegraded in the closed-bottle test. Metronidazole, one of the most important nitroimidazoles, is toxic to algae and daphnids in low milligram per liter concentrations and is effective against anaerobic bacteria. Strong binding to soil is one reason for the poor degradation of some of the antibiotics from contaminated terrain. For example, cyclosporine degraded very slowly after some months in moist samples of garden soils, even though several microorganisms capable of its degradation have been isolated from soils. Sarafloxacin, a fluoroquinolone used against poultry diseases, was mineralized by less than 1% in various soils in 80 days. Virginiamycin, an antibiotic food additive for livestock, was found to biodegrade in various soils with a half-life of 87 to 173 days. Bacteriostatic sulfonamide (sulfa) drugs are used in the treatment of infections in livestock and in the treatment of human infections such as bronchitis and urinary tract infection. Eighty percent of the drug given to livestock is excreted in urine and subsequently dispersed with the sewage on fields and can reach groundwater. Sulfamethoxazole was found in the surface waters in Germany in concentrations of 30 to 85 ng/L (Hartig et al, 1999).

Pharmaceutical Industry Effluent

Wastewaters from pharmaceutical manufacturing contain high levels of suspended solids and soluble recalcitrant organics. Since pharmaceutical plants operate under batch mode, changes in production schedules lead to variability in the effluent flow rate, the principal constituents, and their relative biodegradability. The pharmaceutical industry produces a wide variety of products using both inorganic and organic chemicals as raw materials. Antibiotics and vitamins are produced by fermentation of complex nutrient solutions of organic matter and inorganic salts by fungi or bacteria. Most of the wastes are toxic to biological life and have high biological oxygen demand (BOD), chemical oxygen demand (COD), and a high BOD to COD ratio; they are either highly alkaline (e.g., manufacture of sulfa drugs and vitamin B12) or acidic (e.g., manufacture of organic intermediates). At times cyanide might also be present. But the main advantage of such effluent is that the pollutants are known; hence treatment could be exactly tailored to remove or degrade it efficiently. Common physical treatment methods include coagulation and precipitation, reverse osmosis, and ultrafiltration; biological treatment includes the activated sludge process. Typical characteristics of effluent from a pharmaceutical bulk drug manufacturing plant would be a COD of 12,500 mg/L, a BOD of 6,000 mg/L, sulfate concentration of 9,000 mg/L, total solids of 36,000 mg/L, a pH of 8, and dissolved solids of 29,000 mg/L (Raj and Anjaneyulu, 2003).

It was believed that thermophilic treatment could lead to rapid biodegradation rates, low growth yields, and reduced cooling costs, but the operation of aerobic pharmaceutical wastewater treatment at elevated temperatures seemed to affect the performance of the process. Soluble COD removal efficiency decreased as the treatment temperature was increased from 30 to 60°C (from 62 to 38%, respectively) (Lapara et al., 2001). Untreated wastewater had soluble COD and BOD of 8,150 and 3,800 mg/L, respectively, and a total ammonia concentration of 220 mg/L. An increase in temperature led to a reduction in the number of different bacterial populations, and probably a decrease in biodiversity. Soluble microbial product (SMP) is defined as the organic compounds released into solution from substrate metabolism and biomass decay, and has its own fraction of recalcitrant and biodegradable fractions. Thermophilic culture produced an SMP with a higher fraction of recalcitrant soluble COD than did the mesophilic culture.

Stable reactor performance requires some stability among the individual populations that comprise the microbial community in the bioreactors, even when there is variation in the influent composition. Also, flexibility is needed for the community to adapt in response to changes in the operating conditions. Flexibility with respect to bacterial community structure leads to more stable process performance. The microbial community could adapt to changing environmental conditions. This was confirmed in full-scale pharmaceutical wastewater treatment studies carried out in a series of seven reactors with the first four bioreactors operated under aerobic and thermophilic temperature conditions ($T > 45°C$), while the last three reactors were operated at 25 to 35°C under biological nitrification and denitrification conditions. The overall treatment efficiencies were greater than 95% (LaPara et al., 2002). Short-term variability in influent wastewater composition brought about a greater community shift than did long-term operation with wastewater of consistent composition. The thermophilic reactors had similar community structures to each other; the same was true for the communities from mesophilic reactors. This study brought out the fact that during biological wastewater treatment, temperature served as a selective factor for bacterial community structure development.

Two-stage chemical and biochemical treatment of waste (with a BOD:COD ratio of 0.45 to 0.57) from a pharmaceutical bulk drug manufacturing plant produced very good results. Chemical coagulation using lime led to a reduction of 44 to 48% of the sulfate. Subsequent aerobic oxidation led to COD and BOD reduction efficiencies of 86 and 80%, respectively, at an MLVSS of 1,500 mg/L, a temperature of 30°C, and a hydraulic retention time (HRT) of 4.5 days (Raj and Anjaneyulu, 2003). The influent stream had a COD of 4,000 mg/L. Pharmaceutical waste aerobic activated sludges are normally dispersed, weak with a high solid content (greater than 5,000 mg/L), and resistant to biological degradation. They have poor bacterial filament development and poor dewaterability characteristics (Bernard and Gray, 2000). On the other hand, domestic waste activated sludges are usually compact and strong, with a low solid content and a thick foam.

Bioaugmentation is a technique adapted to maintain sufficient biomass in the medium when enough nutrients and carbon substrates are unavailable. Cells are grown externally (either under different operating conditions or with different substrates) and added to the main reactor from time to time to induce degradation. Manufacture of Cephradine (a main constituent of an antiosmotic drug) generates effluent containing the drug, acetic acid, and ammonia. A mixture containing this effluent and municipal sewage waste was treated in an anaerobic fluidized bed reactor to achieve 89% COD reduction at an HRT of 3 to12 h by bioaugmentation through periodic addition of 30 to 70 g of acclimated cells from an external enrichment reactor every 2 days (Saravanane et al., 2001a). Activated carbon was used as the carrier for the cells. The combined effluent had a COD of 12 to 15,000 mg/L, a BOD of 2,000 mg/L, TSS of 6,000 mg/L, dissolved solids (DS) of 11,000 to 18,500 mg/L, NH_3 of 15 to 40 mg/L, and a pH of 3 to 4. The enrichment reactor was operated as a sequencing batch reactor. A similar bioaugmentation approach was adopted for the treatment of effluent containing Cephalexin (a drug used for the treatment of bronchitis and other lung diseases) in a fluidized bed reactor to achieve an 89% COD removal efficiency at an HRT of 3 to 12 days. The influent had a COD of 12 to 15,000 mg/L. Every 2 days, 30 to 70 g of acclimated cells were added from an offline reactor (Saravanane et al., 2001b).

More than 90% anaerobic degradation of waste fermentation broth from clavulanic acid production (total organic carbon, 50,000mg/L, pH 5, N total = 3,600 mg/L) was achieved in a batch mode of operation when the inoculum was waste sludge from a plant that treated a mixture of domestic and industrial wastewater. Biogas production increased from 92 to 3,067 mg/L and methane content from 54 to 70% when initial total organic carbon (TOC) was increased from 0.05 to 1.7 g TOC/g volatile suspended solids (VSS). Further TOC increases inhibited the anaerobic biodegradation process (Stergar and Konèan, 2002).

Conclusions

Drugs, antibiotics, growth promoters, animal feed supplements, and other pharmaceutical products have been found in marine, underground, and surface waters in many developed countries around the world. Most of these drugs do not get degraded in municipal waste treatment plants. Degradation in contaminated soil is poor, since the drugs bind strongly to the earth, although microbes available for their degradation are found in the nearby proximity. Serious research should be directed toward identifying ways to degrade all these drugs, which have already contaminated the environment, as well as to prevent their entry into the environment and subsequently into the food chain.

References

Backhaus, T., and L. H. Grimme. 1999. The toxicity of antibiotic agents to the luminescent bacterium *Vibrio fischeri*. *Chemosphere* 38(14): 3291–3301.

Bernard, S., and N. F. Gray. 2000. Aerobic digestion of pharmaceutical and domestic wastewater sludges at ambient temperature. *Water Res.* 34(3): 725–734.

Diaz-Cruz, M. S., M. J. Lopez de Alda, and D. Barcelo. 2003. Environmental behavior and analysis of veterinary and human drugs in soils, sediments and sludge. *Trends Anal. Chem.* 22(6):

Forni, C., A. Cascone, M. Fiori, and L. Migliore. 2002. Sulphadimethoxine and *Azolla filiculoides* Lam.: a model for drug remediation. *Water Res.* 36:3398–3403.

Hartig, C., T. Storm, and M. Jekel. 1999. Detection and identification of sulphonamide drugs in municipal waste water by liquid chromatography coupled with electrospray ionisation tandem mass spectrometry. *J. Chromatog. A* 854:163–173.

Kummerer, K., A. Al-Ahmad, and V. Mersch-Sundermann. 2000. Biodegradability of some antibiotics, elimination of the genotoxicity and affection of wastewater bacteria in a simple test. *Chemosphere* 40:701–710.

Lapara, T. M., C. H. Nakatsu, L. M. Pantea, and J. E. Alleman. 2001. Aerobic biological treatment of a pharmaceutical wastewater: effect of temperature on COD removal and bacterial community development. *Water Res.* 35(18): 4417–4425.

LaPara, T. M., C. H. Nakatsu, L. M. Pantea, and J. E. Alleman. 2002. Stability of the bacterial communities supported by a seven-stage biological process treating pharmaceutical wastewater as revealed by PCR-DGGE. *Water Res.* 36:638–646.

Raj, S., and Y. Anjaneyulu. 2005. Evaluation of biokinetic parameters for pharmaceutical wastewaters using aerobic oxidation integrated with chemical treatment. *Process Biochem.* 40(1):165–175.

Saravanane, R., D. V. S. Murthy, and K. Krishnaiah. 2001a. Treatment of anti-osmotic drug based pharmaceutical effluent in an upflow anaerobic fluidized bed system. *Waste Management* 21(6): 563–568.

Saravanane, R., D. V. S. Murthy, and K. Krishnaiah. 2001b. Bioaugmentation and treatment of Cephalexin drug-based pharmaceutical effluent in an upflow anaerobic fluidized bed system. *Bioresource Tech.* 76(3):279–281.

Smith, C. A. 2002. Managing pharmaceutical waste. *J. Pharm. Soc Wisconsin* (Nov/Dec). 17–22.

Stergar, V., and J. Zagorc Konèan. 2002. The determination of anaerobic biodegradability of pharmaceutical waste using advanced bioassay technique. *Chem. Biochem. Eng. Quarterly* 16(1):17–24.

Ternes, T. A. 1998. Occurrence of drugs in German sewage treatment plants and rivers. *Water Res.* 32(11): 3245–3260.

Zuccato, E., D. Calamari, M. Natangelo, and R. Fanelli. 2000. Presence of therapeutic drugs in the environment. *The Lancet* 355(May 20): 1789–1790.

Zwiener, C., and F. H. Frimmel. 2003. Short-term tests with a pilot sewage plant and biofilm reactors for the biological degradation of the pharmaceutical compounds clofibric acid, ibuprofen, and diclofenac. *Sci. Total Environ.* 309:201–211.

Bibliography

Ingerslev, F., L. Torang, M. L. Loke, B. H. Sorensen, and N. Nyholm. 2001. Primary biodegradation of veterinary antibiotics in aerobic and anaerobic surface water simulation studies. *Chemosphere* 44:865–872.

Migliore, L., G. Brambilla, S. Cozzolino, and L. Gaudio. 1995. Effect on plants of sulphadimethoxine used in intensive farming (*Panicum miliaceum, Pisum sativum* and *Zea mays*). *Agri. Ecosystems Environ.* 52:103–110.

Pomati, F., A. G. Netting, D. Calamari, and B. A. Neilan. 2004. Effects of erythromycin, tetracycline and ibuprofen on the growth of *Synechocystis* sp. and *Lemna minor. Aquatic Toxicol.* 67:387–396.

CHAPTER 22

Hospital Waste Treatment

Waste from a hospital, a healthcare facility, a medical center, or a laboratory is considered medical waste, hospital waste, or infectious waste. It could be generated in the diagnosis, treatment, or immunization of human beings or animals, in related research, biological production, or testing. This solid or liquid waste will contain large quantities of multidrug-resistant enterobacteria (0.58 to 40%), which could be simultaneously resistant to 10 or more routinely used antibiotics and enteric pathogens. It has been reported that the detection rate of *Salmonella* and *Shigella* species in hospital wastewater effluents has been as high as 33 and 15%, respectively (Chitnis et al., 2004). This waste could pose a serious threat to the community. The health impact of direct and indirect exposure to hazardous wastes includes carcinogenic, mutagenic, and teratogenic effects, reproductive system damage, as well as respiratory and central nervous system effects. A major problem in many third world countries is scavengers or waste pickers who rummage at the dumpsites, get infected from the waste, and act as carriers for various diseases. In addition, there are several unregulated small clinics and health centers that indiscriminately dump their wastes in community disposal sites without proper precautions.

Hospital solid waste can be divided into (1) nonbiodegradable glass materials, mainly bottles; (2) nonbiodegradable plastic materials, including bottles, syringes, catheters, and blood bags; and (3) biodegradable cellulosic and food materials, including cotton, bandages, pads, amputated organs, and leftover food materials. [In Indian hospitals approximately 47, 32, and 270 g of waste per patient per day is produced under each category, respectively (Ghosh et al., 2000)]. In the United States alone, 3.5 million tons of medical waste per year is produced (1994 figures). The generation rate of medical waste in the United States, Italy, and Thailand is 5 to 7, 3 to 5, and 1 kg/bed day, respectively (Lee et al., 2004). Brazilian hospitals generate 2.63 kg/bed day, of which about 15 to 20% is infectious and biological in nature, and Indian hospitals about 0.5 to 2 kg/bed day, out of which 30 to 35% is infectious (Patil and Shekdar, 2001). Medical waste can be classified into two types based on its toxicity: general waste (or nonregulated) and special (or regulated and hazardous) waste. The latter is considered as a potential health hazard requiring special handling, treatment, and disposal. The solid and liquid contaminants from hospitals and medical centers could flow to the environment through landfill or dump sites and municipal sewage treatment plants.

Enterococci are gram-positive, facultative anaerobic bacteria that live in the intestinal tracts of animals and humans. Escherichia *faecalis* and *E. faecium* are the most important ones in this group that cause infections in humans. They have become resistant to many antibiotics such as penicillins, aminoglycosides, glycopeptides, lincosamides, and tetracyclines. The main causes for their resistance are indiscriminate use of antibiotics in hospitals and commercial animal husbandry, which uses antibiotics as food additives and growth promoters (Klare et al., 2003). In 1998 to 1999 streptogramin-resistant *E. faecium* was isolated in Germany from the wastewater of sewage treatment plants, from fecal samples and meat products of animals, from stools of humans in the community, and from clinical samples.

Liquid waste can be broadly classified as infectious, pathological, radioactive, and general. Most hospitals around the world release their wastewater into the public sewer system, leading to a significant contamination of sewage with all kinds of pharmaceuticals in the milligram per liter range, such as lipid-lowering agents, analgesic agents, x-ray media, and antibiotics. Hospitals release adsorbable, organically bound halogens (AOs) into the aquatic environment. They are persistent in the environment, accumulate in the food chain, and are toxic to humans and other organisms. AOs were found in the effluents of several German hospitals (highest concentration of 0.12 to 17 mg/L) (Kummerer et al., 1997, 1998).

Physical and Chemical Methods

Solid Waste

Landfills and incineration are two methods that are generally used for disposal of solid hospital waste. Studies indicated that 59 to 60% of regulated medical wastes are treated through incineration, 20 to 37% by steam sterilization, and 4 to 5% by other treatment methods. Because of strict regulations concerning onsite incineration, 84% of incineration is through offsite treatment (2000 data) (Askarian et al., 2004). In the United States, it was estimated that there were approximately 2,300 medical incinerators in operation in 1996 (Lee et al., 2002). Dioxins, furans, HCl, SO_2, CO_2, NO_x, and heavy metals including Cd, Hg, Cr, Zn, Ni, Cu, and Pb are produced when solid medical waste is incinerated. Air emission control and disposal of the resulting ash are other serious issues that need to be addressed. Incineration processes that are capable of reaching 900°C in the primary chamber are suitable for the treatment of clinical wastes, but not for cytotoxic wastes. Temperatures on the order of 1,100°C in the secondary chamber with a waste retention time of 1 s is needed for cytotoxic wastes.

Other solid disposal methods that are being considered are microwaving, autoclaving, radiowave and electron-beam irradiation, pulverization, electrothermal deactivation, pyrolysis, oxidation, steam sterilization, steam detoxification, and hydroclaving. Microwaving and autoclaving appear to be economical and competitive with incineration, and can be operated in continuous or batch mode. In the former, wastes are microwaved for 30 min at about 95°C, but it may be difficult to achieve sterilization temperatures above 120°C. Also, some spores may be activated and survive this method. In the latter method, steam, dry heat, or radiation is used to achieve temperatures on the order of 140°C for at least 30 min. This method destroys spores and is ideal for recyclable plastic items (plastic constitutes 30% by weight of all medical wastes). Neither of these methods is suitable for pathological, radioactive, laboratory, and chemotherapy wastes (Lee et al., 2004). The electron beam method utilizes the sterilization or sanitation power of ionizing radiation. The radiation inhibits the action of DNA or RNA molecules of pathogenic organisms. The unit cost of the electron beam method, including depreciation for a 0.5 ton/yr plant, is $0.18/kg as against a cost of $0.28/kg for microwave treatment. The plant for the electron beam process costs 20% more than one for a microwave process (1995 Italy data) (Tata and Beone, 1995). The corresponding cost for treating hospital waste using incineration is about $1.20/kg. Hydroclave is a technique that involves introduction of live steam to the waste and leads to the hydrolysis of the organic matter followed by dehydration. The treatment cost is about $0.06/kg. Chemical methods involve either the use of hydrogen peroxide and lime, or sodium hypochlorite to disinfect the shredded and mashed waste. This method is suitable for clinical waste only. Human body parts must be incinerated or treated by chemical disinfection processes using peroxide and lime, and shredded before disposal to landfill.

The cost of disposal of regulated hospital medical waste is $0.55 to $0.66/kg, which is very high when compared to the $0.035 to $0.066/kg for the disposal of municipal waste (1995 data) (Goldberg et al., 1996). The typical cost for onsite incineration is $1.21 to 15.6/kg, and for microwaving is $0.16/kg. Costs for offsite disposal of infectious waste is $0.79/kg, radioactive waste is $2.87/kg, liquid chemical waste from research labs is $ 3.44/kg, and hospital general solid waste is $0.12/kg (Lee et al., 2004). Although microwaving appears to be cheaper, the inadequate sterilization this method leads to requires that onsite incineration or offsite disposal must also be used. Effective sorting and segregation of hazardous medical waste from general waste could lead to large reductions in solid waste disposal costs.

Liquid Waste

Since hospital wastewater is to a large degree diluted in the municipal treatment plant, its genotoxic activity is no longer detectable. This does not mean that the activity is quelled; it may still be accumulated and could create long-term ecological effects (Giuliani et al., 1996). In addition, waste would also seep out through damaged or leaky municipal sewage pipes. For example, in Germany 17% of public sewers are leaky and another 14% are damaged, and in Great Britain, 23% of the sewers are in critical condition (Hua et al., 2003).

Hospital staff or family members handling excreta from patients treated with antineoplastic drugs or equipment contaminated by these excreta may be exposed to the drugs or their metabolites, which are carcinogenic, mutagenic, teratogenic, and fetotoxic. Spills and disposal of these expired medicines also pose serious problems to the hospital staff. Ifosfamide is a widely used antitumour agent, and concentrations measured in the hospital effluent, sewage treatment plant influent, and effluent were almost the same, indicating that no adsorption, biodegradation, or elimination of it took place (Kümmerer et al., 1998). Cyclophosphamide is another antineoplastic agent used in cancer chemotherapy, 20% of which is released unmetabolized into sewage water by cancer patient excretion; concentrations ranging from 7 to 143 ng/L were detected in the hospital's aerobic sewage water and in the influent and effluent of the communal sewage treatment plant. Chemical oxidation methods are quite effective for the destruction of these highly dangerous drugs. More than 98% destruction of four anticancer drugs Amsacrine, Azathioprine, Asparaginase, and Thiotepa in solution was achieved using sodium hypochlorite or Fenton reagent (HCl and $FeCl_2 \cdot 2H_2O$) within 1 h of treatment. Hydrogen peroxide as the oxidation agent destroyed more than 99% of Asparaginase and Thiotepa in 1 h, but degradation of Amsacrine (28% after 16 h) and Azathioprine (53% degradation in 4 h) were poor (Barek et al., 1998).

A change in the ratio of biological oxygen demand to chemical oxygen demand (BOD/COD) of the effluent after treatment does not reflect the presence or absence of pathogens or multiple-drug–resistant bacteria. Hence bacterial monitoring needs to be included in the effluent parameters. Bacteriological monitoring of a hospital liquid effluent treatment plant indicated that the total viable count (which included coliforms, fecal enterococci, staphylococci, and pseudomonads) decreased by an order of magnitude (from 9×10^6 to 9×10^5 cfu/mL; cfu = colony forming units) in the aeration tank of the activated sludge plant. The bacterial count in the clarifier tank decreased by two orders of magnitude. Interestingly a large increase in the bacterial count was noted in the sun-dried sludge, indicating that the bacteria in the hospital effluent remained firmly adhered to solid particles in the sludge. Multiple-drug–resistant, gram-negative bacteria were present at all stages, and they were inactivated only after chlorine treatment. Ultraviolet rays from solar radiation and temperatures on the order of 30°C did not reduce the microbial count on the solid bed. Sludge from a hospital waste treatment plant is a potential source of infectious organisms. More hypochlorite was needed to disinfect the sludge than was required for ordinary wastewater. The disinfection efficiency of hypochlorite was greater against settled sludge than against thickened sludge (Chitnis et al., 2004). Chlorine dioxide is also used to disinfect the sludge from the activated sludge plant.

Biodegradation Techniques

Composting is an ideal technique for treating solid waste. Most human pathogens are mesophilic so when compost reaches thermophilic conditions, the pathogens are killed. Enteric pathogenic organisms are also destroyed above 50°C. Composting also needs moisture and food in addition to increased temperature. Studies carried out with a 20-kg composting stack maintained at 45% moisture (w/w) by the addition of water and supplemented with cow manure, autoclaved cow manure, horse manure, or food waste to achieve an initial C/N ratio around 27 showed mixed results. Composting without inoculum and supplement showed little degradation activity. Addition of previously composted hospital solid waste to supply a cellulolytic population did not improve degradation. Composting with horse and cow manure achieved maturation within 16 days, with the latter proving to be a source of both nitrogen and microbial population. Composting with autoclaved cow manure took twice as long as nonautoclaved cow manure (Ghosh et al., 2000). The cellulose-degrading bacterial populations (*Bacillus* sp.) increased while the fungal population drastically decreased during the composting process, probably because of the high temperature reached during composting.

A submerged hollow fiber membrane bioreactor had several advantages, including complete solid removal from effluent, effluent disinfection, high loading rate capability, low sludge production, rapid startup, compact size, and lower energy consumption. An aerobic submerged hollow fiber membrane bioreactor of 6 m^3 volume with 96 m^2 surface area was able to treat hospital wastewater effectively—achieving COD, NH_4^+–N, turbidity, and *Escherichia coli* removal efficiencies of 80, 93, 83, and 98%, respectively, at a hydraulic retention time (HRT) of 7.2 h. The treatment removed the effluent color and odor. The hospital wastewater had a COD of 48 to 277.5 mg/L, a BOD of 20 to 55 mg/L, a NH_4^+–N of 10.1 to 23.7 mg/L, a turbidity of 6.1 to 27.9 nephelometric turbidity units (NTU), a pH of 6.2 to 7.1, 9.9 $\times 10^3$ bacteria/L, and an *E. coli* count of over 1,600/100 mL (Wen et al., 2004). The membrane used was made of polyethylene with a pore size of 0.4 μm.

A 125-cm sand column degraded 79 and 67% of total and soluble COD of a pharmaceutical effluent trickling down at a rate of 2 L/day (influent total and soluble COD values were 640 and 300 mg/L, respectively).

The reduction in NH4–N was 94% (Hua et al., 2003). While the effluent trickled through the sand column, 99% of aerobic and anaerobic bacteria, coliforms, and fecal coliforms were also eliminated. All the ibuprofen and naproxen present (about 60 ng/L) in the influent were eliminated, while benzatibrate (300 ng/L) and diclofenac (900 ng/L) were eliminated to a lesser extent (65 to 75%). X-ray contrast media such as iopromide, iomeprol, amidotrizoic acid, iohexol, and iotalamic acid (about 80 ng/L) were poorly removed (less than 30%) from the waste. Several authors have reported use of a sand filter for removal of bacteria and coliforms. Ternes (1998) reported 70 to 90% elimination of drugs such as ibuprofen, Benzafibrate, and Diclofenac from the effluent treated in a domestic sewage treatment plant.

Conclusions

Hospitals generate general and hazardous waste; the majority of the former (about 80%), if properly segregated, can be sterilized and recycled. In Germany waste segregation has lead to a tenfold decrease in the quantity of hazardous waste over a 10-year period. Most of the solid waste at present is either incinerated or landfilled after treatment. Ash from the incinerator is also toxic and needs to be detoxified before disposal. No research is being carried out at present toward the detoxification of solid waste using biological means. Proper attention needs to be given to the handling and disposal of the hazardous liquid waste. Chemical oxidation effectively detoxifies the waste. Generally the effluent is mixed with domestic waste and is treated in the sewage plant. Much of the toxic waste and drugs are not biodegraded in the municipal sewage treatment plant and hence pass through unaffected, contaminating the surface and groundwater. The drug-resistant bacteria and pathogenic organisms found in this effluent should be treated so that they are destroyed. Very little work is reported pertaining to microbial destruction of toxins and drugs from hospital waste. Figure 22-1 gives an approach for handling and treating hospital waste.

References

Askarian, M., M. Vakili, and G. Kabir. 2004. Results of a hospital waste survey in private hospitals in Fars province, Iran. *Int. J. Environ. Health Research* 14(4):295–305.

Barek, J., J. Cvacka, J. Zima, M. De Meo, M. Lagett, J. Michelonx, and M. Castegnaros. 1998. Chemical degradation of wastes of antineoplastic agents amsacrine, azathioprine, asparaginase and thiotepa. *Ann. Occupational Health* 42(4):259–266.

Chitnis, V., S. Chitnis, K. Vaidya, S. Ravikant, S. Patil, and D. S. Chitnis. 2004. Bacterial population changes in hospital effluent treatment plant in central India. *Water Res.* 38:441–447.

Ghosh, S., B. P. Kapadnis, and N. B. Singh. 2000. Composting of cellulosic hospital solid waste: a potentially novel approach. *Int. Biodeterioration Biodegradation* 45:89–92.

Giuliani, F., T. Koller, F. E. Wurgler, and R. M. Widmer. 1996. Detection of genotoxic activity in native hospital waste water by the umuC test. *Mutation Res.* 368:49–57.

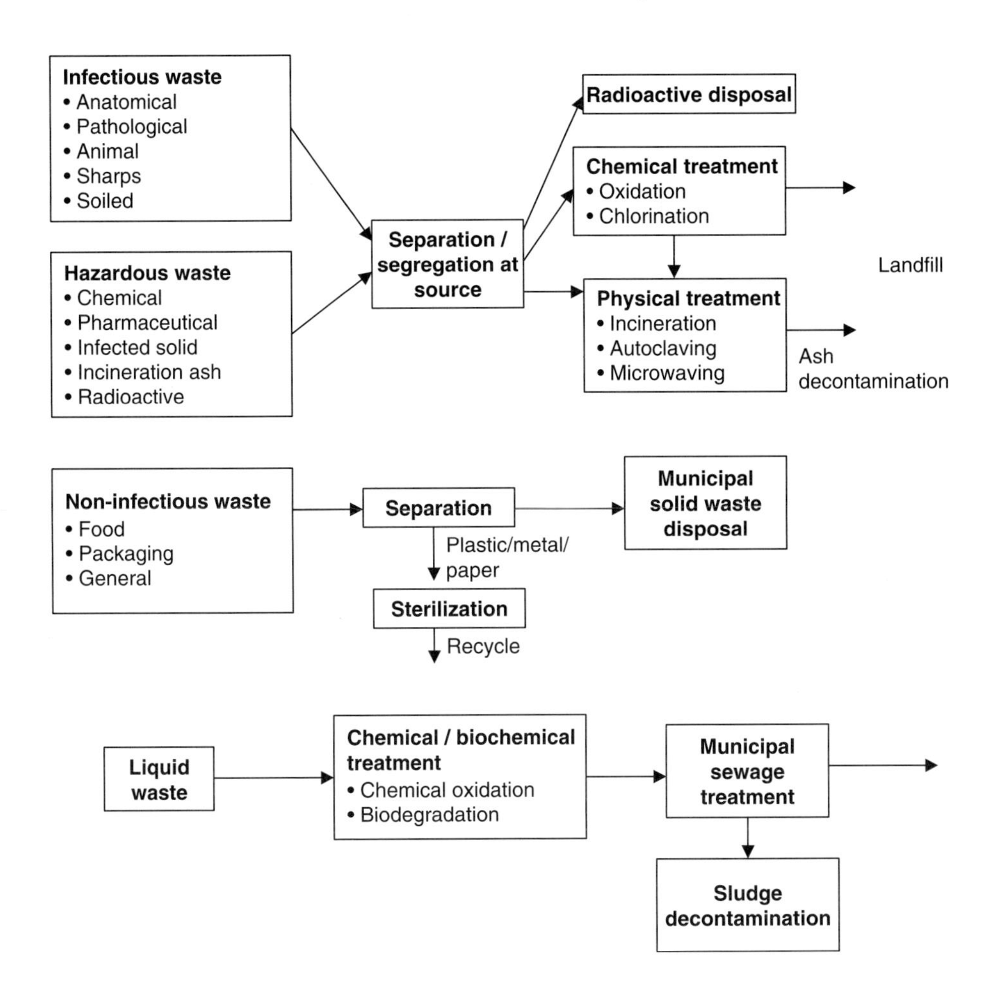

FIGURE 22-1. Suggested treatment of hospital waste.

Goldberg, M. E., D. Vekeman, M. C. Torjman, J. L. Seltzer, and T. Kynes. 1996. Medical waste in the environment: do anesthesia personnel have a role to play? *J. Clin. Anesth.* 8(9):475–479.

Hua, J., P. An, J. Winter, and C. Gallert. 2003. Elimination of COD, microorganisms and pharmaceuticals from sewage by trickling through sandy soil below leaking sewers. *Water Res.* 37:4395–4404.

Klare, I., C. Konstabel, D. Badstubner, G. Werner, and W. Witte. 2003. Occurrence and spread of antibiotic resistances in *Enterococcus faecium. Int. J. Food Microbiol.* 88:269–290.

Kümmerer, K., T. Steger-Hartmann, and M. Meyer. 1997. Biodegradability of the anti-tumour agent ifosfamide and its occurrence in hospital effluents and communal sewage. *Water Research* 31(11):2705–2710.

Kummerer, K., T. Erbe, S. Gartiser, and L. Brinker. 1998b. AOX-emissions from hospitals into municipal waste water. *Chemosphere* 36(11):2437–2445.

Lee, B. K., M. J. Ellenbecker, and R. Moure-Eraso. 2002. Analyses of the recycling potential of medical plastic wastes. *Waste Management* 22:461–470.

Lee, B. K., M. J. Ellenbecker, and R. Moure-Ersaso. 2004. Alternatives for treatment and disposal cost reduction of regulated medical wastes. *Waste Manag.* 24(2):143–151.
Patil, A. D., and A. V. Shekdar. 2001. Health-care waste management in India. *J. Environ. Manage.* 63:211–220.
Tata, A., and F. Beone. 1995. Hospital waste sterilization: A technical and economic comparison between radiation and microwaves treatments. *Radiation. Phys. Chem.* 46(4–6):1153–1157.
Ternes, T. 1998. Occurrence of drugs in German sewage treatment plants and rivers. *Water Res.* 11:3245–3260.
Wen, X., H. Ding, X. Huang, and R. Liu, 2004. Treatment of hospital wastewater using a submerged membrane bioreactor. *Process Biochem.*

Bibliography

Hartmann, T. S., K. Kümmerer, and A. Hartmann. 1997. Biological degradation of cyclophosphamide and its occurrence in sewage water. *Ecotoxicol. Environ. Safety* 36:174–179.

CHAPTER 23

Treatment of Waste from Explosives Industries

Introduction

A chemical explosive may be defined as a compound or mixture of compounds that reacts very rapidly to produce relatively large amounts of gas and heat. The rate of detonation is very high. Exothermic oxidation-reduction reactions provide the energy released during detonation. It is the nearly instantaneous formation of gases plus their rapid expansion due to pressure and heat that results in the destructive force or useful work. Large amounts of explosives are used annually, more for constructive commercial purposes than for military, combat, or terror purposes. The discovery of explosives must be considered as one of the greatest milestones in the development of modern society. Whether it is for mining, excavation of tunnels, construction of roads and pipelines, or rock quarrying, explosives are needed. Explosives contain oxidizers and fuel. Molecular explosives contain both of these within the same molecule (2,4,6-trinitrotoluene, pentaerythritol tetranitrate, and nitroglycerin), while in composite explosives the two portions come from different molecules (ammonium nitrate and liquid fuel oil). Explosives are categorized as three groups, based on their sensitivity to detonation, as follows:

- Primary explosives—most sensitive (get readily initiated)
- Secondary explosives—less sensitive (less hazardous)
- Tertiary explosives—least sensitive

Some of the commonly used explosives are listed in Table 23-1. Of all the known explosives, the most widely known are the ones having a –N=O group. This includes nitro groups (both aromatic and aliphatic), nitrate esters, nitrate salts, nitramines, and nitrosamines. Prominent

TABLE 23-1
Commonly Used Explosives

Compound name	*Symbol*	*Composition*
Primary explosives		
Mercury fulminate	—	$Hg(CNO)_2$
Lead azide	—	$Pb(N_3)_2$
Silver azide	—	AgN_3
Mannitol hexanitrate	MHN	$C_6H_8(ONO_2)_6$
Diazodinitro phenol	DDNP	$C_6H_2N_4O_5$
Secondary explosives		
Nitroglycerin	NG	$C_3H_5(ONO_2)_3$
Pentaerythritol tetranitrate	PETN	$C(CH_2ONO_2)_4$
Trinitrotoluene	TNT	$CH_3C_6H_2(NO_2)_3$
Ethyleneglycol dinitrate	EGDN	$C_2H_4(ONO_2)_2$
Cyclotrimethylene trinitramine	RDX	$C_3H_6N_3(NO_2)_3$
Cyclotetramethylene tetranitramine	HMX	$C_4H_8N_4(NO_2)_4$
Nitroguanidine	NQ	$CH_4N_3NO_2$
Nitromethane	NM	CH_3NO_2
Nitrocellulose	NC	Variable
Ethylenedinitrate	EDDN	$C_2H_{10}N_4O_6$
Prilled ammonium nitrate–fuel oil	ANFO	94/6—AN/FO
Water gels	—	Variable mixtures of oxidizers, fuels, and water
Tertiary explosives		
Mononitro toluene	MNT	$CH_3C_6H_4NO_2$
Ammonium perchlorate	AP	NH_4ClO_4
Ammonium nitrate	AN	NH_4NO_3

examples are nitromethane, 2,4,6-trinitrotoluene (TNT), nitroglycerin (NG), pentaerythritol tetranitrate (PETN), ethylenediamine dinitrate (EDDN), hexahydro-1,3,5–trinitro-1,3,5-triazine (RDX), cyclotetramethylene tetranitramine (HMX), and ammonium nitrate (Fig. 23-1). *The synthesis and use of these explosives contaminates the environment with high amounts of nitrate compounds. The industrial effluent from these industries has low pH value and is usually high in nitrates.*

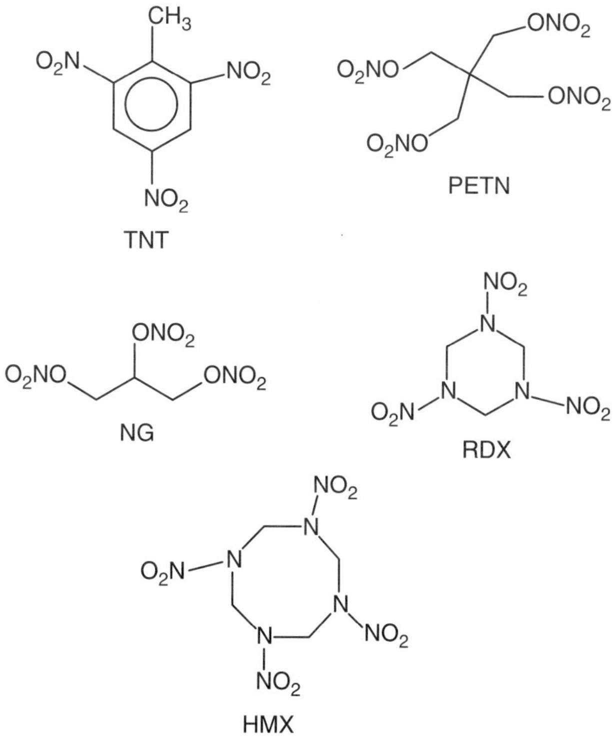

FIGURE 23-1. Commonly used explosives.

Toxicity and Occurrence

The toxicity of nitroorganics, inorganic nitrates, and nitrites is widely known. Some of the common symptoms are irritation of digestive tract, methemoglobinemia, disturbed heart function, kidney trouble, dysfunction of the vascular system, and severe jaundice (Kanekar et al., 2003). The nitroaromatic explosives are toxic, but their environmental transformation products, including arylamines, arylhydroxylamines, and condensed products such as azoxy- and azo- compounds, are equally or more toxic as the parent nitroaromatic. TNT is on the list of U.S. Environmental Protection Agency priority pollutants. RDX is a class C possible human carcinogen and has adverse effects on the central nervous system in mammals.

Aromatic nitro compounds are resistant to chemical or biological oxidation and to hydrolysis because of the electron withdrawing nitro groups (Rodgers et al., 2001). The hydrophilic lipophilic balance (HLB) of these compounds favors lipid solubility, thereby reducing their mobility in the environment. Thus, because of the lipophilic character and deactivated aromatic ring, these compounds accumulate in the environment. Activities associated with manufacturing, training, waste disposal, and closure of military bases have resulted in severe soil and groundwater contamination with explosives (Fournier et al., 2002). These wastewaters are contaminated with explosives as well as the raw materials used for the production of explosives. The nitro aromatic compound TNT is introduced into soil and water ecosystems mainly by military activities like the manufacture, loading, and disposal of explosives and propellants. This problem of contamination may increase in the future because of the demilitarization and disposal of unwanted weapon systems. The disposal of obsolete explosives is a problem for the military and associated industries because of the polluting effect of explosives in the environment (Wyman et al., 1979).

Bioremediation

Past methods of disposing of munitions wastes have included dumping in deep sea, dumping at specified landfill areas, and incineration when quantities were small. All of these cause serious harm to the ecosystem. For example, incineration causes air pollution, and disposal on land leads to soil and groundwater pollution. Other than these, methods such as resin adsorption, surfactant complexing, and liquid–liquid extraction have been used. These methods only transfer the explosive from soil or water into another medium, which then needs further treatment. Chemical methods of oxidation do not yield the necessary products, and the unreacted toxic intermediates still remain. Thus, the biofriendly treatment is bioremediation.

Microorganisms are known for their versatile metabolic activity and have evolved diverse pathways that allow them to mineralize specific nitro compounds. Despite this, relatively few microorganisms have been described as being able to use nitro aromatic compounds as nitrogen, carbon, and energy sources because nitro groups deactivate the aromatic ring to electrophilic attack by oxygenase or other enzymes. Be that as it may, biological degradation is one of the primary routes by which nitro aromatic compounds are broken down in the environment. There has been considerable interest in the past 30 years in the microbial transformation of these compounds.

Both aerobic and anaerobic degradation of nitro aromatics has been reported. Aerobic microorganisms use diverse biochemical reactions to initiate the degradation of nitro aromatic compounds. Reactions that attack the nitro substituent can be grouped into two general categories: oxidative or reductive (Rieger and Knackmuss, 1995). With mono- or di-nitro substituted aromatic compounds, the preferred route for their initial degradation is hydroxylation carried out by mono- or di-oxygenases. These reactions normally result in replacement of the nitro group by a hydroxy group with nitrite release. When the number of nitro substituents on the aromatic ring is greater than two, the predominant initial reactions become reductive. These reactions reduce the nitro (NO_2) substituent first to nitroso (NO), then to hydroxylamino (NHOH), followed by an amino (NH_2) derivative prior to further processing with the release of ammonium ion. In some *Rhodococcus* (Lenke and Knackmuss, 1992) and *Mycobacterium* (Vorbeck et al., 1994) strains, the aromatic ring, rather than the nitro group, may be reduced first to generate a hydride–Meisenheimer complex. On protonation and rearomatization, the nitro group is replaced by a proton and nitrite is released. Most aerobic microorganisms reduce TNT to the corresponding amino derivatives via the formation of nitroso and hydroxylamine intermediates. However, condensation of the latter compounds yields highly recalcitrant azoxytetranitro toluenes. Certain strains of *Pseudomonas* use TNT as the nitrogen source through the removal of nitrogen as nitrite (Esteve-Nunez et al., 2001). *Phanerochaete chrysosporium* mineralizes TNT under lignolytic conditions. Because the manufacturing processes for RDX and HMX are the same, each is present as an impurity in the other. Because of the copresence of RDX and HMX in contaminated waters or at contaminated sites, degradation of both in each other's presence becomes important. *P. chrysosporium* degraded this mixture to carbon dioxide and nitrous oxide (Hawari et al., 2000). In a study of RDX degradation by *Rhodococcus sp.*, nitrite formation was observed with RDX disappearance.

Ecological observations suggest that sulfate-reducing and methanogenic bacteria might metabolize nitroaromatic compounds under anaerobic conditions if appropriate electron donors and electron acceptors are present in the environment. The successful demonstration of the degradation of RDX by sewage sludge under anaerobic conditions (McCormick et al., 1978) further indicated the usefulness of anaerobes in explosive waste treatment. Under anaerobic conditions, the sulfate-reducing bacteria *Desulfovibrio* sp. (B strain) metabolized TNT. Of all the metabolites produced, the formation of toluene was significant (Boopathy and Kulpa, 1992). Most *Desulfovibrio* sp. have nitrite reductase enzymes that reduce nitrate to ammonia. Figure 23-2 elaborates a general pathway for the transformation of TNT that involves the initial reduction of aromatic nitro groups to aromatic amines. Boopathy and Kulpa (1994) isolated a methanogen, *Methanococcus* sp., that transformed TNT to 2,4-diaminonitro toluene. The observations of sulfate reducers and methanogenic bacteria by many workers suggest that these organisms could be exploited for bioremediation of explosives under anaerobic conditions by supplying proper electron donors and electron acceptors. The first step in the metabolism of nitoaromatics is reduction. This step is followed by reductive deamination, which removes all of the nitro groups present in the ring, leaving the ring intact and forming toluene and ammonia as end products. The toluene can be further degraded by toluene-degrading organisms.

As discussed earlier, aerobic transformations of TNT have shown the production of dead-end products like amino derivatives or azoxy compounds.

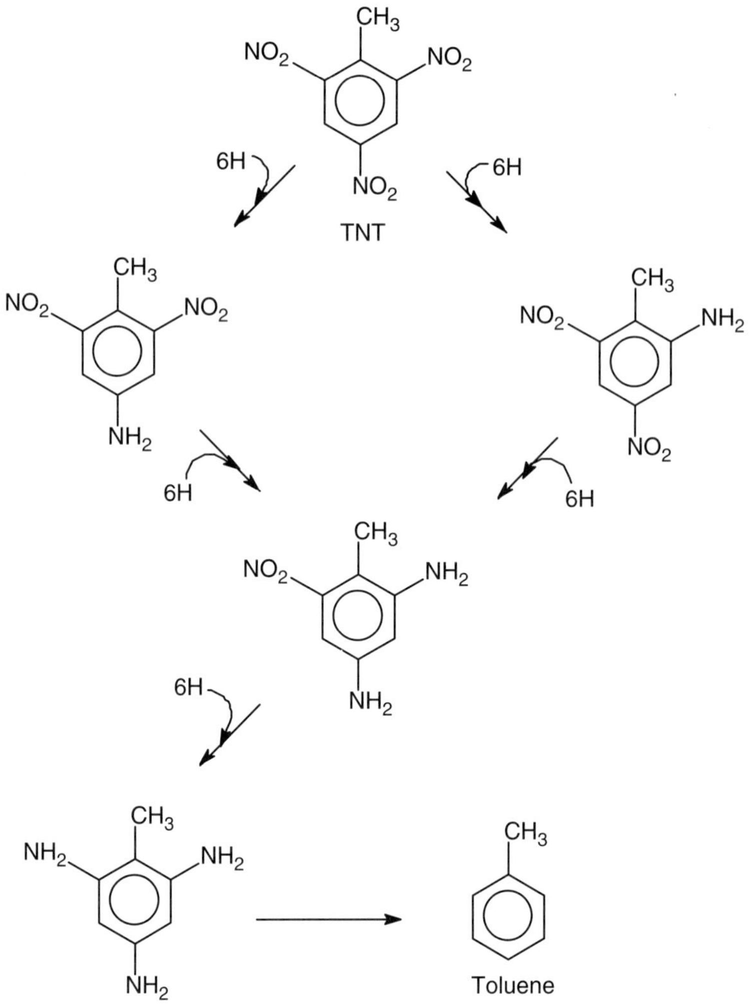

FIGURE 23-2. Degradation of TNT by *Desulfovibrio* sp.

Therefore, the applicability of aerobes in bioremediation of sites contaminated with nitroaromatics is doubtful at present. However, the use of anaerobes like sulfate-reducing bacteria may prove useful in decontaminating sites polluted by nitro compounds.

References

Boopathy, R., and C. F. Kulpa. 1992. Trinitrotoulene (TNT) as a sole nitrogen source for a sulfate-reducing bacterium *Desulfovibrio sp.* (B strain) isolated from an anaerobic digester. *Curr. Microbiol.* 25:235–241.

Boopathy, R. and C. F. Kulpa. 1994. Biotransformation of 2,4,6-trinitrotoulene (TNT) by a Methanococcus sp. (B strain) isolated from a lake sediment. *Can. J. Microbiol.* 40:273–278.
Esteve-Nunez, A, A. Caballerno and J. L. Ramos. 2001. Biological degradation of 2,4,6-trinitrotoulene. *Microbiol. Mol. Biol. Rev.* 65(3):335–352.
Fournier, D, A. Halasz, J. Spain, P. Fiurasek, and J. Hawari. 2002. Determination of key metabolites during biodegradation of hexadhydro-1,3,5-trinito-1,3,5-triazine with *Rhodococcus sp.* Strain DN22. *Appl. Environ. Microbiol.* 68:166–172.
Hawari, J., S. Beaudet, A. Halasz, S. Thiboutot. and G. Ampleman. 2000. Microbial degradation of explosives: biotransformation versus mineralization. *Appl. Microbiol. Biotechnol.* 54(5):605–618.
Lenke, H., and H. J. Knackmuss. 1992. Initial hydrogenation during catabolism of picric acid by *Rhodococcus erythropolis* HL24-2. *J. Bacteriol.* 58(9):2933–2937.
McCormick, N. G., J. H. Cornell, and A. M. Kaplan. 1978. Identification of biotransformation products from ,4-dinitrotoluene. *Appl. Environ. Microbiol.* 35(5):945–948.
Rieger P. G., and H. J. Knackmuss. 1995. Basic knowledge and perspectives on biodegradation of 2,4,6-trinitrotoluene and related nitroaromatic compounds in contaminated soil. In *Biodegradation of nitroaromatic compounds,* J. C. Spain (ed.), pp. 1–18. New York: Plenum Press.
Vorbeck, C., H. Lenke, P. Fischer, and H. J. Knackmuss. 1994. Identification of hybrid-Meisenheimer complex as a metabolite of 2,4,6-trinitrotoluene by a Mycobacterium strain. *J. Bacteriology.* 176: 932–934.
Wyman, J. F, H. E. Guard, W. D. Won. and J. H. Quay. 1979. Conversion of 2,4,6-trinitrophenol to a mutagen by *Pseudomonas aeruginosa. Appl. Environ. Microbiol.* 37:222–226.

Bibliography

Kanekar, P., P. Dautpure, and S. Sarnaik. 2003. Biodegradation of nitro-explosives. *Indian. J. Expt. Biol.* 41(September): 991–1001.
Rodgers, J. D., and N. J. Bunce. 2001. Treament methods for the remediation of nitroaromatic explosives. *Water Res.* 35(9):2101–2111.

CHAPTER 24

Petroleum Hydrocarbon Pollution

Crude oil is unrefined liquid petroleum; it contains predominantly carbon and hydrogen in the form of alkanes (saturated hydrocarbons), alkenes and alkynes (both unsaturated), and aromatic hydrocarbons. The other components present in oil are sulfur, nitrogen, oxygen, trace amounts of iron, silicon, and aluminum. Large amounts of hydrocarbon contaminants are spilled into the environment as a result of various human activities. Major accidental spills from oil exploration sites, oil tankers, pipelines (underwater and underground), spent marine lubricants, and storage tanks have become a common occurrence. Petroleum refineries also generate sludge and other oily effluents. It is estimated that more than 2.5 million tonnes of used lubricating oil is unaccounted for in the United States alone, and the estimated annual oil influx into the ocean is about 5 to 10 million tonnes.

Physical Methods

Oil spills cause short-term as well as long-term damage to the environment (soil, water, aquatic flora, fauna, and animals). Remediation of the affected sites helps to reduce the damage caused to the environment and aid in its recovery. Several physical and chemical techniques for decontamination have been developed and used. The in situ methods include washing with detergent; extraction of topsoil using vacuum, steam, or hot air stripping; soil solidification (binding hydrocarbon to soil); flooding (raising the oil to the surface above the water table), etc. The ex situ methods include excavating the contaminated soil or liquid and subjecting it to chemical oxidation, solvent extraction, adsorption, etc., and later returning the treated soil or liquid back to its original place. Although these techniques are well matured and developed, they are expensive. Ultraviolet illumination on thin oil films can degrade aromatic compounds: the effect is more pronounced for larger polycyclic compounds and more alkylated forms.

Bioremediation

Bioremediation includes stimulating the native microbial populations or introducing microorganisms from external sources that have been known to degrade a particular contaminant, or have been engineered to do so. The environment necessary for the growth of these microorganisms must be created. The in situ treatment procedures include biostimulation, bioventing, bioaugmentation, and addition of a nitrogen-phosphorous-potassium fertilizer. Bioremediation techniques have more advantages than the chemical and physical methods, including treatment cost. For example, the cost to physically wash a marine oil spill is estimated to be about \$1.1 million per meter of the oil-contaminated shoreline, while the cost of biostimulation through fertilizer addition is estimated at \$0.005 per meter. The estimated cost of excavation followed by offsite disposal of a petroleum-contaminated site is around \$3 million, while the cost of onsite bioventing is about \$0.2 million (Atlas and Unterman, 1999). Contrary to the belief of some, after the San Jacinto River flood and oil spill in southeast Texas, intrinsic bioremediation achieved a 95% reduction in hydrocarbon concentration within 150 days (Mills et al., 2003). During this period, ammonium concentration in the sediment decreased from 43 to 4.8 ppm N.

Aerobic

The microorganisms make use of hydrocarbons as their carbon and/or energy sources and degrade the hydrocarbons to carbon dioxide and water. Since the crude oil contains paraffinic, simple aromatic, and polyaromatic hydrocarbons (PAHs), its biodegradation involves the interaction of many different microorganisms. The common hydrocarbon-degrading organisms in the marine environment are *Pseudomonas, Acinetobacter, Nocardia, Vibro,* and *Achromobacter* (Floodgate, 1984; Salleh et al., 2003). Oxygen is essential for in situ degradation of hydrocarbons. Since injecting oxygen gas is expensive, other soluble electron acceptors such as nitrates or sulfates are also used, but these acceptors slow down the reaction.

Straight chain alkanes are easily and rapidly degraded by several microorganisms, including *Acinetobacter* sp., *Actinomycetes, Arthrobacter, Bacillus* sp., *Candida* sp., *Micrococcus* sp., *Planococcus, Pseudomonas* sp., *Calcoaceticus,* and *Streptomyces* (Surzhko et al., 1995). Although microorganisms degrade *n*-alkanes up to a chain length of 40 carbon atoms, the solubility of long chained alkanes in water is poor; therefore the availability of the alkanes decreases, leading to reduced biodegradation. The general degradation pathway is via the oxidation of the terminal methyl group to its corresponding carboxylic acid, possibly through various intermediates (Fig. 24-1), which finally get mineralized. But in some cases, the preterminal carbon is also oxidized. Anaerobic biodegradation of crude oil using seawater and sediment as inocula produced a two orders of magnitude

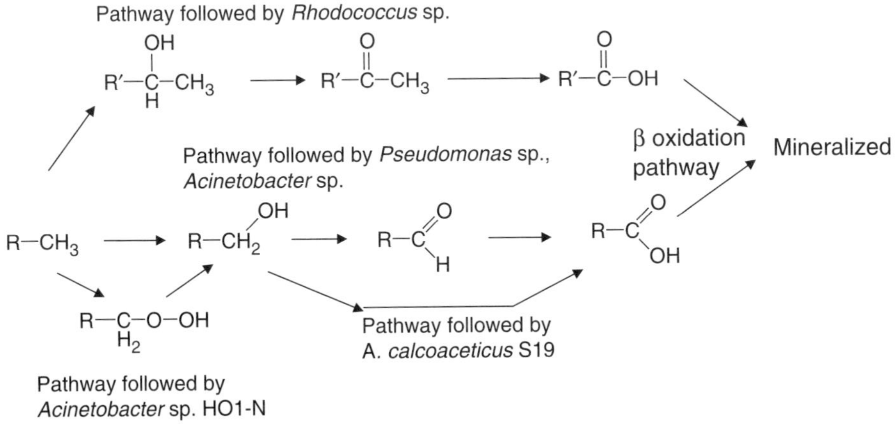

FIGURE 24-1. Aerobic degradation of hydrocarbon.

increase in the degradation of C_{10} to C_{20} carboxylic acids in 5 days, which were further degraded, leaving behind higher (greater than C_{20}) molecular weight cyclic and branched carboxylic acids as recalcitrant material (Watson et al., 2002). An *Acinetobacter* sp. isolated from soil was able to mineralize long-chain *n*-paraffins (C_{16-36} chain) in car engine oil (Koma et al., 2001). Long chain *n*-paraffins were metabolized via the terminal oxidation pathway of *n*-alkane, which was confirmed from the products of degradation, namely *n*-hexadecane, 1-hexadecanol, and 1-hexadecanoic acid.

Pseudomonas sp., *Ralstonia* sp., *Rhodococcus* sp,. and *Sphingomonas* sp. are some of the microorganisms that are known to oxidatively degrade monoaromatics like benzene, toluene, and xylenes (BTEX) as shown in Fig. 24-2 (Lee and Lee, 2001; Parales et al., 2000). Toluene aerobically degrades more rapidly than other BTEX compounds in a wide variety of strains (*Pseudomonas putida* mt-2 and P., *P. mendocina*, *R. picketti* PKO1 etc.), either through the formation of substituent groups on the benzene ring or on the methyl group. The products could be cresols, benzyl alcohol, or dihyrol. A *Pseudomonas* sp. oxidizes xylenes at the methyl group, similar to the degradation of toluene, forming several intermediates.

Polyaromatics (PAHs) persist in soil and sediment because of their low water solubility and high stability (because of the presence of multiple fused aromatic rings); their half-life is directly proportional to the number of fused rings. Motor vehicle exhausts, lubricating oils, paint solvents, and greases contribute to PAHs, and many of them are carcinogenic. *Burkholderia cepacia* F297 degrades a variety of polycyclic aromatic compounds, including fluorene, methyl naphthalene, phenanthrene, anthracene, and dibenzothiophene (Harayama, 1997). Several microorganisms have been reported to degrade PAHs, and they include *Rhodococcus* sp., *Alteromonas* sp.,

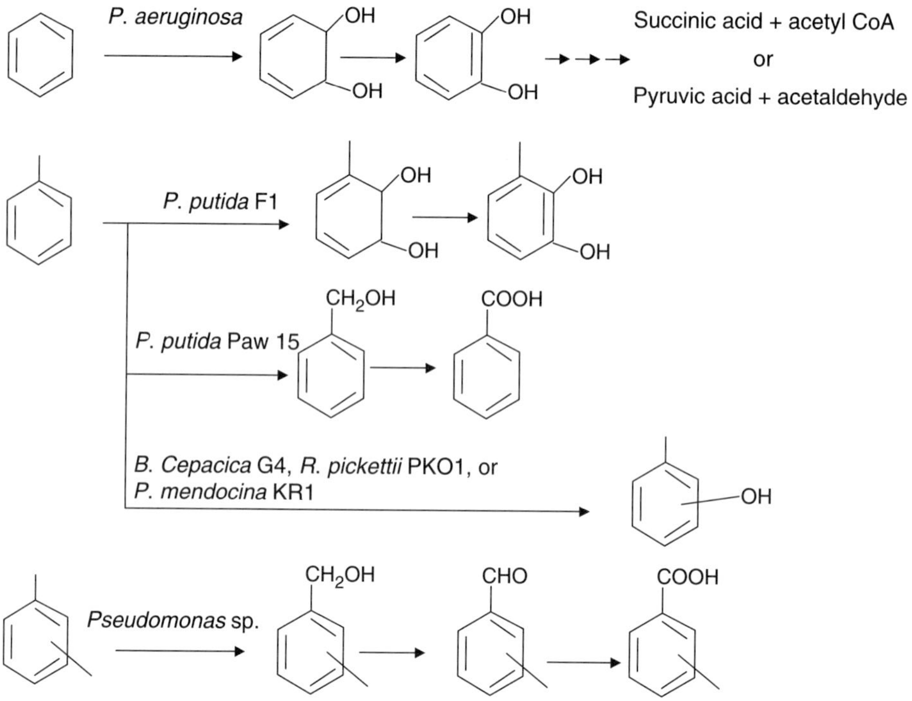

FIGURE 24-2. Aerobic biodegradation pathway of aromatics.

Arthrobacter, Bacillus, Mycobacterium sp., *Pseudomonas* sp., and *Phanaerochaete chrysporium* (Barclay et al., 1995). Other microorganisms, including bacteria and fungi, that are specific for a substrate include (Juhasz and Naidu, 2000; Aitken et al., 1998):

- Naphthalene—*Mycobacter calcoaceticus, Pseudomonas paucimobillis, Pseudomonas putida, Pseudomonas fluorescens, Sphingomonas paucimobilis*
- Acenaphthene—*Beijernickia* sp., *P. putida, P. fluorescens,* and other *Pseudomonas* sp., *Burkholderia cepacia*
- Anthracene—*Beijernickia* sp., *Mycobacterium* sp., *Pseudomonas paucimobilis, Cycloclasticus pugeti, Ulocladium chartarum, Absidia cylindrospora*
- Phenanthrene—*Aeromonas* sp., *Alcaligenes faecalis, Achromobacter denitrificans, Bacillus cerus, A. faecalis*

- Fluoranthene—*Mycobacterium* sp., *P. putida, Sp. paucimobilis, P. paucinobilis*
- Pyrene and chrysene—*Sphingomonas* sp.
- Pyrene—*Caenorhabditis elegans, Phanerochaete chrysosporium, Penicillium* sp., *Penicillium janthinellum*
- Chrysene—*P. janthinellum, Syncephalastrum racemosus, Penicillium* sp.
- Benz[a]anthracene—*C. elegans, Trametes versicolor, Phanerochaete laevis, P. janthinellum*
- Dibenz[a,h]anthracene—*Trametes versicolor, P. janthinellum*

Most degradative mechanisms reported for fungi are cometabolic, where an alternate carbon source is utilized for energy and growth, while as a consequence PAH is transformed into other products. White-rot fungus, *Phanerochaete chrysosporium*, has been reported to mineralize phenanthrene, fluorene, fluoranthene, anthracene, and pyrene in nutrient-limited cultures. Fungal metabolism of several low molecular weight PAHs has been reported in literature. They include:

- Naphthalene—*Absida glauca, Aspergillus niger, Basidiobolus ranarum, Candida utilis, Choanephora campincta, Circinella* sp.
- Acenaphthene by *C. elegans, T. versicolor*
- Phenanthrene—*C. elegans, P. chrysosporium, P. laevis, Pleurotus ostreatus, T. versicolor*
- Anthracene—*Bjerkandera* sp., *Bjerkandera adjusta, C. elegans, P. chrysosporium, P. laevis, Ramaria* sp., *Rhizoctonia solani, T. versicolor, Pleurotus ostreatus*
- Fluoranthene—*C. elegans, C. blackesleeana, C. echinulata, Bjerkandera adjusta, Pleurotus ostreatus*
- Pyrene—*C. elegans, P. chrysosporium, Penicillium sp., P. janthinellum, P. glabrum, P. ostreatus*
- Benz[a]anthracene—*C. elegans, T. versicolor, P. laevis*
- Chrysene—*P. janthinellum, Syncephalastrum racemosus, Penicillium* sp.

Algae and cyanobacteria also oxidize naphthalene (*Oscillatoria* sp., *Microcoleus chthonoplastes, Nostoc* sp.) and phenanthrene (*Oscillatoria* sp., *Agmenellum quadruplicatum*).

Salicylate, a central intermediate in the metabolism of naphthalene, undergoes oxidative decarboxylation to yield catechol; it also acts as an inducer for degradation in the presence of gram-negative bacteria like *Pseudomonas* (Gibson and Subramanian 1984). Whereas salicylate does not act as an inducer, it is hydroxylated to gentisate in the presence of gram-positive bacteria such as members of the *Rhodococcus* sp. (Grund et al. 1992).

Benzo[a]pyrene (BaP), a five-ring fused compound, is known to degrade via the formation of 4,5 or 7,8 or 9,10 dihydrols, followed by the formation of carboxylic acids in the presence of bacterial species that include *Rhodococcus* sp. strain UW1, *Burkholderia cepacia*, *Mycobacterium*, *S. maltophilia*, as well as a mixed culture containing *Pseudomonas* and *Flavobacterium* (Juhasz and Naidu, 2000). In addition, fungal isolates that include *Phanerochaete chrysosporium*, *Trametes versicolor*, and *Pycnoporus cinnabarinus* grown on an alternate carbon source can remove more than 90% of BaP in 30 h, producing about 15% carbon dioxide, indicating mineralization. Fungal BaP oxidation is mediated by cytochrome P-450, leading to the formation of *trans*-dihydrol via the formation of epoxide. The green alga *Selanastum capricornutum* oxidizes BaP to 4,5 or 7,8 or 9,10 or 11,12 dihydrodiols. The bioavailablity of BaP in contaminated soils could be increased by the use of surfactants, which could increase its dissolution and hence enhance the mass transfer rates. Bacterial-fungal cocultures can lead to peroxidation of BaP by fungus, which could lead to an increase in the rate of BaP mineralization by bacteria. Similar behavior was observed in the case of pyrene.

Naphthalene dioxygenase is induced by naphthalene, salicylate, and succinate, and is isolated in gram-negative bacteria (mainly *Pseudomonas).* The enzyme helps to incorporate molecular oxygen into the substrate to produce *cis*-dihydrodiol, which is the intermediate degradation component. *P. putida* was able to grow on naphthalene as a sole carbon source, synthesizing the enzyme naphthalene-dioxygenase when activated initially on salicylate.

Operating Conditions

The rate of microbial degradation depends on several operating factors that include ambient temperature, pH, salinity, oxygen availability, amount of nutrients available, chemical composition of the petroleum, its physical state and concentration in the contaminated area, and adaptation of the microorganism to the contaminated site.

Higher temperatures lead to increased rates of degradation, as well as decreased viscosity of the oil, which in turn increases its availability for the organism in the aqueous phase. Biodegradation of petroleum has been reported in Arctic and Antarctic seawater. Strains have been known to degrade diesel oil at 0 to 10°C. Below 10°C, some of the long chain hydrocarbons also solidify, reducing their availability to the microbes. A temperature-dependent diffusion barrier in the thin layer of unfrozen water limited metabolic activity (Rivkina et al., 2000). Studies carried out by Rike et al. (2003) in winter months at an Arctic site have shown that cold-adapted microorganisms are capable of in situ biodegradation. Although degradation of crude oil has been observed even at 60°C, at higher temperatures the membrane toxicity of hydrocarbons is increased, hindering biodegradation. A neutral pH is favored by most of the strains, although degradation of hydrocarbons has been reported in acidic as well as in alkaline pH conditions.

Organisms found in seawater are able to degrade oil at salt concentrations that vary from 0.1 to 2.0. *M. Pseudomonas* sp., enterobacteria, and a few gram-negative aerobes are known to work under saline conditions. Aerobic degradation requires 3.1 mg oxygen to degrade 1 mg hydrocarbon. Although the amount of oxygen dissolved in aqueous medium is good, it decreases sharply with the depth of the water. Addition of urea and ammonia-based fertilizers used for oil spills can exert an oxygen demand that results from biological oxidation of ammonia. Also on fine sediment beaches, mass transfer of oxygen may not be sufficient. Hence aerobic biodegradation is restricted to a small layer floating on top of the water layer. Oil slicks and globules of tar that sink below persist for a long time because of the absence of oxygen. Under oxygen-limited conditions, anaerobic degradation occurs in the presence of sulfate-reducing bacteria, metal-reducing bacteria, methanogens, and nitrifiers.

For sustained microbial activity, the C:N:P ratio must be maintained at 120:10:1. During oil spills, the carbon amount increases, which disturbs the nutrient balance and hence microbial growth, causing biodegradation to slow down. Organic (fertilizers) as well as inorganic sources (salts) for N and P have been added and found to be very effective (Rosenberg et al., 1992). Oleophilic fertilizer was found to be very effective in degrading oil after the Exxon Valdez spill (Pritchard and Costa, 1991). The fertilizer preferably is added in slow-release form to have a maximum effect; it also cannot exceed the toxic concentrations of ammonia and/or nitrate so that the nutrient addition does not limit the microbial population. A field study conducted on the shoreline contaminated during the *Sea Empress* incident showed that addition of N and P led to significant decomposition of aliphatic hydrocarbons, but biodegradation of aromatics was not affected (Maki et al., 2003).

Petroleum has different compositions depending upon its source; hence its rate of biodegradability varies. Generally *n*-alkanes are easily susceptible, followed by branched alkanes, low molecular weight aromatics, and finally cyclic alkanes. Also biodegradation rates from highest to lowest are saturated compounds, light aromatics, heavy aromatics, and finally polar compounds, which are recalcitrant. The physical state of the oil has an effect on the degradation rate; emulsified spills degrade faster than tar balls because of the availability of the spill's large surface area. An increase in oil concentration can lead to an increase in membrane toxicity or can upset the C:N:P balance. Oxygen limitations due to the presence of a thick oil fraction can also affect the activity of the microorganisms. Surprisingly, the percentage degradation of naphthalenes and fluorenes was greater than that of alkanes, dibenzothiophenes, and phenanthrenes in contaminated soils. There are probably two reasons for this: (1) The low molecular weight aromatic compounds have a higher solubility in water than the high molecular weight aromatics and alkanes, and (2) the water solubility, and thus the availability, of alkanes is reduced by their high adsorption dry sand. The latter could be addressed by using suitable surfactants to solubilize the alkanes into the aqueous phase. Oil spills at sea are exposed to solar radiation, which could be hostile to microbial growth. Jezequel et al. (2003) have observed that alkanes in oil spills that have little exposure to sunlight but that are damp degrade faster.

A mixture of *Acinetobacter* sp. and *Pseudomonas putida* PB4 degraded a light crude oil efficiently, with the degradation taking place in a sequential manner. The *Acinetobacter* sp. degraded the alkanes and other hydrocarbons and formed metabolites; the P. putida PB4 formed aromatic compounds by growing on the metabolites (Nakamura et al., 1996).

Anaerobic Degradation

Petroleum hydrocarbons can serve as electron donors and as a carbon source for bacteria under a variety of redox conditions. The *Azoarcus/Thauera* group was found to be the major bacterial group responsible for the anaerobic degradation of alkylbenzenes and *n*-alkanes, and a methanogenic consortium composed of two archaeal species related to the genera *Methanosaeta* and *Methanospirillum*, and a bacterial species related to the *Methanospirillum* was responsible for toluene degradation (Watanabe, 2001).

Alkanes are very inactive compounds, and during aerobic degradation, oxygen (which is absent during anaerobic degradation) is available to activate them. Sulfate-reducing and denitrifying bacteria that completely oxidize alkanes with 6 to 20 carbon atoms have been isolated. The sulfate reducers are able to produce the corrosive and toxic gas hydrogen sulfide with crude oil as a substrate (Holliger and Zehndner, 1996). Similar to toluene, which gets added to fumarate, a common cell metabolite, via a radical mechanism, *n*-alkanes also get activated via radical mechanism and are added to fumarate. However, the *n*-alkanes were not activated at the terminal carbon but at C2, as was the case with *n*-hexane (Wilkes et al., 2003). The proposed pathway for anaerobic degradation is that fumarate reacts with the C2 of the alkane through a radical mechanism and forms (1-methyl-alkyl)-succinate. It is activated by coenzyme A (HSCoA), several rearrangements follow, and then β oxidation occurs. The final end product is CO_2 (see Fig. 24-3). The metabolites formed during anaerobic biodegradation are various alkylsuccinates with alkyl chains (linked at C2) that had a carbon chain length of 4 to 8.

Under anaerobic conditions, aromatic compounds are transformed into a few intermediates [namely, to benzoate (or benzoyl-CoA) and, to a lesser extent, resorcinol and phloroglucinol], followed by the cleavage of the rings by hydrolysis, resulting in the formation of noncyclic compounds, which are then converted into metabolites by β oxidation (Fuchs, 1994). Two examples of activation reactions are:

- Hydroxylation of benzene ring to form phenol
- Methyl hydroxylation of toluene to form benzyl alcohol

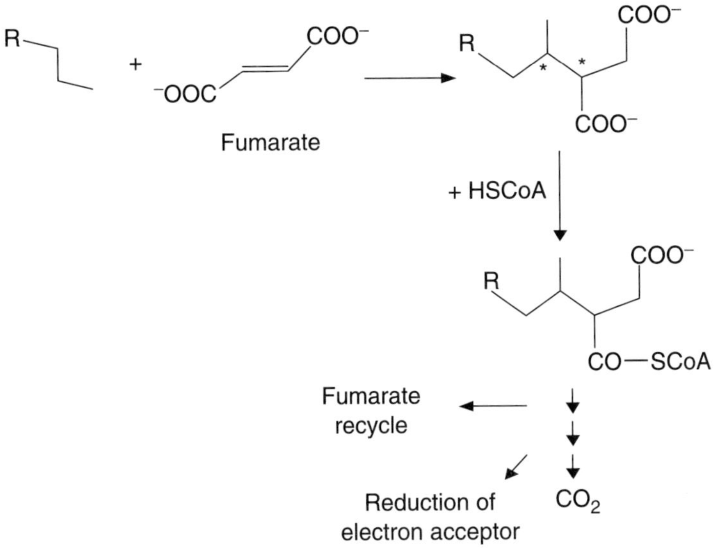

FIGURE 24-3. Anaerobic biodegradation.

Two examples of ring cleavage reactions are:

- Hydrolytic cleavagc
- Reduction of an aromatic ring to an alicyclic ring

Benzene is transformed to phenol in the presence of methanogenic cultures and to *p*-hydroxybenzoate in the presence of denitrifying bacteria and finally to the central intermediate benzoate. Pure cultures of denitrifying, iron-reducing, and sulfate-reducing bacteria (under the genera *Thauera* and *Azoarcus)* utilize toluene as a carbon and energy source. A sulfate-reducing bacterium that oxidizes toluene has been isolated and found to belong to the *Desulfobacula toluolica* genus/species. Toluene degrades via benzoyl-CoA. The oxidation of the methyl group occurs by the formation of benzyl alcohol, going to benzaldehyde, and finally to benzoate. Ethyl benzene is stable under anaerobic conditions. Denitrifying and methanogenic bacteria degrade the three isomers of xylene. Except for naphthalene, none of the PAHs have been known to degrade under anaerobic conditions.

Phytoremediation

Phytoremediation is a technique by which plants and the associated rhizosphere microorganisms are utilized to remove, transform, or contain toxic chemicals located in soils, sediments, groundwater, surface water, and the atmosphere. Phytostimulation involves the stimulation of the microorganisms in the location by using plants that have been tested for the destruction of PAH, BTEX, and other petroleum hydrocarbons.

Phytoextraction, which involves removal of a contaminant from the site using plants, has been adopted in the decontamination of soil and groundwater affected by PAHs using alfalfa (*Medicago sativa*) and hybrid poplar trees. Rhizofiltration (use of microorganisms around the zone near the roots to filter contaminants) and phytodegradation (use of plants for the degradation of the contaminants) using grasses and clover (*Trifolium* spp.) have been adopted for the treatment of a PAH-contaminated site (Susarla, 2002).

Typha latifolia, *T. angustifolia*, *Phragmites communis*, *Scirpus lacustris*, *Juncus* spp., different algae, and microflora consisting of different heterotrophic and autotrophic microorganisms, including different oil-degrading bacteria and fungi present in an artificially made wetland, were able to efficiently decontaminate water consisting of crude oil and heavy metals (namely cadmium, copper, iron, lead, and manganese) (Groudeva et al., 2001). Paraffins and napthenes were more easily degraded than other hydrocarbons, and low molecular weight PAHs degraded more easily than high molecular weight PAHs.

Reactors

Anaerobic bioremediation of soil contaminated with No. 2 diesel fuel (550 mg petroleum hydrocarbon/kg of soil) in a slurry reactor at a pH of 6.5 led to 81, 55, 50, and 40% biodegradation in 290 days, with mixed electron acceptor, sulfate-reducing, nitrate-reducing, and methanogenic conditions (Boopathy, 2003). A fibrous-bed bioreactor, constructed by winding a porous wire cloth, to which the cells are attached and entrapped, provides a suitable, novel cell immobilization support (Shim and Yang, 1999). Such a bioreactor containing immobilized *Pseudomonas putida* and *P. fluorescens* degraded 10, 20, 20, and 12% of benzene, toluene, ethylbenzene, and o-xylene, respectively, under hypoxic conditions. Immobilized cells tolerated higher concentrations (greater than 1,000 mg/L) when compared with the free cells. Cells in the bioreactor were relatively insensitive to benzene toxicity. Substrate inhibition was observed for all substrates.

A continuous stirred tank reactor (CSTR) and a soil slurry-sequencing stirred batch reactor (SS-SBR) were tested for the degradation of a diesel fuel–contaminated soil under aerobic conditions and with added nutrients (C:N:P ratios ~60:2:1) (Cassidy et al., 2000). The diesel fuel removal efficiency was higher in the SS-SBR than in the CSTR (96 and 75%, respectively). Microbial growth was approximately 25% greater in the SS-SBR than the CSTR, probably because of the variety of environments faced by the organisms and because the induction or acclimatization of the bacteria is favored under dynamic conditions. Significant amounts of biosurfactant were produced in the SS-SBR, which was not observed in the CSTR. Periodic aeration and venting strategy was found to be better in treating soil contaminated by diesel fuel in a SS-SBR (Cassidy and Irvine, 1997). A combination of SS-SBR followed by a solid phase bioreactor (biopile) was found to be cost effective in treating soil contaminated (2.5% oil) with car diesel fuel or *n*-decane (achieving 80% degradation). Addition of an anionic surfactant increased the degradation rate. Improved porosity of the soil led to enhancement of the contaminant removal rate (Nano et al., 2003).

An effluent mixture containing brewery and petroleum wastes (1:2) was treated in a fluidized bed reactor using a mixed culture obtained from a petroleum refinery waste separation pond. The culture was supported on low density polyethylene (LDPE) particles (Ochieng et al., 2003). There were 36 and 64% decreases in COD for petroleum-only and mixed wastes, respectively. Addition of extra nutrients to the mixed waste increased the reduction in COD to 90%.

Conclusions

Petroleum or crude oil contains a large number of hydrocarbons, aromatics, and fused ring structures; identifying microbes or microbial communities that could degrade all of them is a challenge. In addition, PAHs are refractive; they are hydrophobic, which decreases their water solubility, making them inaccessible to the microorganisms. Thus they have a tendency to be adsorbed to the soil matrix. Nitrogen and sulfur compounds present in the petroleum may also be toxic to the microorganisms. A large number of microorganisms, fungi, and algae have been reported to degrade hydrocarbons under aerobic and anaerobic conditions. The white rot fungi *Phanerochaete chrysosporium* and *Pleurotus ostreatus* appear to be general-purpose organisms capable of degrading a wide range of hydrocarbons and PAHs. Addition of extra nutrient helps degradation but adds to the operating cost. Bioaugmentation appears to be a good method for enhancing degradation if the microorganism population at the contaminated site is not sufficient.

References

Aitken M. D. 1998. Characteristics of phenanthrene degrading bacteria isolated from soils contaminated with PAHs. *Can. J. Microbiol.* 44:743–752.

Atlas, R. M., and R. Unterman. 1999. Bioremediation. In *Manual of industrial microbiology and biotechnology*. Eds. A. C. Demain and J. E. Davis, pp. 666–681, 2nd edition. Washington, DC: ASM Press.

Barclay, C. D. 1995. Biodegradation and sorption of polyaromatic hydrocarbons by *Phanaerochaete chrysporium. Appl. Microbiol. Biotech.* 62:1188–1196.

Boopathy, R. 2003. Use of anaerobic soil slurry reactors for the removal of petroleum hydrocarbons in soil. *Int. Biodeterioration Biodegradation* 52:161–166.

Cassidy, D. P., and R. L. Irvine. 1997. Biological treatment of a soil contaminated with diesel fuel using periodically operated slurry and solid phase reactors. *Water Sci. Technol.* 35(1): 185–192.

Cassidy, D. P. S, Efendiev, and D. M. White. 2000. A comparison of cstr and sbr bioslurry reactor performance. *Water. Res.* 34(18): 4333–4342.

Floodgate, G. D. 1984. The fate of petroleum in marine ecosystems. In *Petroleum microbiology*. Ed. R. M. Atlas, pp. 355–397. New York: MacMillan.

Fuchs G., M. E. S. Mohamed, U. Altenschmidt, J. Koch, A. Lack, R. Brackmann, C. Lochmeyer, and B. Oswald. 1994. Biochemistry of anaerobic biodegradation of aromatic compounds. In *Biochemistry of microbial degradation.* ed. C. Ratledge, 513–553. Dordrecht, The Netherlands: Kluwer Academic Publishers.

Gibson, D. T., and V. Subramanian. 1984. Microbial degradation of aromatic hydrocarbons. In *Microbial degradation of organic compounds*. Eds. D. T. Gibson and M. Dekker, pp. 181–252. New York.

Giraud, F., P. Guiraud, M. Kadri, G. Blake, and R. Steiman, 2001. Biodegradation of anthracene and fluoranthene by fungi isolated from an experimental constructed wetland for wastewater treatment. *Water Res.* 35(17): 4126–4136.

Groudeva, V. I., S. N. Groudev, and A. S. Doycheva. 2001. Bioremediation of waters contaminated with crude oil and toxic heavy metals. *Int. J. Mineral Process.* 62:293–299.

Grund, E., B. Denecke, and R. Eichenlaub. 1992. Naphthalene degradation via salicylate and gentisate by *Rhodococcus* sp. strain B4. *Appl. Environ. Microbiol.* 58(6):1874–1877.

Harayama, S. 1997. Polycyclic aromatic hydrocarbon bioremediation design. *Current Opinion Biotech.* 8:268–273.

Holliger, C., and A. J. B. Zehnder. 1996. Anaerobic biodegradation of hydrocarbons. *Current Opinion Biotech.* 7:326–330.

Jezequel, R., L. Menot, F.-X. Merlin, and R. C. Prince. 2003. Natural cleanup of heavy fuel oil on rocks: an in situ experiment. *Marine Pollution Bull.* 46:983–990

Juhasz, A. L., and R. Naidu. 2000. Bioremediation of high molecular weight polycyclic aromatic hydrocarbons: a review of the microbial degradation of benzo[a]pyrene. *Int. Biodeterioration Biodegradation* 45:57–88.

Koma, D., F. Hasumi, E. Yamamoto, T. Ohta, S.-Y. Chun, and M. Kubo. 2001. Biodegradation of long-chain n-paraffins from waste oil of car engine by *A cinetobacter* sp. *J. Biosci. Bioeng.* 91(1): 94–96.

Lee, S. K., and S. B.V. Lee. 2001. Isolation and characterization of a thermo tolerant bacterium *Ralstonia* sp. strain PHS1 that degrades benzene, toluene, ethyl benzene, and o-xylene, *Appl. Microbiol. Biotech.* 56:270–275.

Maki, H., M. Utsumi, H. Koshikawa, T. Hiwatari, K. Kohata, H. Uchiyama, M. Suzuki, T. Noguchi, T. Yamasaki, M. Furuki, and M. Watanabe. 2003. Instrinsic biodegradation of heavy oil from Nakhodka and the effect of exogenous fertilization at a coastal area of the Sea of Japan. *Water, Air, and Soil Pollution* 145(1):123–138.

Mills, M. A., J. S. Bonner, T. J. McDonald, C. A. Page, and R. L. Autenrieth. 2003. Intrinsic bioremediation of a petroleum-impacted wetland. *Marine Pollution Bull.* 46:887–899.

Nakamura, K. 1996. Construction of bacterial consortia that degrade Arabian light crude oil, *J. Ferment. Biotech.* 48:677–686.

Nano, G., A. Borroni, and R. Rota. 2003. Combined slurry and solid-phase bioremediation of diesel contaminated soils. *J. Hazardous Mat. B* 100:79–94.

Ochieng, A., J. O. Odiyo, and M. Mutsago. 2003. Biological treatment of mixed industrial wastewaters in a fluidized bed reactor. *J. Hazardous Mat. B* 96:79–90.

Pritchard, H. P., and C. F. Costa. 1991. EPA's Alaska oil spill bioremediation report. *Environment. Sci. Tech.* 25:372–379.

Rike, A. G., K. Braathen Haugen, M. Børresen, B. Engene, and P. Kolstad, 2003. In situ biodegradation of petroleum hydrocarbons in frozen Arctic soils. *Cold Regions Sci. Tech.* 37:97–120.

Rivkina, E. M., E. I. Friedmann, C. P. McKay, and D. A. Gilichinsky. 2000. Metabolic activity of permafrost bacteria below the freezing point. *Appl. Environ. Microbiol.* 66(8): 3230–3233.

Rosenberg, E. 1992. Petroleum biodegradation—A multi phase problem. *Biodegradation.* 3:337–350.
Salleh, A. B., F. M. Ghazali, R. N. Zaliha, A. Rahman, and M. Basri. 2003. Bioremediation of hydrocarbon pollution. *Indian J. Biotech.* 2:411–425.
Shim, H., and S.-T. Yang. 1999. Biodegradation of benzene, toluene, ethylbenzene, and o-xylene by a coculture of *Pseudomonas putida* and *Pseudomonas fluorescens* immobilized in a fibrous-bed bioreactor. *J. Biotech.* 67:99–112.
Surzhko L. F. 1995. Utilization of oil in soil and water by microbial cells. *Microbiol.* 64:330–334,
Watson, J. S., D. M. Jones, R. P. J. Swannell, and A. C. T. van Duin. 2002. Formation of carboxylic acids during aerobic biodegradation of crude oil and evidence of microbial oxidation of hopanes. *Org. Geochem.* 33:1153–1169.
Wilkes, H., S. Kuhner, C. Bolm, T. Fischer, A. Classen, F. Widdel, and R. Rabus. 2003. Formation of *n*-alkane- and cycloalkane-derived organic acids during anaerobic growth of a denitrifying bacterium with crude oil. *Org. Geochem.* 34:1313–1323.

Bibliography

Head, I. M., and R. P. J. Swannell. 1999. Bioremediation of petroleum hydrocarbon contaminants in marine habitats. *Current Opinion Biotech.* 10:234–239.
Sridhar, S., V. F. Medina, and S. C. McCutcheon. 2002. Phytoremediation: An ecological solution to organic chemical contamination. *Ecol. Eng.* 18:647–658.
Watanabe, K. 2001. Microorganisms relevant to bioremediation. *Current Opinion Biotech.* 12:237–241.

CHAPTER 25

Biodesulfurization

The crude oil being produced around the world is showing a higher content of organic sulfur [world figures: 1990 output: 70,800 thousand barrels a day (tbd) with 1.13 wt% sulfur; 2010 projected output: 83,450 tbd with 1.27 wt% sulfur], so refineries now have to deal with severely impure feedstocks. Sulfur levels in crude oil range from 1,000 to 30,000 ppm. Diesel sulfur levels are much higher and are on the order of 5,000 ppm. Currently acceptable sulfur levels are on the order of 500 ppm and are soon to be reduced to 10 to 15 ppm. Sulfur oxides obtained from the combustion of gasoline poison the catalytic converters in automobile exhaust systems.

It is estimated that around 80 million barrels of oil are pumped from the earth every day. Most of these hydrocarbons are burned for energy. Most liquid and solid reserves are contaminated with sulfur, so direct combustion will release large amounts of sulfur oxides into the atmosphere, the natural consequence of which is acid rain. To avoid such adverse environmental effects, the following measures have been adopted.

- Limit sulfur emissions from power plants by using low sulfur fuels and postcombustion wet gas scrubbing
- Impose increasingly stringent restrictions on sulfur levels in transportation fuels and home heating oil (targets: European Union, heating oil—1,000 ppm by 1999 and diesel fuel—less than 100 ppm by 2005; United States, gasoline—50 to 100 ppm).

Recently, it has been found that for any refiner, the major source of sulfur is cat-cracked naphtha. The various options other than biodesulfurization available for reducing sulfur content in the gasoline pool are (1) the blending or sale of high-sulfur components, (2) the use of fluidized catalytic cracker unit (FCCU) catalyst additive, (3) extractive caustic treatments, (4) hydrodesulfurization, and (5) catalytic-cracker-feed desulfurization.

Hydrodesulfurization

Hydrodesulfurization is a high-pressure (150 to 250 psig) and high-temperature (200 to 425°C) process that uses hydrogen gas to reduce the sulfur in petroleum fractions (particularly diesel) to hydrogen sulfide, which is then readily separated from the fuel. Hydrodesulfurization units are expensive to build and operate. In addition, this chemistry does not work well on certain sulfur molecules in oil, particularly the polyaromatic sulfur heterocycles (PASHs) found in heavier fractions.

It is believed that refining industries will spend about $37 billion on new desulfurization equipment and an additional $10 billion on annual operating expenses over the next 10 years to meet the new sulfur regulations. In addition to this opportunity in the refinery, there is also a large potential in the desulfurization of crude oil itself. Approximately half of the 60 million barrels of crude produced each day is considered "high sulfur" (greater than 1%). The partial desulfurization of this material represents a significant chance to "upgrade" the crude and its value.

Biodesulfurization

Biodesulfurization of fossil fuel is an old concept that has recently been given renewed importance. The goal of biodesulfurization is to develop an economical and feasible process to reduce the sulfur content of fluid catalytic cracking (FCC) gasoline from 1,000 ppm to less than 200 ppm using biochemical means. The advantages of bio over hydro desulfurization are given in Table 25-1. The challenges here are isolating and characterizing bacterial strains that produce enzymes that utilize the sulfur in thiophenes and benzothiophenes as the sole sulfur source for growth. Then they must be isolated and cloned, and the genes responsible for biodesulfurization overexpressed.

TABLE 25-1
Comparison of Hydro- and Biodesulfurization Technologies

Hydrodesulfurization	*Biodesulfurization*
Higher production cost.	Lower production and capital cost (by about 50%).
Energy-intensive process.	Energy savings.
High temperature and pressure required.	Ambient temperature and pressure required.
Hydrogen is required for reducing sulfur, hence needs a hydrogen generation plant.	Hydrogen plant not needed.
Toxic byproducts.	Nontoxic byproducts.
Amount of saturated olefins increases, which lowers octane rating.	No saturation of olefins, so product quality not lowered.

These genes must also be engineered into a gasoline-tolerant host organism. Points to be kept in mind while developing a biocatalytic process are that the quality of the gasoline should not be affected, the environment for the individuals operating the bioreactor should be safe, and the product recovery system should be suitable and economical. The current research in this area includes the following:

- Elucidation of the desulfurization pathway including the isolation, identification, and quantification of the pathway intermediates
- Enhancement of solvent tolerance of the catalyst
- Definition of the basis for required genetic improvements
- Determination of the rate and extent of gasoline desulfurization

Biodesulfurization could be ideally suited for refineries that process high-sulfur crudes but that lack residue upgrading capabilities. It can be used instead of hydrodesulfurization. When applied to FCC gasoline, biodesulfurization can reduce sulfur content to less than 200 ppm.

Mechanism of Biodesulfurization

Alkylbenzothiophene, alkyl dibenzothiopehene, and benzothiophenes with ethylpropyl and butyl groups are some of the sulfur-containing compounds in fuels. Diobenzothiophene (DBT) is the model compound chosen by researchers, and its desulfurization is studied as a basis of removal of all sulfur-containing compounds from fuels because it has a hindered sulfur atom that reacts with difficulty. If degradation as opposed to desulfurization is chosen, then the calorific value of the fuel would be altered, which is not desirable. Gram-positive bacteria, such as *rhodococcal* strains, *mycobacterial* strains, and *Puenibacillus* sp. have been mainly used for biodesulfurization research.

The metabolism of DBT follows one of the two pathways, namely:

- The ring destructive, or Kodama, pathway followed by the DBT degrading microorganisms
- The hydrocarbon conserving pathway, or 4S pathway, followed by DBT desulfurizing microorganisms

The 4S pathway is favored as it leaves the hydrocarbon fraction of the fuel unaltered (see Fig. 25-1) (Monticello, 2000). The first step in desulfurization of these molecules is the transfer of the molecules from the oil phase into the cells, which appears to happen in *Rhodococcus* spp. This is likely because *Rhodococcus* spp. and other bacteria have been shown to metabolize many "insoluble" molecules in this fashion. The desulfurization (*dsz*) genes have been transferred to other organisms, such as *Escherichia coli* and *Pseudomonas putida*. In these cells, the PASH appear to partition to the water phase before being brought into the cell. In *Rhodococcus*, the *dsz*

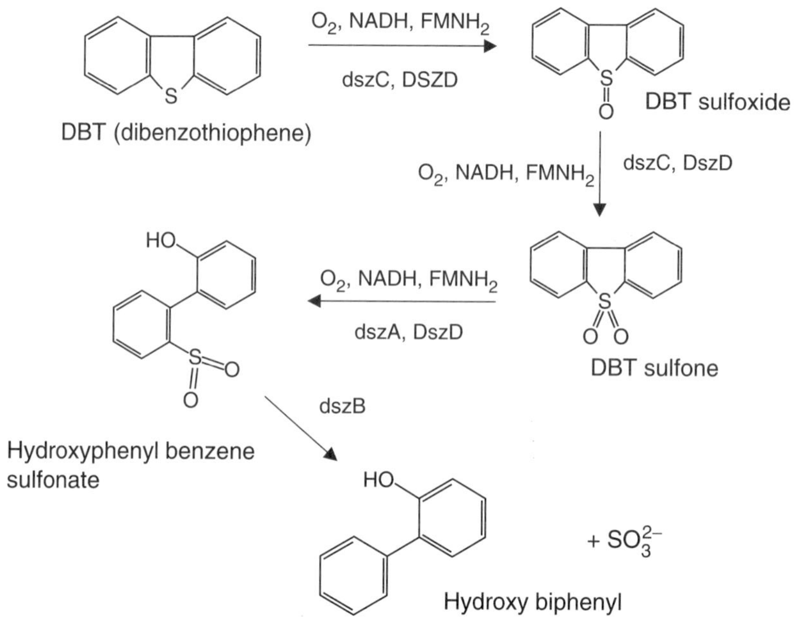

FIGURE 25-1. The pathway of biological desulfurization of the model compound dibenzothiophene relies on biocatalysts for specificity. NADH is reduced nicotinamide adenosine dinucleotide; FMN is flavin mononucleotide; DSZA, DSZB, DSZC, and DSZD are the catalytic gene products of *dszA*, *dszB*, *dszC*, and *dszD*, respectively (Monticello, 2000, 540).

enzymes are soluble and probably found in the cytoplasm. Once the Cx-DBT molecules find their way into the cell, they are then subjected to a series of oxidations.

The *Pseudomonas aeruginosa* gene was isolated, characterized, and evaluated for its capacity for the uptake of DBT in *n*-tetradecane by Noda (2003). Here, a transposon vector was used to transfer the *dsz* desulfurization gene cluster from *Rhodococcus erythropolis* KA2-5-1 into the chromosome of *Pseudomonas aeruginosa* NCIMB9571. All of the recombinant strains were able to completely desulfurize 1 m*M* DBT in *n*-tetradecane (*n*-TD) except for one named PARM1. This could desulfurize DBT in water but not in *n*-tetradecane, although the *n*-alkane utilization ability, the biosurfactant production, and the fatty acid composition of cells in strain PARMl were the same as those of the other recombinants. One can conclude that *P. aeruginosa* NCIMB9571 has a specific system of transporting hydrophobic compounds such as DBT in oil, and the recombinant PARM1 obtained was a mutant deficient in a DBT-transport system operational in *n*-TD.

The second step in the process is the conversion of DBT to the sulfone. The enzyme directly responsible for the first two oxidations has been isolated and characterized in some detail, and the gene for this enzyme (*dszC*) has been cloned and sequenced. The enzyme is named DBT monooxygenase ($FMNH_2$:DBT oxidoreductase) to reflect the reaction it catalyzes: the transfer of an electron from flavin mononucleotide ($FMNH_2$) to DBT to produce oxidized FMN ($FMNH_2$), DBT sulfoxide (DBTO), and DBT sulfone ($DBTO_2$). DBT monooxygenase catalyzes the oxidation of DBT to the sulfoxide and also the oxidation of the sulfoxide to the sulfone. The enzyme appears to operate as a tetramer in the cell.

The third step in the 4S process is the cleavage of the first C—S linkage. The first cleavage of the carbon-sulfur bonds is catalyzed by DBT sulfone monooxygenase ($FMNH_2$:$DBTO_2$ oxidoreductase, which transfers another electron from $FMNH_2$ to $DBTO_2$). This enzyme and its gene *dszA* have also been characterized. It appears to operate in the cell as a dimer.

The final step in the reaction pathway is the liberation of inorganic sulfur, production of sulfite, and an intact hydrocarbon molecule. This is catalyzed by a "desulfinase" (coded as *dszB* gene) and leads to the release of the sulfur as sulfite and the production of the oil-soluble product, 2-hydroxy biphenyl (HBP), which finds its way back into the petroleum fraction while retaining the fuel value (Fig. 25-2). It is unclear how many separate steps are involved in the transfer of Cx-DBT molecules from the oil to the first enzyme. Experimental results suggest that the reaction is not limited by the mass transfer from oil-to-water and later from water-to-cell. It is also unclear how Cx-HBP or Cx-HPBS exit the cells.

Other desulfurizing microorganisms that follow the 4S pathway include *Rhodococcus erythropolis* IGTS8, *R. erythropolis* D-l, *R. erythropolis* KA2-5-l, and *Mycobacterium* strain G3. The end-product of the desulfurization of DBT in all these cases was HBP (Okada, 2002).

The desulfurization ability of *R. erythropolis* KA2-5-1 using asymmetrically alkyl-substituted DBTs as substrates suggested that resting-cell reactions of KA2-5-1 with (alkylated dibenzothiophenes) Cx-DBTs occur through a specific carbon-sulfur (C—S) bond-targeted cleavage, yielding their alkylated hydroxybiphenyls. The attack on the DBT skeleton is found to be affected by the position, as well as the number and length of the alkyl substituents (Onaka, 2001); *Paenibacillus* sp. strain All-2 also utilized DBT and its alkylated derivatives (Cx-DBTs) as the sole sulfur source, producing the degraded product alkylated mono-hydroxybiphenyls. The substrate shape had a marked effect on selectivity for the first C—S bond cleavage by these desulfurizing microbes.

Rhodococcus erythropolis KA2-5-l desulfurizes DBT via a sulfur-specific pathway in which DBT is converted to the end product 2-hydroxybiphenyl by releasing sulfite via DBT-sulfone and 2-(2′-hydroxyphenyl) benzene sulfinate. Compared with glucose or glycerol, ethanol was found to be a better carbon source for obtaining high specific activity of desulfurization (50 vs 135.5 mmol 3HBP/kg-dry cell weight/h, respectively). It was postulated that NADH that is produced by the

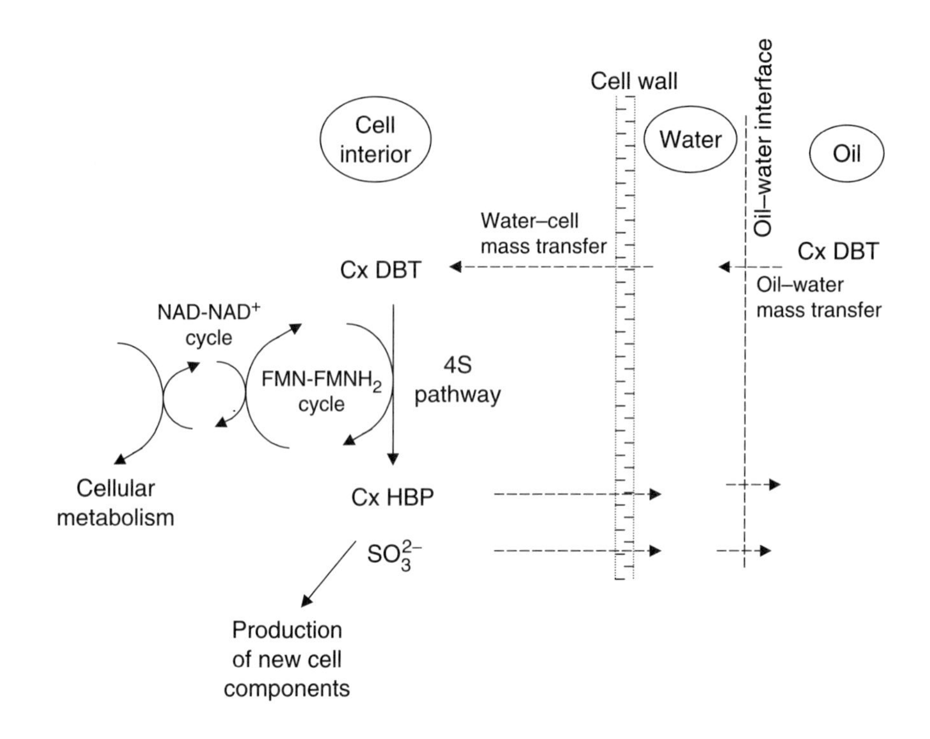

FIGURE 25-2. A conceptual diagram of some of the steps in the desulfurization of oil.

biochemical reaction of NAD with ethanol, which is catalyzed by alcohol dehydrogenase, might contribute to the conversion of FMN to $FMNH_2$, which is a coenzyme for the activities of desulfurization enzymes (Yan et al., 2000).

It was difficult to desulfurize alkyl DBTs with molecular weights higher than that of C5-DBT using *R. erythropolis* I-19, a derivative of IGTS8. Desulfurization of highly alkyl DBTs was achieved when the bacterial cell walls were destroyed (cell-free reaction system). This is attributed to the limitations of mass transfer from the oil phase to the cell interior. *Mycobacterium* sp. G3 has the ability to take up higher molecular weight alkyl DBTs like 4,6-dipropyl DBT and 4,6-dipentyl DBT. Also, the strain G3 was able to desulfurize diesel oil, thereby reducing the concentration of sulfur in diesel oil from 116 to 48 mg/L within 24 h (Okada et. al, 2002).

R. erythropolis KA2-5-1 was found to retain high desulfurization activity for extended periods (Kobayashi, 2000). PCR cloning and DNA sequencing of a KA2-5-1 genomic DNA fragment showed that it was practically identical with *dsz* ABC genes from *Rhodococcus* sp. IGTS8. This was a representative bacteria that targeted the carbon-sulfur bond of DBT. KA2-5-1 also desulfurized alkyl benzothiophenes. The purified monooxygenase, encoded by *dszC* of KA2-5-1, converted benzothiophene and dibenzothiophene into corresponding sulfones with the aid of an NADH-dependent oxidoreductase.

The thermophilic DBT-desulfurizing bacterium *Mycobacterium phlei* WU-F1, which grew in a medium with hydrodesulfurized light gasoline oil (LGO) as the sole source of sulfur, exhibited high desulfurizing activity toward LGO between 30 and 50°C. When WU-F1 was cultivated at 45°C with B-LGO (390 ppm S), F-LGO (120 ppm S), or X-LGO (34 ppm S) as the sole sulfur source, biodesulfurization was around 60 to 70% for all three types of hydrodesulfurized LGOs. When resting cells were incubated at 45°C with hydrodesulfurized LGOs in the reaction mixtures containing 50% (v/v) oils, biodesulfurization reduced the sulfur content from 390 to 100 ppm of B-LGO, from 120 to 42 ppm of F-LGO, and from 34 to 15 ppm of X-LGO (Furuyaa, 2003).

Process Development

Biodesulfurization technological issues include good reactor design, product recovery, and oil-water separations. Generally, batch–stirred tank reactors have been used because of the absence of immobilization technologies. Multistaged airlift reactors can also be used to overcome poor reaction kinetics at low sulfur concentrations and to reduce mixing costs. This would enhance the concept of continuous growth and regeneration of the biocatalyst in the reaction system rather than in separate, external tanks.

A typical process consists of charging the biocatalyst, oil, air, and a small amount of water into a batch reactor (Fig. 25-3). In the reactor, as the PASHs are oxidized to water-soluble products, the sulfur segregates into the aqueous phase. The oil–water–biocatalyst–sulfur–by-product emulsion from the reactor effluent is separated into two streams, namely, the oil (which is further processed and returned to the refinery) and the water–biocatalyst–sulfur–by-product stream. A second separation is needed to achieve this and allow most of the water and biocatalyst to return to the reactor for reuse.

Mass-transfer issues strongly influence the process design. A *Pseudomonas* system with naphthalene dioxygenase-like ring cleavage enzymes has been used to show that the transfer of DBT from the oil to the water phase and then from the water to the cells can limit the rate of DBT metabolism. Similar limitations are observed when *dsz* genes are cloned into *Pseudomonas* hosts. However, the case differs for *R. erythropolis* IGTS8. In an oil-water system, these bacteria adhere to the oil-water interface. When a cell-water-oil suspension is allowed to settle, the bacteria are found to be associated with oil droplets at the interface, and the aqueous phase remains clear. On the other hand, when dealing with *Pseudomonas* systems, the cells remain suspended in the aqueous phase. The explanation for this is that *Rhodococcus* "drinks from the oil" whereas *Pseudomonas* "drinks from the water" (Monticello, 2000).

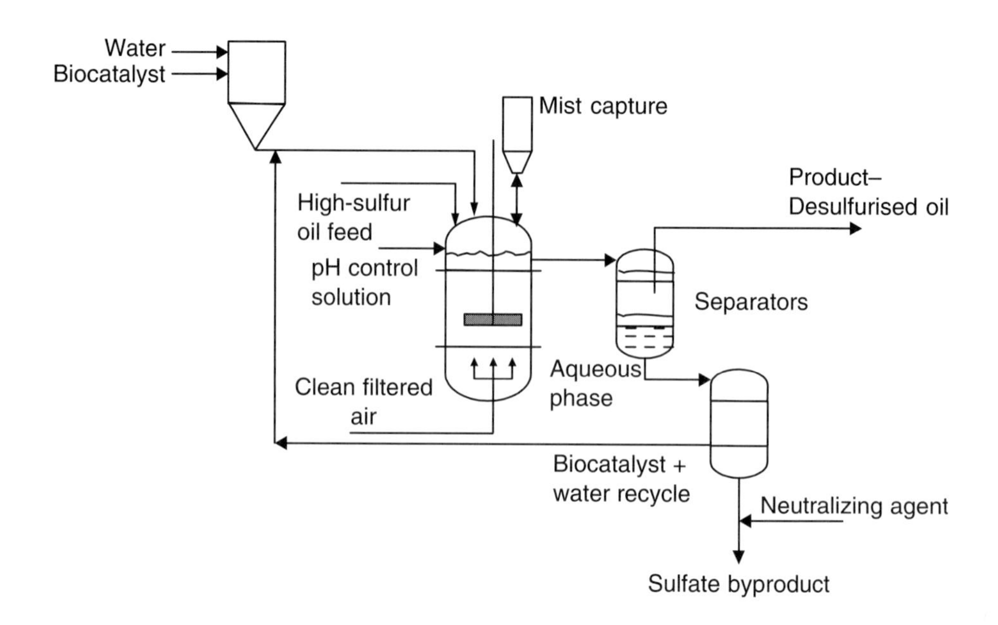

FIGURE 25-3. Process flowsheet.

Effective oil-cell-water contact and mixing is essential for good mass transfer. Unfortunately, a tight emulsion is usually formed, and it must be broken in order to recover the desulfurized oil, recycle the cells, and separate the byproducts. The phases are usually separated by liquid–liquid hydroclones. Another approach is to separate two immiscible liquids of varying densities by using a settling tank, where the liquid mixture is given enough residence time for them to form two layers, which are then drained (U.S. DOE).

The bacteria usually partition to the oil-water interface and move with the discontinuous phase into a two-phase emulsion. In a water-in-oil emulsion, cells associate with the water droplets. A small amount of fresh oil can then be added to create an oil-in-water emulsion so that the cells will stick to the oil droplets. Passage of this emulsion through a hydrocyclone will yield a clean water phase and a concentrated cell and oil mixture that can be recycled to the reactor. By manipulating the nature of these emulsions, relatively clean oil and water can be separated from the mixture without resorting to high energy separations. This reduces the capital and operational costs tremendously.

A spin-off from the biodesulfurization process is the production from the oil of a "biopetrochemical," Cx-HPBS (hydroxy phenyl benzene sulfonate), which is the penultimate product of the 4S pathway. The HPBS molecule has useful hydrotrope properties and is an effective detergent when derivatized with a long chain alkyl side chain. The feedstock for this material is the low-value high-sulfur refinery stream. Also, the reaction is carried out at low temperatures and pressures, which is economical.

Inverse Phase Transfer Biocatalysis

The problems involved in the biodesulfurization of fuels are inhibition of the biocatalyst by the byproducts and slow diffusion between the organic and aqueous phases. Inverse phase transfer biocatalysis (IPTB) uses supramolecular receptors (modified cyclodextrine-like hydroxy propyl-β-cyclodextrine), which selectively pick up the sulfur aromatic compounds in the organic phase and transfers them into the water phase, which contains the biocatalyst. The IPTB approach can increase mass transfer of water insoluble substrates between aqueous and organic phases, and eliminate or reduce feedback inhibition of the biocatalyst due to accumulation of the byproducts in the water phase.

It has been reported that 2-hydroxybiphenyl (HBP) inhibits *Rhodococcus rhodochrous* IGTS8. Conversion of DBT drops from 100 to 20% when the amount of 2-HBP is increased from 0 to 50 ppm. In the presence of 10 m*M* cyclodextrine, the growth of IGTS8 falls from 100 to only 80%. This indicates that cyclodextrine probably picks up the HBP in solution as well as directly from the interphase of the cellular biomembrane, thus protecting the microorganisms from the irreversible inhibition effect of this phenol (Setti et al., 2003).

It is also observed (Setti et al., 2003) that hydroxypropyl cyclodextrine can improve the mass transfer of water insoluble substrates such as DBT in *n*-hexadecane and aqueous phase. The specific rate of the DBT converted by IGTS8 (i.e., the parts per million of DBT converted per hour per gram of dry cell) at a DBT concentration of 120 ppm in *n*-hexadecane increases from 2.9 to 4.3 in the absence and presence of 3.14 m*M* of hydroxypropyl cyclodextrine.

Coal Biodesulfurization

Compared to the biodesulfurization of oil, that of coal is more difficult, since the highly polymeric material does not penetrate bacterial cells very well. The efficiency of microbial oxidation of coal depends on several parameters, including particle size of the powdered coal, nutrient media composition, pH, temperature, aeration, and reactor design. As of now, there are no commercial processes available. Sulfur-reducing bacteria were also reported to desulfurize sulfur compounds in coal to hydrogen sulfide gas. The ability to remove both inorganic and organic sulfur has been found in *Rhodococcus* species. Desulfurizing *Rhodococcus* species include *R. erythropolis* IGTS8, *R. erythropolis* D-1, *R. erythropolis* H-2. Among them, *R. erythropolis* IGTS8 is the most widely studied. It could remove 55.2% sulfate sulfur, 20% pyritic sulfur, 23.5% organic sulfur, and 30.2% total sulfur from Mengen lignite in 96 h (Prayuenyong, 2002).

Although microbial metabolism of inorganic sulfur in coal has been known for some time, degradation of organically bound sulfur is inefficient and hence prevents the development of a viable technology for the microbial desulfurization of coal. Recently, a mixed culture IGTS7 was found to remove about 91% of organic sulfur from coal. However, it is incapable of growing at acidic pH, which is typical of conditions conducive to the microbial metabolism of inorganic sulfur (Kilbane, 1990). *Agrobacterium* MC501 and the mixed culture composed of *Agrobacterium* MC501, *Xanthomonas* MC701, *Corynebacterium* sp. MC401, and *Corynebacterium* sp. MC402, isolated from a coal mine, were able to desulfurize DBT sulfone by using it as a sole source of sulfur for growth. The sulfone was metabolized to 2-hydroxybiphenyl and sulfate. These two cultures could also utilize a wide range of organic and inorganic sulfur compounds such as DBT, thianthrene, diphenylsulfide, thiophene-2-carboxylate, dibutilsulfide, methionine, cysteine, sulfate, and sulfite as sources of sulfur (Constantía, 1996).

Conclusions

Because of stringent environmental regulations and apparent cost advantages, the biodesulfurization process is continuously being studied. The mechanism of degradation by strains such as *Rhodococcus* of sulfur heterocycles dissolved in oil is well understood now. The physical properties of the cells are important for good mass transfer, effective separations, and good product recovery. Inverse phase transfer biocatalysis technology using supramolecular receptors holds plenty of promise for improving the mass transfer between oil and water and between water and cell phases; it also decreases biocatalyst deactivation due to byproduct inhibition. Factors like biocatalyst specificity, stability, activity, and bioreactor design affect conversion. The volumetric ratio between the oil phase and the aqueous medium is the limiting factor in industrial scale applications.

Other possibilities in petroleum biorefining or petroleum refining using biotechnology could be the denitrogenation of fuels, removal of heavy metals, transformation of heavy crudes into light crudes, as well as depolymerization of asphaltenes.

References

Constantía, M., J. Giraltb, and A. Bordonsa. 1996. Degradation and desulfurization of dibenzothiophene sulfone and other sulfur compounds by *Agrobacterium* MC501 and a mixed culture. *Enzyme Microb. Tech.* 19(3): 214–219.

Furuyaa, T., Y. Ishiia, K.-I. Nodab, K. Kinoa, and K. Kirimura. 2003. Thermophilic biodesulfurization of hydrodesulfurized light gas oils by *Mycobacterium phlei* WU-F1. *FEMS Microbiol. Letters* 221(1): 137–142.
Kilbane, J. J. 1990. Sulfur-specific microbial metabolism of organic compounds. *Resources, Conservation and Recycling* 3(2–3): 69–79.
Monticello, D. J. 1998. Riding the fossil fuel biodesulfurization wave. *Chemtech.* 28(7): 38–45.
Monticello, D. J. 2000. Biodesulphurisation and the upgrading of petroleum distillates. *Current Opin. Biotech.* 11(6): 540–545.
Noda, K., K. Watanabe, and K. Maruhashi. 2003. Isolation and characterization of a transposon mutant of *Pseudomonas aeruginosa* affecting uptake of dibenzothiophene in *n*-tetradecane. *Letters Appl. Microbiol.*, 37(2), 95–99.
Okada, H., N. Nomura, T. Nakahara, and K. Maruhashi, 2002. Analysis of dibenzothiophene metabolic pathway in *Mycobacterium* strain G3. *J. Biosci. Bioeng.* 93(5), 491–497.
Onaka, T., Jin Konishi, Yoshitaka Ishii, I , Kenji Maruhashi. 2001. Desulfurization characteristics of thermophilic paenibacillus sp. strain a 11-2 against asymmetrically alkylated dibenzothiophenes. *J. Biosci. Bioeng.* 92(2): 193–196.
Pienkos, P. T. 1999. Choosing the best platform for the biotransformation of hydrophobic molecules. In *Microbial ecology in industry, Microbial Biosystems*: New Frontiers Proceedings of the 8th International Symposium on Microbial Ecology, eds. C. R. Bell, M. Brylinsky, and P. Johnson-Green. Halifax, Canada: Atlantic Canada Society for Microbial Ecology.
Praxyuenyong, P. 2002. Coal biodesulfurization process. *Songklanakarin J. Sci. Technol.* 24(3):493–507.
Setti, L., S. Bondi, E. Badiali, and S. Guiliani. 2003. Inverse phase biocatalysis for a biodesulfurization process for middle distillate. BECTH. MOCK. YH-TA. CEP.2.XNMNR. 44(1):80–83.
U.S. DOE. 2003. Gasoline Biodesulfurization, Office of Energy Efficiency and Renewable Energy, May. U. S. Department of Energy, www.eere.energy.gov/industrial.
Yan, H., M. Kishimoto, T. Omasa, Y. Katakura, K.-I. Suga, K. Okumura, and O. Yoshikawa. 2000. Increase in desulfurization activity of *Rhodococcus erythropolis* KA2-5-l using ethanol feeding. *J. Biosci. Bioeng.* 89(4): 361–366.

Bibliography

Borgne, S. L., and R. Quintero. 2003. Biotechnological processes for the refining of petroleum. *Fuel Processing Tech.* 81(2): 155–169.
Denome, S. A., E. S. Olson, and K. D. Young. 1993. Identification and cloning of genes involved in specific desulfurization of dibenzothiophene by *Rhodococcus* sp. strain IGTS8. *Appl Environ. Microbiol.* 59:2837–2843.
Gray, K. A., O. S. Pogrebinsky, G. T. Mrachko, L. Xi, D. J. Monticello, and C. H. Squires. 1996. Molecular mechanisms of biocatalytic desulfurization of fossil fuels. *Nature Biotechnol.* 14:1705–1709.
Kobayashi, M., T. Onaka, Y. Ishii, J. Konishi, M. Takaki, H.Okada, Y. Ohta, K. Koizumi, and M. Suzuki. 2000. Desulfurization of alkylated forms of both dibenzothiophene and benzothiophene by a single bacterial strain. *FEMS Microbiol. Letters* 187(2): 123–126.
Mcfarland, B. Biodesulphurisation. 1999. *Curr. Opin. Microbiol.* 2:257–262.
Monticello, D. J., and W. R. Finnerty. 1985. Microbial desulfurization of fossil fuels. *Annu. Rev. Microbiol.* 39:371–389.
Noda, K. I., K. Watanabe, and K. Maruhashi. 2003. Isolation of the *Pseudomonas aeruginosa* gene affecting uptake of dibenzothiophene in *n*-tetradecane. *J. Biosci. Bioeng.* 95(5): 504–511.
Ohshiro, T., T. Hirata, and Y. Izumi. 1996. Desulfurization of dibenzothiophene derivatives by whole cells of *Rhodococcus erythropolis* H-2. *FEMS Microbiol. Letters* 142:65–70.
Ohshiro, T., and Y. Izumi. 1999. Microbial desulphurisation of organic sulphur compounds in petroleum. *Biosci. Biotech. Biochem.* 63:1–11.

CHAPTER 26

Treatment of Solid Waste

Introduction

Solid waste is defined as waste that is collected and transported by a means other than water. Solid waste can be classified into different types depending on the source:

- Household waste, also called municipal waste
- Industrial waste
- Hospital or biomedical waste.

Municipal solid waste consists of household waste, construction and demolition debris, sanitation residue, and waste from streets. This garbage is generated mainly from residential and commercial complexes. Garbage itself can be classified into four categories:

- Organic waste: kitchen waste, vegetables, flowers, leaves, fruits.
- Toxic waste: old medicines, paints, chemicals, bulbs, spray cans, fertilizer and pesticide containers, batteries
- Recyclable: paper, glass, metals, plastics
- Soiled: waste from first aid, cleaning vehicles and other machine parts

Over the last few years, the consumer market has grown rapidly, leading to products being packed in cans, aluminum foil, plastics, and other such nonbiodegradable items. Industrial solid waste includes metals, chemicals, paper, pesticides, dyes, rubber, and plastics. Hospital waste is generated during the diagnosis, treatment, or immunization of human beings or animals, in research activities in these fields, and in the production or testing of biologicals. These are in the form of disposables, swabs, bandages, etc. This waste is highly infectious and can be a serious threat to human health

TABLE 26-1
The Type of Litter We Generate and Approximate Time It Takes to Degenerate (Untreated)

Type of litter	*Approximate time it takes to degenerate*
Organic waste (vegetable, fruit peels, leftover foodstuff, etc.)	A week or two
Paper	10–30 days
Cotton cloth	2–5 months
Wood	10 to 15 years
Woolen items	1 year
Plastic bags	Undetermined (many years)
Glass bottles	Undetermined

if not managed in a scientific and discriminate manner. These different categories of waste each take their own time to degenerate if left untreated (as illustrated in the Table 26-1).

Bioremediation

Solid waste management and treatment calls for a multipronged approach; ideally it should involve all the four Rs of waste management, alongside judiciously planned biotreatment. Biotreatment, if planned, is the most suitable because it would generate methane gas, which can be used for energy purposes (biogas), while ensuring that detoxification is achieved.

The need for a biological approach to improve environmental conditions directly relates to the increasing size of the human population on a planet of finite dimensions. In 1996 earth's estimated population was 6 billion people, but by the year 2100 that number is expected to almost double (Ashford and Noble, 1996). As populations grow in size, increases in a variety of adverse human health and ecological effects (and associated costs such as healthcare expenses) are also expected. The U.S. EPA's Toxic Substances Control Act Chemical Inventory includes more than 72,000 chemicals, with approximately 2,300 new chemicals submitted to the U.S. Environmental Protection Agency every year (Hoffmann, 1982). Along with population increases, the number of different chemicals and the total amount of chemicals produced are also bound to increase in the future. In 1990, the total release of toxicants into the environment by U.S. manufacturers was approximately 4.8 billion pounds (Ember, 2000). In addition, large quantities of a number of toxic products are released into the environment by end users in more or less unaltered form. These products include those designed for household use, as well as industrial materials such as fuels, detergents, fertilizers, dielectric fluids, preservatives, flavorings, flame retardants, heat transfer fluids, lubricants, protective coatings, propellants, pesticides, refrigerants, and many other chemicals. Such materials or their breakdown products often accumulate in soil and aquifers near landfills and dumps, in surface lakes and streams, and in sediment. These pollutants are present not only in concentrated waste sites but are widely distributed throughout the environment, although in many cases at levels too low to trigger regulatory action. The kinds and amounts of these chemicals are also likely to increase as human populations swell.

There are a number of excellent reviews on bioremediation of solid wastes. Composition-based remediation methods are covered in some way in other chapters. Hence, the scope of the present chapter will be to give an overview of newer technologies emerging in this field. Innovative alternate technologies will be given attention.

Landfill

The main method used to dispose of municipal solid waste (MSW) is to place it in a "landfill"—also called a "garbage dump" or a "rubbish tip"—85 to 90% of domestic waste and commercial waste is disposed of in this way. If the landfill is suitably aerated and if it has sufficient amounts of organic waste, aerobic degradation naturally sets in. Depending on the components of the landfill, i.e., if it has sufficient amounts of organic matter with no toxic chemicals, then both aerobic and anaerobic degradation set in. Initially anaerobic degradation produces volatile carboxylic acids and esters, which dissolve in the water that is present. In the next stage of decomposition, significant quantities of methane gas (biogas) are released as these acids and esters are degraded to methane and carbon dioxide. The presence of heavy metals and polyhalogenated aromatics dampen the growth of microorganisms. Care must be taken to ensure that these pollutants are pretreated before being dumped into the landfill. Another way to overcome the presence of these growth retardants is to inoculate the landfill with microorganisms adapted to high concentrations of these toxins. One of the major problems of landfills is the *leachate*—water seepage from the landfill. This leachate contains organic, inorganic, and microbial contaminants extracted from solid waste, which may contaminate the groundwater. Aerobic degradation is the typical treatment for rapidly decreasing the biological oxygen demand (BOD) of the *leachate*. In the past, landfills were often simply "holes in the ground" that had been created by mineral extraction. Modern municipal landfills are much more highly designed and engineered. Anaerobic digestion is gaining more acceptance in the treatment of solid wastes. The high solids reactor concept for anaerobic digestion can handle more than 30% dry solids in the feed material and achieve a high conversion of organics to methane (Rivard, 1993).

Compost Treatment

A new compost technology, known as compost bioremediation, is currently being used to restore contaminated soils. Compost bioremediation refers to the use of a biological system of microorganisms in a mature, cured compost to sequester or break down contaminants in soil. Microorganisms digest, metabolize, and transform contaminants in soil and ground into humus and inert byproducts, such as carbon dioxide, water, and salts. Compost bioremediation has proven effective in degrading or altering many types of contaminants such as chlorinated and nonchlorinated hydrocarbons, wood-preserving chemicals, solvents, heavy metals, pesticides, petroleum products, and explosives. The compost used in bioremediation is referred to as "tailored" or "designed" compost in that it is specially made to treat specific contaminants at specific sites. In addition to reducing contaminant levels, compost advances this goal by facilitating plant growth. In this role, compost provides soil conditioning and also provides nutrients to a wide variety of vegetation. In 1979, at a denuded site near the Burle Palmerton zinc smelter facility in Palmerton, PA (United States), a remediation project was started to revitalize 4 square miles of barren soil that had been contaminated with heavy metals. Researchers planted Merlin Red Fescue, a metal-tolerant grass, in lime fertilizer and compost made from a mixture of municipal wastewater treatment sludge and coal fly ash. The remediation effort was successful, and the area now supports a growth of Merlin Red Fescue and Kentucky Bluegrass (Chaney, 1994). A similar success story was observed for the remediation of soil contaminated with petroleum hydrocarbons (Fordham, 1995).

Use of Enzymes

There is a growing recognition that enzymes can be used in many remediation processes to target specific pollutants for treatment. Recent biotechnological advances have allowed the production of cheaper and more readily available enzymes through better isolation and purification procedures (Karam and Nicell, 1997). Improvement in the useful life of the enzyme, and thereby a reduction in treatment cost, has been accomplished through different methodologies, and one of the most promising was enzyme immobilization (Nicell et al., 1993). The effect of immobilized horseradish peroxidase (HRP) (on activated alumina) and hydrogen peroxide concentration on the removal efficiency of phenol showed that one molecule of HRP was needed to remove approximately 1,100 molecules of phenol when the reaction was conducted at pH 8.0 and at room temperature. Both tyrosinase and birnessite were able to catalyze the transformation of phenolic compounds through oxidative polymerization, a process that leads to humification. Bollag (2003) suggested that it is possible to enhance the natural process of xenobiotic binding and incorporation into the humus by adding laccase to the soil. Chlorinated phenols and anilines were transformed in

TABLE 26-2
Enzymes and Their Potential Applications in Biodegradation

Enzymes	*Source*	*Applications*
Peroxidases	Horseradish	Phenol, chlorophenol, aniline degradation, dewatering of slimes
	Artromyces ramosus	Phenol, PAH, herbicide degradation, polymerization of humic acid
	Plant material	Water decontamination
Chloroperoxidase	*Caldariomyces funago*	Phenol degradation
Lignin peroxidase	*Phanerochaete chrysosporium*	Aromatic compounds, phenols degradation.
Manganese peroxidase	*Phanerochate chrysopsorium*	Phenols, lignins, pentachlorophenol, dyes degradation
	Nematolona frowardie	Lignin degradation
Tyrosinase	*Agaricus bisporus*	Catechol degradation
Laccase	*Trametes hispida*	Dye degradation
	Pyricularia oryzae	Azo-dye degradation
	Trametes versicolor	Chlorophenol, urea derivative degradation
Catechol dioxygenases	*Pseudomonas pseudoalacaligenes*	Polychlorinated biphenyls, chlorothanes
Phenoloxidase	*Phanerochate chrysopsorium*	Chlorinated compounds

soil by oxidative and detoxified coupling reactions mediated by laccase, peroxidase, or metal oxides such as birnessite. The potential applications of enzymes in biodegradations are listed in Table 26-2 (Duran and Esposito, 2000). Oxidative enzymes play an important role in the decontamination of soils. At present, however, the commercial use of enzymes is still not realized because of the high cost of their isolation, purification, and production. Immobilization will play an extremely important role in cost reduction.

Phytoremediation

Phytoremediation is also an innovative technology that is gaining recognition as a cost-effective and aesthetically pleasing method of remediating contaminated soils. There are several categories of phytoremediation:

- Phytoextraction: Plants are often capable of the uptake and storage of significant concentrations of some heavy metals and other compounds in

their roots, shoots, and leaves. This method is ideally suitable for soil contaminated with heavy metals.
- Phytotransformation: Plants metabolize some compounds and render them less toxic. This method is suitable for soil contaminated with organic pollutants.
- Phytostabilization: Plant root exudates (enzymes and other chemicals) chelate with some contaminants and reduce their migration through the soil. This process effectively reduces the bioavailabilty of harmful contaminants.
- Phytostimulation: At the soil-root interface, known as the rhizosphere, there is a very large and active microbial population. Often the plant and microbial populations provide needed organic and inorganic compounds for one another. The rhizosphere environment is high in microbial abundance and rich in microbial metabolic activity, which has the potential to enhance the rate of biodegradation of contaminants by the microorganisms. Generally, the plant is not directly involved in the biodegradation process. It serves as a catalyst for increasing microbial growth and activity, which subsequently increases the biodegradation potential.

According to preliminary studies, enhanced degradation of pesticides (atrazine, metolachlor, and trifluralin) was observed in contaminated soils where plants of the *Kochia* sp. have been planted. Many plants and bacteria have evolved various means of extracting essential nutrients, including metals, from their environment. In the course of prospecting for minerals, unusually tolerant species have been observed in the vicinity of metal-rich deposits. In some cases, these tolerant organisms concentrate metals several thousandfold over ambient concentrations. Zajic (1969), Baker and Brooks (1989), Shann (1995), and other authors point out that such organisms may provide the opportunity to return waste material to useful products rather than merely transform them to innocuous substances. However, a practical phytoremedial technology remains to be developed, although progress has been made with transgenic *Arabidopsis thaliana* expressing *merApe9* (Rugh et al., 1996). Grown on medium containing $HgCl_2$, at concentrations of 25 to 100 *M* (5 to 20 ppm), these transgenic *merApe9* seedlings evolved considerable amounts of Hg^0 relative to control plants. However, the transformation of ionic mercury to the metallic elemental form, which then volatilizes to become an air pollutant, is a less than ideal remedial solution.

Vermicomposting

Municipal solid waste (MSW) is highly organic in nature, so vermicomposting has become an appropriate alternative for safe, hygienic, and cost effective disposal. Earthworms feed on the organics and convert material into castings (ejected matter) rich in plant nutrients. The chemical analyses of cast show 2 times the available magnesium, 15 times the available nitrogen, and 7 times the available potassium compared with the surrounding soil.

The action of earthworms in the process of vermicomposting of waste is physical and biochemical. The physical process includes substrate aeration, mixing as well as actual grinding, while the biochemical process is influenced by microbial decomposition of substrate in the intestine of earthworms (Hand et al., 1988). Various studies have shown that vermicomposting of organic waste accelerates organic matter stabilization (Neuhauser et al., 1998) and provides chelating and phytohormonal elements that have a high microbial matter content and stabilized humic substances. A number of references are available on the potential of earthworms in the vermicomposting of solid waste, particularly household waste (Edwards, 1980). Advanced systems for vermicomposting are based on top feeding and bottom discharge of a raised reactor, thus providing stability and control over key areas of temperature, moisture, and aeration. Price and Phillips (1990) have developed an improved mechanical separator, having a novel combining action, for removing live earthworms from vermicomposts. Vermicomposting provides other advantages, too; some earthworms (*Lempito mauritii*) can also be used for specific wastes such as those from medical facilities (Hori et al., 1974) and those with high concentrations of protein or pig feed (Mekada et al., 1979), as well as in nematode control (Dash et al., 1980).

Conclusion

Solid waste management is a necessary prerequisite for healthy living. Given the growth in population and industry, solid waste is increasing geometrically year after year. Unless there is a concerted, focused effort in dealing with this waste, both at the level of the individual and the community, waste will become a major health hazard. Bioremediation is the most suitable and economical method for degrading this waste. Many newer processes are being developed; of these, the most promising are (as discussed previously):

- Landfill
- Use of enzymes
- Composting
- Phytoremediation
- Vermicomposting

Rather than adopting any single method of remediation, it is advisable that a combination of two or more of these methods be adopted. This would ensure faster degradation of the waste while producing biomass (sludge) that can be used for a variety of commercial purposes.

References

Ashford, L. S., and J. A. Noble. 1996. Population policy: consensus and challenges. *Consequences* 2(2):25–36.

Baker, A. J. M., and R. R. Brooks. 1989. Terrestrial higher plants which hyperaccumulate metallic elements–a review of their distribution, ecology, and phytochemistry. *Biorecovery* 1:81–126.

Bollag, J. M., H.-L. Chu, M. A. Rao, and L. Gianfreda. (2003). Enzymatic oxidative transformation of chlorophenol mixtures. *J. Environ. Qual.* 32:63–69.

Chaney, R. L. 1994. Phytoremediation potential of *Thlaspi caerulescens* and *Bladder campion* for zinc. *J. Environ. Qual.* 23:1151–1157.

Dash, M. C., B. K. Senapati, and C. C. Mishra. 1980. Nematode feeding by tropical earthworms. *Trop. Ecol.* 20:10–12.

Duran, N., and E. Esposito. 2000. Potential applications of oxidative enzymes and phenoloxidase-like compounds in wastewater and soil treatment: a review. *Appl. Catalysis B: Environ.* 28:83–99.

Edwards, C. A. 1981. Earthworms, soil fertility, and plant growth. In: *Workshop on the Role of Earthworms in Stabilization of Organic Residues*, vol. 1, M. Appelhof (ed.), pp. 61–86 Kalamazoo, Michigan: Beech Leaf Press.

Ember, L. 2000. Reclassifying chemical relics of the Cold War. *Chem. Eng. News* 78(3):44.

Fordham, W. 1995. Yard trimmings composting in the Air Force. *Biocycle* 36:44.

Hand, P., W. A. Hayes, J. C. Frankland, and J. E. Satchell. 1998. The vermicomposting of cow slurry. *Pedobiologia,* 31:199–209.

Hoffmann, G. R. 1982. Mutagenicity testing in environmental toxicology. *Environ. Sci. Technol.* 16:560–573.

Hori, M., K. Kondo, T. Yosita, E. Konsihi, and S. Minami. 1974. Studies of antipyretic components in the Japanese earthworm. *Biochem. Pharmacol.* 23(11):1583–1590.

Karam, J., and J. A. Nicell.1997. Potential applications of enzymes in waste treatment. *J. Chem. Technol. Biotechnol.* 69:141–153.

Mekada, H., N. Hayashi, H. Yokota, and J. Okumura. 1979. Performance of growing and laying chickens fed diets containing earthworms*(Eisenia foetida). Jpn. Poult. Sci.*, 16:293–297.

Neuhauser, E. F., R. C. Loehr, and M. R. Malecki. 1998. *Earthworms in waste and environmental management*, The Hague: SPB, Academic Publishing.

Price, J. S., and V. R. Phillips. 1990. An improved mechanical separator for removing live worms from worm-worked organic wastes. *Biol. Waste.* 33(1):25–37.

Rivard, C. J. and N. J. Nagle. 1993. Anaerobic biodegradation of sewage-derived fat, oil, and grease (FOG) at mesophilic and thermophilic temperatures. In: *Proceedings of the 1994 food industry environmental conference*, p.71, Atlanta, GA: Georgia Tech Research Institute.

Rugh, C. L., H. D. Wilde, N. M. Stack, D. M. Thompson, A. O. Summers, and R. B. Meagher. 1996. Mercuric ion reduction and resistance in transgenic *Arabidopsis thaliana* plants expressing a modified bacterial merA gene. *Proc. Natl. Acad. Sci. USA* 93:3182–3187.

Shann, J. R. 1995. The role of plants and plant/microbial systems in the reduction of exposure *Environ. Health. Perspect.* 103(5):13 15.

Zajic, J. E. 1969. *Microbial biogeochemistry*. New York: Academic Press.

CHAPTER 27

Treatment of Municipal Waste

Introduction

The term "sewage" refers to the wastewater produced by a community, which may originate from three different sources:

- Domestic wastewater
- Industrial wastewater
- Rainwater

Domestic wastewater is usually the main component of sewage and is often used as a synonym. The sewage flow rate and composition vary considerably from place to place, basically depending on economic aspects, social behavior, climatic conditions, water consumption, type and conditions of the sewer systems, and so forth. It is not uncommon for water polluted by organic substances associated with animal or food waste or sewage to have an oxygen demand that exceeds the maximum equilibrium solubility of dissolved oxygen. Under such circumstances, unless the water is continuously aerated, it will soon be depleted of its oxygen, and fish living in the water will die. The average composition of sewage is given in Table 27-1.

Improved bioremediation of biological wastes is envisioned as a necessary first step in breaking the chain of events associated with microbial pathogenesis. In England, the recent outbreak of bovine spongiform encephalopathy (mad cow disease), which is believed to be associated with Creutzfeldt-Jakob disease in humans, has increased concern over disease transmission from food animals to humans (Narang, 1996). In fact, a great many microbial diseases (zoonotic diseases) can and often do cross over to affect humans. Diseases that can pass to humans from swine, for example, include bacterial infections, such as anthrax (*Bacillus antracis*), brucellosis (*Brucellosis suis*), ampylobacteriosis (*Campsylobacter jejuni*), erysipeloid (*Erysipelothrix rhusiopathiae*); viral infections, such as encephalomyocarditis (*Cardiovirus*), influenza (*Influenzavirus*), Japanese

TABLE 27-1
Average Composition of Sewage

Constituents	*Amount* (mg/L)
TSS	330
VSS	200
BOD	180
COD	550
NH3	30
Total phosphorous	7
Sulfates	10
Chlorides	78
Alkalinity	280
Calcium	110
Magnesium	100
Microorganisms	
E. coli	4×10^7 (no. in 100 mL)
Viruses	—
Emerging contaminants	
Antibacterial agents	—
Acidic pesticides	—
Surfactant metabolites	—

B encephalitis [*Flavivirus (gp A)*], and vesicular stomatitis (*Vesiculovirus*); nematode infections, such as ascariasis (*Ascaris suum*) and trichinosis (*Trichinella spp.*); protozoan infections, such as balantidiasis (*Balantidium coli*), toxoplasmosis (*Toxoplasma gondii*), amoebic dysentary/amebiasis (*Entamoeba polecki*) and sarcocystosis (*Sarcocystis suihominis*); and spirochetal infections, such as leptospirosis (*Leptospira interrogans*) (Beran, 1994).

Although the advent and continued development of antibiotics have kept infectious disease in developed countries under control for many years, there is growing evidence that this may not be effective indefinitely because increasingly virulent and antibiotic-resistant strains continue to evolve (Tenover, 1995). Hence, proper treatment of the sewage becomes essential for maintaining a healthy environment.

Treatment

Wastewater purification is the clearest paradigm of environmentally friendly technologies. Some negative aspects of development and urbanization can be diminished, or even eliminated, through a comprehensive treatment of domestic and industrial wastewater, directly and immediately enhancing the quality of the environment.

Bioremediation is not new to the human race, although new approaches that stem from advances in molecular biology and process engineering are emerging. An important, long-standing, and increasingly problematic bioremediation area is processing biological nitrogen waste (feces and urine) produced by humans and the animals that humans depend on for food. As human population size, industrial production, and chemical use have increased, so have populations of farm animals. Much of the waste ends up in river waters and estuaries, where it causes enormous problems, with secondary contributions to air and groundwater pollution (Culotta, 1992). It is no wonder that, worldwide, the effects of poor water quality are second only to malnutrition in the total disease burden and cause of death of human beings (Murray and Lopez, 1996).

Direct discharge to the environment is still the most common way of dealing with sewage. Yet several technological options are available today in the field of sewage treatment, including conventional aerobic treatment in ponds, trickling filters, and activated sludge plants, direct anaerobic treatment (upflow anaerobic sludge blanket [USAB] and expanded granular sludge bed [EGSB] reactors) (Seghezzo et al., 1998), and a combination of aerobic and anaerobic treatments. Sewage treatment can be broadly classified into three categories, viz:

- Septic tank
- Artificial marshes
- Sanitary sewer systems

Biotreatment is an integral part of all these types of sewage treatment. Anaerobic treatment is increasingly recognized as the core method of an advanced technology for environmental protection and resource preservation, and it represents, combined with other proper methods, a sustainable and appropriate wastewater treatment system for developing countries. Anaerobic treatment of sewage is increasingly attracting the attention of sanitary engineers and decision makers.

Septic Tank Treatment

Raw sewage is treated by one of the following methods depending on the size and the economic status of the community. In many rural and small communities, septic tanks are used to decontaminate sewage, since central sewage facilities are not available. These concrete or open underground tanks often receive the wastewater from only one home. The solids settle to the bottom, and the bacteria in the wastewater feed on the organic matter, liquefying the waste. Since the conditions are anoxic, most of the processes are anaerobic degradation, although a small portion of aerobic degradation does occur. This small aerobic degradation converts most of the nitrogen compounds to nitrates. A lack of denitrifying organisms will lead to the water being contaminated by nitrates. Around 1860, a French engineer, Louis H. Mouras, built a closed chamber with a water seal in which all "excrementitious matter" was rapidly transformed. This invention named "Mouras Automatic Scavenger" was enthusiastically defined at that time as "the most simple, the most beautiful, and perhaps the grandest of modern inventions" (McCarty, 1981). Septic tanks are another large and imperfect bioremedial system that contributes nitrogen and other waste to the impairment of water quality, particularly to groundwater. U.S. Environmental Protection Agency studies (U.S. EPA, 1980) indicate that about one-third of all septic tanks operate improperly; as a result, septic tanks are the primary source of groundwater contamination in many parts of the country. This contamination leads to nitrates, chemicals, and pathogens in the well water that some people drink.

Artificial Marshes

An alterative to the processing of wastewater through a conventional treatment plant in small communities is biological treatment in an artificial marsh, also called a "constructed wetland." Here along with bioremediation, phytoremediation takes place. Phytoremediation, the use of vegetation for the in situ decontamination of soils and sediments of heavy metals and organic pollutants, is a low-cost, nonobtrusive method of remediation. Certain plants are hyper accumulators of metals and organic compounds. They absorb high levels of these heavy metals and some organic compounds through their roots. The organic compounds are stored or sometimes metabolized. The plants can then be harvested and burnt to get ash, which has high concentrations of these heavy metals. Some plants also ooze root exudates (enzymes) that chelate and thereby again reduce the toxicity of these metals. These wetlands commonly have plants such as bull rushes, reeds, and cattails, which take up metal ions and organic compounds through their root systems. The microbes (aerobic and anaerobic) that live among the plants' roots and rhizomes also degrade the organic matter. The plant growth uses up the pollutants and increases the pH, which serves to destroy some harmful microorganisms. The greatest advantage of this type of decontamination is that great amounts of sludge are not generated, unlike in the conventional methods. Thus artificial marshes (wetlands) are one of the best and most convenient methods of sewage decontamination.

Sanitary Sewer Systems

General Aspects Sewer systems consist of three stages of wastewater treatment (Fig. 27-1). Primary treatment is a purely mechanical treatment; secondary treatment is a bioremediation step, while tertiary treatment is a chemical treatment.

In the primary treatment stage, the larger particles (including sand and silt) are removed by allowing the water to flow across screens and then slowly along a lagoon. Fats, oils, waxes, and the products of the reaction of soap and calcium and magnesium, normally termed "liquid grease," float on the

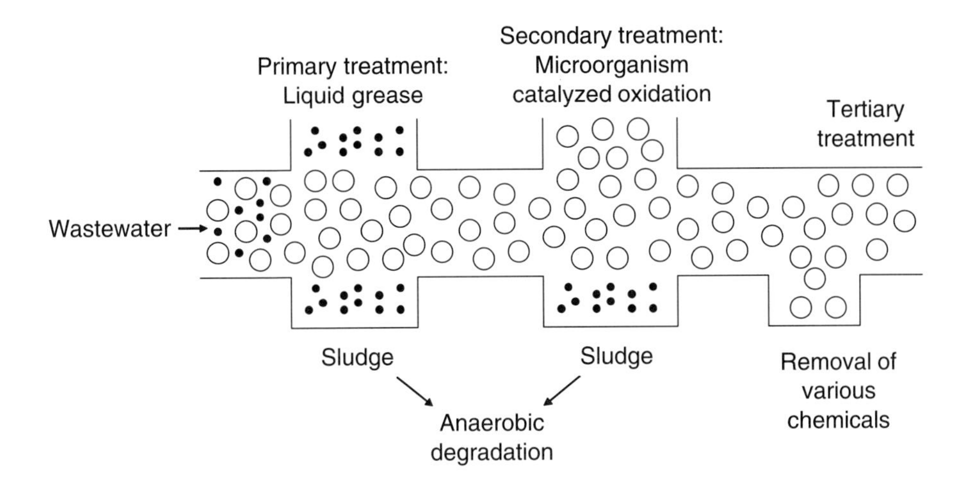

FIGURE 27-1. The common stages of treatment of sewage.

water's surface. This is skimmed off. The sludge of insoluble particles (predominantly organic matter) that forms at the bottom of the lagoon is digested anaerobically by microbes. The water now cleaned of the liquid grease and sludge still has very high biological oxygen demand (BOD), which is due mainly to the organic colloidal particles.

In the secondary treatment stage, most of this suspended organic matter, as well as that which is actually dissolved in the water, is oxidized by microorganisms. Additional sludge may be produced in this process and can be easily separated from water. The biological oxidation in the secondary treatment stage is predominantly by aerobic organisms because in this stage air is pumped through the water, providing sufficient oxygen for the organisms to thrive. Anaerobic treatment is being preferred now because the amount of sludge produced is much less. Biological treatment involves the transformation of dissolved and suspended organic contaminants to biomass and evolved gases (CO_2, CH_4, N_2, and SO_4^{2-}). The activated sludge process is the most widely used biological wastewater treatment in the world for domestic and industrial plants.

The treated water from the secondary stage now has a relatively low BOD. It is further purified in the tertiary treatment stage by various chemical means—alum treatment, activated charcoal, lime addition, etc.—before final release into rivers or other bodies of water. In some cases the water from the tertiary treatment stage is further purified by reverse osmosis or pollutants are removed by electrodialysis. The water thus treated is suitable even for reuse. Several authors have shown that particles represent the major part, up to 85%, of the total chemical oxygen demand (COD) in domestic sewage (Levine et al., 1985). The size of particles in domestic sewage affects both biological and physical processes. Anaerobic treatment reduces these colloidal particles and improves the degradability of sewage by aerobic systems. It was observed that the presence of surfactants (detergents) in these wastewaters enhanced the biodegradability of particles (Elmitwali et al., 2001).

Technological Aspects Because of the importance of clean water to human health, sewage treatment plants (STPs) constitute the largest and most important bioremediation enterprise in the world. There are approximately 16,000 municipal STPs in the United States (Laws, 1993). The major components of raw sewage are suspended solids, organic matter, nitrogen, phosphorus, pathogenic microorganisms, and chemicals (e.g., pesticides and heavy metals), and even the most rudimentary STPs make some reductions in most of these factors. Several methods are used for sewage treatment.

Generally, primary treatment consists of a screening device to remove the large trash and debris (usually hauled away to landfills), a settling tank where coarse grit and sand particles are removed, and a primary clarifier (essentially a large tank from which floating solids and settled sludge are removed after the sewage has resided in the tank for a brief period, usually a few hours). The limited time in the primary clarifier means that microorganisms living in the tank do not have the opportunity to consume a large amount of the nutrient material contained in the sewage. The floating solids and the sludge are then pumped to an anaerobic digester. The liquid effluent is disinfected, usually with chlorine, before its release into the environment. Alternatively, additional processes, referred to as secondary- and tertiary-level sewage treatments, may be applied to further reduce the levels of nutrients, pathogens, and chemicals. The anaerobic digester contains microorganisms adapted to grow and multiply in the absence of oxygen at elevated temperatures. In this process, nutrients are converted primarily to microbial biomass, methane, and carbon dioxide, and thus are consumed. The liberated methane is used to heat the digester. The objectionable qualities (less odor as well as reduced numbers of pathogens) of the sludge coming out of the anaerobic digester are reduced considerably. The sludge is typically transported to a landfill or applied to the land as fertilizer.

Secondary sewage treatment consists of two main types: trickling filters and activated sludge. Trickling filters are cylindrical tanks containing loosely packed rocks that range in size from 2 to 10 cm. Effluent enters through the top; air is introduced from the bottom. Distributed throughout the column is a variety of organisms that are attached to the surfaces of the rocks and fill the intervening spaces. Bacteria and fungi are the first to consume the organic constituents, and in turn the bacteria and fungi are consumed by higher trophic level organisms, including protozoa, rotifers, nematodes, worms, and insects. Activated sludge systems consist of a series of tanks. Effluent is introduced at one end, and it exits at the other.

In between, the sewage is mixed and aerated vigorously. Bacteria are the main decomposing organisms in the activated sludge system, but protozoans, rotifers, and nematodes are also present. All the various life forms tend to occur together in flocculant masses.

Both activated sludge and trickling filter secondary STP systems can be effective, but there are advantages and disadvantages to each. Trickling filters seem to be more tolerant of industrial chemicals, perhaps because of greater species and metabolic diversity. However, trickling filters require more space, cost more to construct, and tend to create more of an odor problem. Activated sludge systems tend to achieve greater reductions in organic nutrients and suspended solids.

Regardless of which secondary process is used, without further (i.e., tertiary) treatment, large amounts of nitrogen and phosphorus remain in secondary STP effluents (Ellis, 1983). These inorganic nutrients in turn encourage algal and phytoplanktonic growth in receiving waters. Ultimately, these organisms die and decompose, which consumes oxygen and thereby promotes hypoxic and anoxic conditions. Fish kills resulting from oxygen deprivation are notable consequences; in extreme cases, millions of fish are killed (Schindler, 1974). The technology to remove both nitrogen and phosphorus (and as a result, counteract these effects) has been available for some time (Eliassen and Tchobanoglous, 1969). Inorganic phosphorus can be precipitated from solution by the addition of calcium (as lime, CaO), aluminum (as alum, aluminum sulfate), or a variety of other relatively inexpensive chemicals. Nitrogen can be removed both chemically and biologically. Most of the nitrogen in secondary sewage effluent occurs as ammonium ion (NH_4^+). The process of ammonia stripping involves the conversion of NH_4^+ to ammonia gas (NH_3) by raising the pH and providing vigorous agitation. However, the liberated ammonia gas then becomes a potential atmospheric pollutant. Biological conversion of nitrogen gas (N_2) by denitrifying bacteria is an alternative approach, although there are other approaches as well (e.g., break point chlorination, reverse osmosis, and distillation) (Pressley et al., 1973). In spite of the available technology, implementation has been limited, and eutrophication, caused in part by the effluent from STPs, still commonly occurs in many coastal regions throughout the world.

The discharge of STP effluent on land rather than in water has been tried many times, often with at least initial success (Allhands and Overman, 1989). The potential advantages of land deposition are that groundwater resources can be recharged and that valuable nutrients become available to assist with crop growth and other vegetation. The disadvantages include possible groundwater contamination with nitrates (NO_3^-), associated with methemoglobinemia in infants, cancer, and birth defects (Xu et al., 1992), and other toxic, possibly carcinogenic, chemicals, including biocides (Garry et al., 1996). Other disadvantages are the increased risk of exposure to disease pathogens and the gradual accumulation of heavy metals in soils such that the growth of crops can eventually become inhibited (McGrath et al., 1995). In spite of these problems, land application of STP effluent has been remarkably useful in many cases (e.g., the reclamation of strip-mined soil) (Sopper and Seaker, 1984).

Conclusion

It is estimated that more than half of the rainwater that falls is converted to wastewater by people, cities, and industry. Although there are many less-than-ideal systems, bioremediation carried out in STPs does a reasonable overall job of cleaning up this huge amount of waste. Agricultural operations, on the other hand, sometimes do not tend to their animal wastes. Sixty percent of water quality impairment is attributed to silt and fertilizer runoff (Outwater, 1996).

Thus, bioremediation forms the basic core around which other processes—chemical and mechanical—function in sewer treatment plants. Anaerobic degradation occurs at the primary treatment stage, while aerobic processes occur at the secondary treatment stage.

References

Allhands, M. N., S. A. Allck, A. R. Overman, W. G. Leseman, and W. Vidak. 1989. *Municipal water reuse at Tallahassee.* Florida, Trans-ASAE, 38:411–418.

Beran, G. W., ed. 1994. *Handbook of Zoonoses,* 2nd ed., Boca Raton, FL: CRC Press.

Culotta, E. 1992. Red menace in the world's oceans. *Science* 257:1476–1477.

Eliassen, R., and G. Tchobanoglous. 1969. The indirect cycle of water reuse. *Environ. Sci. Tech.* 3:536–541.

Ellis, K. V. 1983. Stabilization ponds—design and operation. *Critical Reviews in Environmental Control* 13(2):69–102.

Elmitwalli, T. A., J. Soellner, A. De Keizer, H. Bruning, G. Zeeman, and G. Lettinga. 2001. Biodegradability and change of physical characteristics of particles during anaerobic digestion of domestic sewage. *Wat. Res.* 35(5):1311–1317.

Garry, V. F., D. Schreinemachers, M. E. Harkins, and J. Griffith. 1996. Pesticide appliers, biocides, and birth defects in rural Minnesota. *Environ. Health Perspect.* 104:394–399.

Laws, E. A. 1993. *Aquatic pollution. an introductory text,* 2nd ed. New York:Wiley & Sons.

Levine, A., G. Tehobanaglous, and T. Asano. 1985. Characterizations of size distribution of contaminants in wastewater: treatment and reuse implications. *J. Water Poll. Control Fed.* 57:805.

McCarty, P. L. 1981. In *Anaerobic digestion,* eds. D. E. Hughes, D.A. Stafford, Badder, and D. Andrew, Amsterdam: Elsevier.

McGrath, S. P., A. M. Chaudri, and K. E. Giller. 1995. Long-term effects of metals in sewage sludge on soils, microorganisms and plants. *J. Ind. Microbiol.* 14:94–104.

Murray, C. J. L., and A. D. Lopez. 1996. Evidence-based health policy lessons from the global burden of disease study. *Science* 274:740–743.

Narang, H. 1996. Origin and implications of bovine spongiform encephalopathy. *Proc. Soc. Exp. Biol. Med.* 211:306–322.

Outwater, A. 1996.*Water: A Natural History.* New York: Basic Books.

Pressley, T. A., D. F. Bishop, A. P. Pinto, and A. F. Cassel. 1973. Ammonia-nitrogen removal by breakpoint chlorination. *US-EPA (670/2-73-058) Workbook.*

Schindler, D. W. 1974. Eutrophication and recover in experimental lakes: implications for lake management. *Science* 164:897–899.
Seghezzo, L., G. Zeeman, J. B Van Lier, and G. Lettinga. 1998. A review: the anaerobic treatment of sewage in UASB and EGSB reactors. *Bioresource Technol.* 65:175–190.
Sopper, W. E., and E. M. Seaker. 1984. Strip mine reclamation with municipal sludge. *US-EPA, Municipal Environ. Res. Lab.* EPA-600/S2-84-035. Workbook.
Tenover, F. C. 1995. The best of times, the worst of times. The global challenge of antimicrobial resistance. *Pharm. World. Sci.* 17(5):149–151.
U.S. EPA. 1980. *Groundwater protection.* Washington, DC: U.S. Environmental Protection Agency.
Xu, G., P. Song, and P. I. Reed. 1992. The relationship between gastric mucosal changes and nitrate intake via drinking water in a high-risk population for gastric cancer in Moping county, China. *Eur. J. Cancer. Prev.* 1:437–443.
Zeeman, G., T. A. Elmitwalli, J. Secllner, A. De Keizer, H. Burning, and G. Lettinga. 2001. *Water Res.* 35(5): 1311–1317.

CHAPTER 28

Groundwater Decontamination and Treatment

Introduction

A large amount of the available freshwater on earth lies underground. As one digs into the ground, first an initial belt of soil moisture is encountered. Below that, soil along with a thin film of water and air is encountered, which is called the aeration or unsaturated zone. Then the saturated zone in which the water has displaced all the air is encountered. We find the water table here. Groundwater is the name given to the freshwater in the saturated zone (see Fig: 28-1). Nearly 30 to 35% of the world's total drinking water supply is from this groundwater.

Pollutants in the Groundwater

Industrialization has brought in its wake the problem of waste disposal. Lack of proper planning in siting of industrial units, inadequate development of infrastructure, and lack of waste management facilities have resulted in contamination of surface water bodies and groundwater aquifers. The collective discharges from industries, municipalities, small industries, and farms is one source of organic pollutants (nitrates, harmful bacteria and viruses, detergents, and household cleaners) in groundwater. Much of the groundwater pollution is blamed on intensive farming practices that depend on increased use of chemical fertilizers and pesticides. It has been estimated that 35 to 45% of chemical fertilizers leach into the groundwater in the form of nitrates. Gasoline enters the soil via surface spills, underground storage tank leaks, and pipeline ruptures. Once they descend to groundwater, the water soluble components are preferentially leached into water and can migrate rapidly in the dissolved state. The insoluble (BTY, MTBE, and chlorinated solvents) components form a plume (a liquid blob). Very slowly—in a

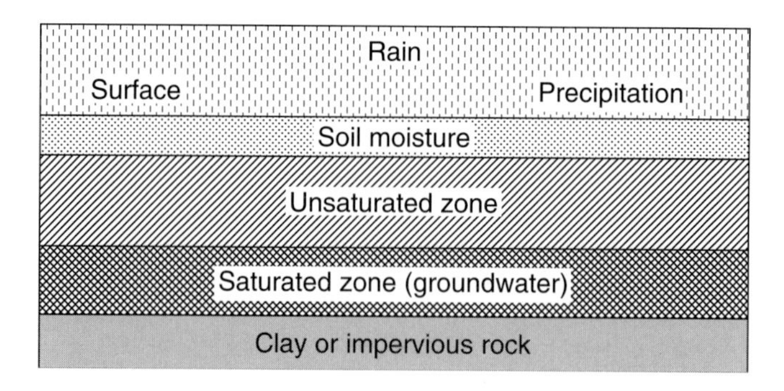

FIGURE 28-1. Regions in the soil.

process that often takes decades or centuries—these poorly soluble compounds gradually dissolve in the water that passes over the blob, and so there is a continuous supply of these organic pollutants to the groundwater. The most common inorganic contaminant in groundwater is the nitrate ion (NO_3^-), which occurs in both rural and suburban aquifers. It has been estimated that 35 to 45% of chemical fertilizers leach into the groundwater in the form of nitrates. Although uncontaminated groundwater generally has nitrogen levels (as nitrate) of less than 2 ppm, shallow aquifers have nitrate levels exceeding 10 ppm. The inorganic contaminants of greatest concern in groundwater are fluoride and arsenic. Although they are not common worldwide, in certain parts of India, Bangladesh, and Africa, they pose a major health hazard. It is proposed that they are inherent in the type of soil and leach into the groundwater from the rocks adjacent to the aquifers. Studies at Cornell University have established that 99.9% of the pesticides sprayed (whether in developed or the developing countries) go into the environment, with only 0.01% of pesticides reaching the target pest. The report by North Carolina Pesticide Board states that a total of 36 chemicals were found in wells. Of those, 31 were pesticides or pesticide breakdown products. The various contaminants of groundwater are summarized in Table 28-1.

Treatment

The type and quantity of pollutants in water (groundwater) vary from place to place. Therefore, the processes used for purification also vary from place to place. Although various methods are in vogue, most of these have common stages, such as:

- Aeration
- Settling and precipitation
- Hardness removal

TABLE 28-1
Contaminants of Groundwater

Pollutant	*Aerobic degradation*	*Anaerobic degradation*
Organic		
Most common		
Chloroform	—	✓
Bromodichloromethane	✓	✓
Dibromochloromethane	✓	✓
Bromoform	✓	✓
Less common		
Trichloroethene	—	✓
Tetrachloroethene	—	✓
1,1,1-Trichloroethene	—	✓
1,2-Dichloroethene	✓	✓
1,1-Dichloroethene	✓	✓
Carbon tetrachloride	—	✓
Dichloroiodomethane	✓	✓
Xylenes	✓	✓
Benzene	✓	
Toluene	✓	✓
Present at wells close to hazardous waste sites		
Methylene chloride	—	✓
Ethylbenzene	✓	✓
Acetone	✓	—
1,1-Dichloroethene	✓	✓
1,2-Dichloroethane	✓	✓
Vinyl chloride	—	✓
Methyl ethyl ketone	✓	—
Chlorobenzene	—	✓
1,1,2-Trichloroethane	✓	✓
Fluorotrichloromethane	—	✓
1,1,2,2-Tetrachloroethane	—	✓
Methyl isobutyl ketone	✓	—
Relatively common inorganic contaminants		
Nitrate ions	—	✓
Fluoride ions	✓	✓
Arsenic	Chemical	✓
Calcium ions	Chemical	—
Magnesium ions	Chemical	—
Ferrous ions	✓	✓
Phosphate ions	Chemical	—
Various metal ions (Hg, Cd, Pb, Sn)	Chemical, ✓	✓
Microorganisms		
Harmful bacteria	Chemical/UV	—
Harmful viruses	Chemical/UV	—

- Disinfection
- Purification

Of these various stages, the different methods vary from one another in adopting different disinfection procedures. For disinfection the common procedures are:

- Chlorination
- Ozone treatment
- Chlorine dioxide treatment
- Ultraviolet light treatment

Biotreatment

Recently, there has been commendable progress reported in using bioremediation to cleanse water contaminated by organic pollutants (gasoline and chlorinated solvents). It is advisable that even when bioremediation is used, the first three stages—aeration, settling, and hardness removal—be done routinely. Both aerobic and anaerobic organisms are effective in cleansing the water. Aerobic organisms are preferred for benzene, toluene, ethylbenzene, PAHs, and other aromatic systems, while anaerobic organisms are preferred for polychlorinated aliphatic compounds (PCE, TCE, DCM, and TCM) and polychlorinated aromatic compounds (dioxins and PCBs). However, the choice of the organism will depend not only on the substrate, but also on the other contaminants present. Since biodegradation is basically oxidation of the substrate, there has to be a terminal electron acceptor (TEA) in the system. Under aerobic conditions it is oxygen; under anaerobic conditions it could be nitrate, sulfate, iron (III), manganese (IV), or carbon dioxide.

Of all the petroleum hydrocarbons, benzene is considered the most problematic because of its high toxicity, relatively high aqueous solubility, and stability under anaerobic conditions. Although recent studies indicate that benzene is degraded anaerobically by a few organisms, its degradation is competitively inhibited by the presence of more readily degraded compounds such as toluene and xylene. Aquifers are often anoxic. In the absence of dissolved oxygen in groundwater, benzene degradation rates decrease or can stop altogether. The aerobic degradation of benzene (via catechol) is well established. The application of oxygen to anoxic soil, sediments, and groundwater is possible by using biopiles: injecting O_2, air, aerated water, or hydrogen peroxide or chlorite. All of these are intrusive and, therefore, relatively expensive measures. Hence, where feasible, monitored natural attenuation (intrinsic bioremediation) is likely to remain the most widespread remediation technique for petroleum-contaminated aquifers. Natural attenuation encompasses a host of physical processes—dispersion, dilution, sorption, and volatilization—as well as chemical and biological degradation. There have been reports of degradation of benzene under anoxic conditions by organisms belonging to:

- Two strains of the genus *Dechloromonas* (with nitrate as the sole electron acceptor)
- A community including members of the genus *Geobacter*

Similarly, members of the genus *Nocardiodes* were reported to degrade phenanthrene and members of the genus *Burkholderia* were reported to degrade dinitro toluene (Johnson et al., 2001). Toluene is readily degraded by both aerobic and anaerobic organisms; some of the organisms reported to degrade toluene are:

- *Pseudomonas putida*
- Nitrate-reducing genera *Azoarcus* and *Thauera*
- Iron-reducing genera *Geobacter metallireducens*.

The generalized biodegradation pathway of benzene and related contaminants by both aerobic and anaerobic pathways is summarized in Fig. 28-2 (Nyer, 1992).

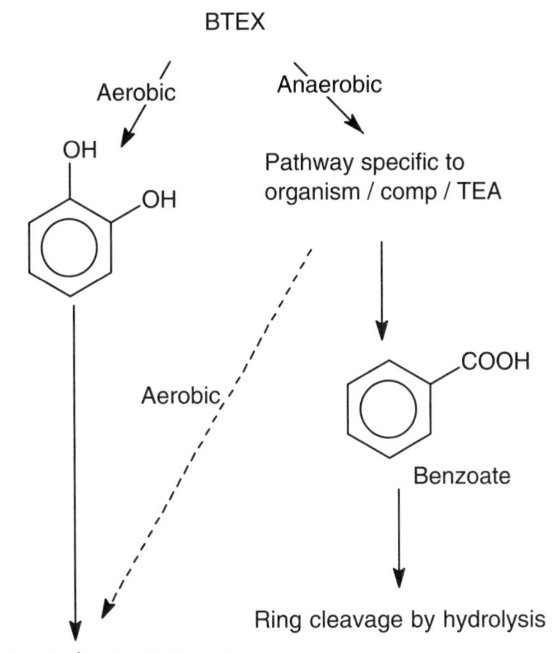

FIGURE 28-2. Generalized benzene toluene ethylbenzene and xylene (BTEX) biodegradation pathway.

Phytoremediation

Plant materials were found useful in the decontamination of water polluted with phenolic compounds. Enzymes exuded by roots of some families of *Fabaceae, Gramineae,* and *Solanaceae* release oxidoreductase, which takes part in the oxidative degradation of certain soil constituents. Horseradish-mediated removal of 2,4-dichlorophenol in a model solution was comparable with that achieved using purified horseradish peroxidase. In addition, horseradish could be reused up to 30 times. Because of the apparent ease of application, the use of plant material may present a breakthrough in the enzyme treatment of contaminated water.

Heavy metals are among the most dangerous substances in the environment because of their high level of durability and harmfulness to living organisms. Mercury has a high propensity to accumulate in organisms; the most harmful are organic compounds of Hg, especially in water (Henry, 2000). Chromium is biologically inactive in a metallic state. Organisms weakly absorb Cr (III), but Cr (VI) is more dangerous because its compounds easily penetrate physiological barriers. Phytoremediation is one of the ways to solve the problem of heavy metal pollution using plants. In the process of phytoremediation, pollutants are collected by plant roots and either decomposed to less harmful forms or accumulated in the plant tissues.

Phytoremediation is used to clean up waters, soils, slimes, and sediments from pesticides, PAHs, fuels, explosives, organic solvents, chemical manures, heavy metals, and radioactive contaminants. There are many plants that can bind heavy metals, and they are called "hyperaccumulators" (Adams et al., 2000). Table 28-2 lists some of the plants that are hyperaccumulators of chromium and mercury. *Azolla caroliniana* has been tested as a biofilter to purify water and to remove nitrogen and phosphorous, elements that cause water eutrophication. It can also remove sulfa drugs (Forni et al., 2001) and metals such as Sr, Cu, Cd, Cr, Ni, Pb, Au, and Pt and even radioactive elements as U (Zhao and Duncan, 1997). Bennicelli et al. (2004) showed that *A. caroliniana* accumulates Hg (II), Cr (III), and Cr (VI).

TABLE 28-2
Hyperaccumulators of Chromium and Mercury

Metal ions	
Cr (III)	*Hg (II)*
Dicoma niccolifera	*Arabidopsis thaliana*
Sutera fodina	*Nicotiana tobacum* (tobacco)
Pearsonia metallifera	*Liriodendro tulipifera*
Berkheya coddii	*Salix* spp.
Solanum elaeagnifolium	*Azolla caroliniana*
Azolla caroliniana	

Contaminated subsurface aquifers frequently accompany soil contamination. Bioremediation of groundwater resources presents unique problems and risks. Among the most obvious of these problems are that groundwater is mobile, whereas soil is generally stationary, and that people and livestock frequently drink untreated groundwater. Thus, there will often be an additional urgency factor associated with groundwater cleanup that may justify more drastic and expensive measures.

The usual approach in remediation of contaminated aquifers is groundwater pumping and surface treatment to eliminate the water-soluble wastes. The treated water is then recharged into the aquifer via one or more injection wells at some point up gradient to the contaminated zone. Pump-and-treat operations can incorporate bioremediation in at least two ways. The most obvious method uses biological (bioreactor) surface treatment, but like any pump-and-treat approach, this method is only able to degrade wastes in the mobile, aqueous phase. It is important to recognize that many organic wastes have low water solubilities, and aquifer-associated soils will often contain larger volumes of organic wastes than the water itself. The conviction that pump-and-treat measures can be effective has led to an appreciable effort in this direction at numerous hazardous waste sites. However, the goal of remediating aquifers to drinking water standards by such techniques may be unrealistic in many, if not most, cases. Contamination levels at remediation sites are typically two to three orders of magnitude above allowable drinking water limits. Based on past experience gained in pumping and treating contaminated aquifers, treatment typically drops pollutant concentrations by a factor of 2 to 10, which then level out with no further decline. Cessation of pumping is often followed by a rebound in aquifer waste concentrations. The problem is largely that sites are typically contaminated with organic wastes that do not readily dissolve in the aqueous phase. The waste either remains adsorbed to the soil matrix, floats on the top, or sinks to the bottom of the water table. Therefore, wastes only slowly seep into the groundwater at a diffusion-limited rate and cannot be significantly changed by groundwater pumping. Pump-and-treat measures may dramatically reduce pollutant concentration in the aqueous phase of the aquifer; when the pumps are switched off, however, pollutants gradually leach out of the soil and the aqueous concentration rises again. Many leading hydrologists have concluded that hundreds to thousands of years of pumping could be required to purge some contaminated aquifers of their organic waste contaminants. The implication is that although pump-and-treat measures may be useful to limit dispersal of a waste plume into the water table, massive excavation of soil is usually required to remove the source of the problem.

A more recently developed bioremediation approach to water treatment is subsurface in situ remediation. The treated water can be nutrient- and oxygen-enriched prior to recharge, stimulating aerobic biodegradation of soil-bound, water-insoluble wastes by indigenous soil microorganisms. The actual oxygen content of the water can be boosted by air pumps, or alternative oxygen sources such as hydrogen peroxide may be added. Surfactants and other organic waste desorbing chemicals can also be added to increase waste bioavailability. If a surface bioreactor is used, some portion of the active microbial biomass can be recharged with the water, providing continuous inoculation of the contaminated aquifer and soil. Although stimulation of aerobic metabolism is the objective of most systems, the reinjected groundwater can be enriched with nitrate to stimulate growth and enhance the biodegradative action of anaerobic denitrifying microbes. Recently, this approach has proved effective in degrading the various organic constituents of gasoline, with toxicity reductions comparable to those seen in aerobic degradation (Carroquino et al., 1992). As with in situ soil treatment, the success of subsurface aquifer bioremediation is largely determined by waste and soil characteristics. Soil permeability is especially important to the success of nutrient enrichment and inoculation efforts.

Over the years, a number of lakes and rivers have become seriously contaminated with various industrial wastes in many parts of the world. In some cases, the sources of pollution have been reduced or even eliminated. Public demand for remediation to a condition safe for fishing and other recreational uses is growing. Unfortunately, technologies for surface water remediation are not nearly as well developed as those for soil or even groundwater. Part of the problem lies in the size of many bodies of water. It is technically and environmentally impractical to divert a large flowing body of water from its course for treatment. Also, as with underground aquifers, the water contamination problem is largely a sediment contamination problem. Many persistent wastes become tightly bound to bottom sediments from which they slowly leach out and thus cannot readily be removed by water treatment. Conventional treatment typically involves dredging and removing bottom sediment in the most polluted areas, but such measures can themselves be environmentally devastating and there is a risk of remobilizing toxicants accumulated over many years. Furthermore, the excavated sediment must still be treated and/or disposed of as toxic waste. Workers are turning to in situ bioremediation almost as a last resort.

Surface-water bioremediation technologies are largely being developed in place. An ongoing example is the General Electric (GE) site in Fort Edward, NY, on the upper Hudson River. For many years GE legally released PCBs into the river from a plant that manufactured capacitors. When PCBs became priority environmental pollutants in the early 1980s, at least 20 miles of the river bottom were found to be contaminated downstream of the plant. GE began looking for remediation options. In 1991, GE conducted an extensive field research program to characterize natural degradation of PCBs at this site (General Electric Company, 1992) and discovered that the indigenous consortia of microorganisms was exceptionally good at degrading PCBs. Presumably, since the PCBs have been present in this site for a significant period (at least 35 years), the indigenous microorganisms have adapted to utilize the material as a food source. Both anaerobic and aerobic biodegradation have been identified as part of the natural process of remediation, which can be slow. Field tests in cylindrical caissons sunk into the river sediment at this site have identified the variables that can be manipulated to enhance in situ biodegradation of PCBs. The addition of inorganic nutrients, the organic cometabolite biphenyl, and oxygen significantly increased PCB degradation rates. Addition of selected PCB-degrading bacterial cultures did not dramatically improve biodegradative efficiency. No more than 60% of the PCBs was degraded in any laboratory or field experiments, a finding attributed to tight sediment adsorption of the least water-soluble PCB compounds (Harkness et al., 1993). More information on degradation rates, products, and variability under natural conditions is required for a realistic evaluation of the role that bioremediation may play in this and other surface water sites contaminated by organic waste.

Conclusion

Groundwater, the most important source of drinking water, must be effectively and efficiently purified to ensure good health. Bioremediation is the most suitable method for the degradation of pollutants because other methods either involve elaborate expensive procedures or give rise to incompletely transformed products. Anaerobic treatment followed by aerobic treatment will ensure complete mineralization of all pollutants. Ideally, aquifers would be inoculated with nonpathogenic bacteria that function under anoxic conditions. An aerobic treatment after pumping the groundwater will clean up the rest of the pollutants and their anaerobic transformed products.

References

Adams, N., D. Carroll, K. Madalinski, S. Rock, T. Wilson, and B. Pivetz. 2000. *Introduction to Phytoremediation*, National Risk Management Research Laboratory, Office of Research and Development, U. S. Environmental Protection Agency, EPA/600/R-99/107.

Bennicelli, R., Z. Stepniewska, A. Benach, K. Szajnocha, and J. Ostrowski. 2004. The ability of *Azolla caroliniana* to remove heavy metals (Hg(II), Cr(III), Cr(VI)) from municipal wastewater. *Chemosphere* 55:141–146.

Carroquino, M. J., R. M. Gersberg, W. J. Dawsey, and M. D. Bradley. 1992. Toxicity reduction associated with bioremediation of gasoline-contaminated groundwaters. *Bull. Environ. Contam. Toxicol.* 49:224–231.

Forni, C., A. Cascone, S. Cozzolino, and L. Migliore. 2001. The duckweed *Lemna minor, L.* is another free-floating aquatic plant known to accumulate heavy metals. *Minerva Biotechnol.* 13:151–152.

General Electric Company. 1992. *A field study on biodegradation of PCBs in Hudson River sediment.* Final Report, Schenectady, NY: General Electric Company.

Harkness, M. R., J. B. McDermott, D. A. Abramowicz, J. J. Salvo, W. P. Flanagan, M. L. Stephens, F. J. Mondello, R. J. May, J. H. Lobos, and K. M. Carroll. 1993. In situ stimulation of aerobic degradation of PCB biodegradation in Hudson River sediments. *Science* 259:503–507.

Henry, J. R. 2000. An overview of the phytoremediation of lead and mercury, *National. Network of Environmental Management Studies* (NNEMS), U. S. EPA, Solid Waste and Emergency Reponse, Technology Innovation Office.

Johnson, S. J., K. J. Woolhouse, H. Prommer, D. A. Barry, and N. Ghristofi. 2001. Contribution of anaerobic microbial activity to natural attenuation of benzene in groundwater. *Eng. Geol.* 70:343–349.

Nyer, E. K. 1992. *Groundwater treatment technology,* 2nd ed. New York: Van Nostrand Reinhold.

Zhao, M., and J. R. Duncan. 1997. Batch removal of sexivalent chromium by *Azolla filiculoides. Biotech. Appl. Biochem.* 26:172-179.

Bibliography

Colin Baird, 1999. *Environmental chemistry,* New York: W. H. Freeman and Company.

CHAPTER 29

Denitrification

Introduction

Nitrogen occurs in natural waters in organic and inorganic forms. There are several environmentally important forms of nitrogen that differ in the extent of oxidation of the nitrogen atom. Nitrate ion (NO_3^-) is the most oxidized form, while ammonia (NH_3) and the ammonium ion (NH_4^+) are the most reduced forms. The common oxidation states of nitrogen occurring in nature are illustrated in Table 29-1.

Nitrate and nitrite ions in drinking water are a potential health hazard because they can result in methemoglobinemia. The nitrite combines with and oxidizes the hemoglobin in blood, thereby leading to respiratory failure. Nitrate ion is also implicated in stomach cancer.

Occurrence

The main source of nitrate ion is the runoff from agricultural land. Fertilizers are the major source of nitrates. Nitrate-bearing wastes result from production of fertilizers, explosives, nitro-organic compounds, and pharmaceuticals (Pinar et al., 1997). Other industries such as nuclear fuel processing use significant amounts of nitric acid. While these industries do not produce nitrogen-containing products, they can generate significant volumes of nitrate-bearing waste streams. In response to this problem, nitrate-containing compounds were added to the U.S. Environmental Protection Agency (U.S. EPA) Toxic Release Inventory in 1995.

Biotreatment

A commonly used method for nitrate removal is ion exchange separation (Kapoor and Viraraghavan, 1997). Regeneration of ion exchange resins with sodium chloride produces brine containing high concentrations of nitrate that can be difficult to remove using standard biological, physical, or chemical technologies. It was observed that *Halomonas campisalis* completely

TABLE 29-1
Common Oxidation States of Nitrogen Occurring in Nature

Oxidation state of N						
−3	0	+1	+2	+3	+4	+5
NH_4^+ (aq)				NO_2^- (aq)		NO_3^- (aq)
NH_3 (aq)						
NH_3 (g)	N_2 (g)	N_2O (g)	NO (g)		NO_2 (g)	

reduced nitrate even at a concentration of 125 g/L NaCl and pH 9 (Peyton et al., 2001). Microorganisms oxidize ammonium ions and nitrogen to nitrate ions, which are suitable for plant uptake; the process is termed "nitrification." In the reverse process, microorganisms catalyze the reduction of nitrate and nitrite to nitrogen—termed "denitrification." Both these processes are important in soils and natural waters. Biological denitrification is the microbe-catalyzed transformation of nitrate to nitrogen gas via several intermediate compounds (NO_2^-, NO, and N_2O). This is an attractive treatment option because the nitrate ion is converted by the denitrifying bacteria to inert nitrogen gas, and the waste product usually contains only biological solids. Denitrification is a respiratory process in which an electron donor is needed as an energy source; the reduction of nitrate to nitrogen involves the transfer of electrons from some source to nitrate ions. Electrons originating from organic matter, reduced sulfur compounds, or molecular hydrogen are transferred to oxidized nitrogen compounds instead of oxygen to build up a proton motive force usable for ATP regeneration. The enzymes involved are nitrate reductase, nitrite reductase, nitric oxide reductase, and finally nitrous oxide reductase. Dinitrogen (nitrogen gas) is the main end product of denitrification, while the nitrogenous gases occur as intermediates at low concentrations. Denitrification also occurs in the presence of oxygen. The range of oxygen concentrations permitting aerobic denitrification is broad and differs from one organism to another. Denitrifying bacteria are usually heterotrophic and need organic carbon for the reduction of nitrate to nitrogen gas. In heterotrophic denitrification, if the wastewater to be treated presents a deficit of electron donors, an exogenous carbon source, such as alcohols, glucose, acetate, or lactate, must be added (Knowles, 1982). The literature suggests that methane can be used as the sole electron donor in denitrification (Islas-Lima et al., 2004). Methane is a low-cost electron donor frequently available from wastewater treatment plants. Carrera et al. (2003) studied the denitrification of real industrial wastewater with a concentration of 5,000 mg N–NH_4^+/L using a two-sludge system. They achieved complete denitrification using two different carbon sources, one containing ethanol and the other one methanol. The maximum denitrification rate (MDR) reached with ethanol (0.64 g N–NOx^- g/VSS day) was about six times higher than the MDR reached with methanol (0.11 g N–NOx$^-$ g/VSS day) (Carrera et al., 2003). The denitrification process is performed by various chemoorganotrophic, lithoautotrophic, and phototrophic bacteria and some fungi, especially under reduced oxygen or anoxic conditions.

Many modifications and processes have been developed and implemented for nitrogen removal from wastewaters. Basically, these processes for nitrogen removal (from wastewaters) can be classified as:

- Suspended sludge
- Fixed-film cultures
- Aerobic granulation

Nitrification is a necessary first step in nitrogen removal when the nitrogen is present in the reduced form (ammonia or ammonium ion). The key enzyme of nitrite oxidizing bacteria is the membrane-bound nitrite oxidoreductase, which oxidizes nitrite with water as the source of oxygen to form nitrate. The electrons released from this reaction are transferred via a- and c-type cytochromes to a cytochrome oxidase of the aa_3-type (Hollocher et al., 1982). The proteobacterial ammonia oxidizers obtain their energy for growth from both aerobic and anaerobic ammonia oxidation. The main products are nitrite under oxic conditions and dinitrogen, nitrite, and nitric oxide under anoxic conditions (Nold et al., 2000). Aerobic and anaerobic ammonia oxidation is initiated by the enzyme ammonia monooxygenase (AMO), which oxidizes ammonia to hydroxylamine. This is further oxidized to nitrite by hydroxylamine oxidoreductase (HAO). The four reducing equivalents derived from this reaction (Fig. 29-1) enter the AMO reaction, the CO_2 assimilation, and the respiratory chain. The reducing equivivalents are transferred to the terminal electron acceptor O_2 under oxic conditions, or nitrite under anoxic conditions. The reduction of nitrite under anoxic conditions leads to the formation of nitrogen (N_2), resulting in a nitrogen loss of about 45±15% (Phillips et al., 2000). Under oxic conditions aerobic nitrifiers convert ammonia to nitrite. At lower concentrations of oxygen (less than 0.8 mg O_2/L), they use small amounts of the nitrite produced as terminal electron acceptors, producing NO, NO_2, and N_2. In the absence of nitrogen oxides, up to 15% of the converted ammonia can be denitrified (Prosser, 1989). *Nitrosomonas eutropha* was shown to nitrify and simultaneously denitrify under

$$NH_3 + O_2 + 2H^+ + 2e^- \longrightarrow NH_2OH + H_2O \; [G^{o\prime} -120 \text{ kJ mol}^{-1}]$$

$$NH_3 + N_2O_4 + 2H^+ + 2e^- \longrightarrow NH_2OH + 2NO + H_2O \; [G^{o\prime} -140 \text{ kJ mol}^{-1}]$$

$$NH_2OH + H_2O \longrightarrow HNO_2 + 4H^+ + 4e^- \; [G^{o\prime} -289 \text{ kJ mol}^{-1}].$$

FIGURE 29-1. AMO catalyzed ammonia oxidation.

fully oxic conditions in the presence of NO_2 and NO. The presence of the nitrogen oxides enhances the conversion of ammonia to nitrogen gas. Influenced by nitrogen oxides, ammonia oxidizers convert ammonia to gaseous dinitrogen (about 60% of the converted ammonia) and nitrite. Some strains of *Nitrobacter* were shown to be denitrifying organisms as well. The anaerobic oxidation of ammonia proceeds via hydrazine, a volatile and toxic intermediate. An enzyme that resembles HAO from aerobic ammonia oxidizers is responsible for the oxidation of hydrazine to nitrogen gas. Today, with newly discovered anaerobic ammonia oxidizing organisms (*planctomycetes*), the preconversion of ammonia to nitrate is not necessary for the denitrification of ammonia. Partial nitrification needs less aeration, so the subsequent denitrification consumes less chemical oxygen demand (COD), since only nitrite and not nitrate must be reduced to molecular nitrogen. This is cost effective if the low C:N ratio of the wastewater necessitates the addition of a synthetic electron donor such as methanol. In this case, the process also emits less CO_2 to the atmosphere.

The anaerobic ammonia oxidation (anammox) process is the denitrification of nitrite with ammonia as the electron donor (Strous et al., 1997). This is the aesthetically most satisfying process because both the nitrogenous pollutants, ammonia and nitrite, are consumed in one process. Anammox needs a preceding partial nitrification step that converts half of the ammonium in wastewater to nitrite. This anammox process is mediated by a group of planctomycete bacteria (*Candidatus* sp.). Conceptually and practically, removal of nitrogen pollutants can be brought about by:

- Complete autotrophic nitrogen removal over nitrite—the two groups of bacteria cooperate and perform two sequential reactions simultaneously. The nitrifiers oxidize ammonia to nitrite, consume oxygen, and so create the anoxic conditions that the anaerobic ammonia oxidation needs (Third et al., 2001).
- Controlling and stimulating the denitrification activity of *Nitrosomonas*-like microorganisms by adding nitrogen oxides offers new possibilities in wastewater treatment. In the presence of NOx, *Nitrosomonas*-like microorganisms nitrify and denitrify simultaneously even under fully oxic conditions, with N_2 as the main product (Poth, 1986). NOx (NO/NO_2) is the regulatory signal inducing the denitrification activity of the ammonia oxidizers, and it is only added in trace amounts (NH_4^+/NO_2 ratio about 1,000 to 5,000 per liter). As a consequence about 50% of the reducing equivalents [H] are transferred to nitrite as the terminal electron acceptor instead of oxygen.

Discovering the group of annamox microorganisms opened new techniques for nitrogen removal. Also the discovery of the versatility of aerobic ammonia oxidizers led to the development of new treatment processes, such as the addition of small amounts of NOx. In the future, the combination of different groups of nitrogen converting microorganisms and the optimization of the process management will improve nitrogen removal. One of the options is complete nitrogen removal by a mixed population of "aerobic" ammonia oxidizers and anammox bacteria under anoxic conditions in the presence of NO_2(Schmidt et al., 2003).

In an unusual study, Volokita et al. (1996) observed biological denitrification of drinking water using newspaper. Microbial denitrification of drinking water was studied in laboratory columns packed with shredded newspapers. Newspaper served as the sole carbon and energy substrate, as well as the only physical support for the microbial population. Complete removal of nitrate (100 mg/L) was readily achieved without accumulation of nitrite (Volokita et al., 1996).

Kesseru et al. (2003) immobilized *Pseudomonas butanovora* on composite beads and used them as fill in the reactor system. Using a continuous-flow bioreactor, they observed a nitrate removal efficiency of nearly 100%.

Conclusion

There have been a number of reports in the past on the various organisms that bring about denitrification; many reports also appeared in literature on the types of processes suitable for a particular type of nitrate waste. All the same, there is no single best process for nitrogen removal from wastewaters. In designing a process, consideration must be given to selection of the reactor type, loading criteria, sludge production, oxygen requirements and transfer, nutrient requirements, control of filamentous organisms, and effluent characteristics. Apart from these issues, the nature of wastewaters and local environmental conditions should also be considered.

References

Carrera, J., J. A. Baeza, T. Vicent, and J. Lafuente. 2003. Biological nitrogen removal of high-strength ammonium industrial wastewater with two-sludge system, *Water Res.* 37(17):4211–4221.

Hollocher, T. C., S. Kumar, and D. J. D. Nicholas. 1982. Respiration-dependent proton translocation in *Nitrosomonas europaea* and its apparent absence in *Nitrobacter agilis* during inorganic oxidations. *J. Bacteriol.* 149(3):1013–1020.

Islas-Lima, S., F. Thalaso, and J. Gomez Hernandez. 2004. Evidence of anoxic methane oxidation coupled to denitrification. *Water Res.* 38:13–16.

Kappor, A., and T. Viraraghavan. 1997. Nitrate removal from drinking water—a review. *J. Environ. Eng.* 123(4):371–380.

Kesseru, P., I. Kiss, Z. Bihari, and B. Polyok. 2003. Biological denitrification in a continuous-flow pilot bioreactor containing immobilized *Pseudomonas butanovora* cells. *Bioresource Technol.* 87(1):75–80.

Knowles, R. 1982. Denitrification. *Microbiol. Rev.* 46:43–70.

Nold, S. C., J. Z. Zhou, A. H. Devol, and J. M. Tiedje. 2000. Pacific northwest marine sediments contain ammonia-oxidizing bacteria in the subdivision of the *Proteobacteria. Appl. Environ. Microbiol.* 66:4532–4535.

Peyton, B. M., M. R. Mormile, and J. N. Petersen. 2001. Nitrate reduction with *Halomonas campisalis:* kinetics of denitrification at pH 9 and 12.5% NaCl. *Water Res.* 35(17):4237–4242.

Phillips, C. J., D. Harris, S. L. Dollhopf, K. L. Gross, J. I. Prosser, and E. A. Paul. 2000. Effects of agronomic treatments on the structure and function of ammonia oxidizing communities. *Appl. Environ. Microbiol.* 66:5410–5418.

Pinar, G., E. Duque, A. Haidour, J. Oliva, L. Sanchez-Barbero, V. Calvo, and J. L. Ramos. 1997. Removal of high concentrations of nitrate from industrial wastewaters by bacteria. *Appl. Environ. Microbiol.* 63:2071–2073.

Poth, M. 1986. Dinitrogen production from nitrite by a *Nitromonas* isolate. *Appl. Environ. Microbiol.* 52:957–959.

Prosser, J. I. 1989. Autotrophic nitrification in bacteria. *Adv. Microbiol. Physiol.* 30:125–181.

Schmidt, I., O. Sliekers, M. Schmid, E. Bock, J. Fuerst, J. G. Kuenen, M. S. M. Jetten, and M. Strous. 2003. New concepts of microbial treatment processes for the nitrogen removal in wastewater. *FEMS Microbiol. Rev.* 27:481–492.

Strous, M., E. Van Gerven, J. G. Kuenen, and M. S. M. Jetten. 1997. Effects of aerobic and microaerobic conditions on anaerobic ammonium-oxidizing (Anammox) sludge. *Appl. Environ. Microbiol.* 63:2446–2448.

Third, K. A., A. O. Sliekers, J. G. Kuenen, and M. M. Jetten. 2001. The CANON system (completely autotrophic nitrogen-removal over nitrate) under ammonium limitation: interaction and competition between three groups of bacteria. *Syst. Appl. Microbiol.* 24:588–596.

Volokita, M., S. Belkin, A. Abeliovich, and M. I. M. Soares. 1996. Biological denitrification of drinking water using newspaper. *Water Res.*, 30:965–971.

CHAPTER 30

Gaseous Pollutants and Volatile Organics

Sulfur dioxide, nitrogen oxides, volatile organic compounds (VOCs), and particulates are the four major components of air pollution and are the main causes of environmental damage and many diseases, including cancer. The sulfur and nitrogen oxides and particulates are the result of burning petroleum fuels, coal, wood, etc. Printing and coating facilities and foundries, as well as the electronics, petrochemical, metal finishing, and paint industries, produce VOCs, which include solvent thinners, degreasers, cleaners, lubricants, and liquid fuels. They originate from breathing and loading losses from storage tanks, venting of process vessels, and leaks from piping and equipment, wastewater streams, and heat exchange systems. A few common VOCs are methane, ethane, tetrachloroethane, methyl chloride, and various chlorohydrocarbons, perfluorocarbons, styrene, and naphthalenes. The European Community emissions limit is 35 g total organic compounds (TOC) per cubic meter of gasoline loaded (35 g TOC/m^3): the U.S. Environmental Protection Agency emission limit is 10 g TOC/m^3. Particulates from air can be removed using several well established physical methods that use a gravity settler, centrifugal collector, wet spray venturi collector, electrostatic precipitator, and fabric filter.

Physical Methods

Two general methods for abatement include recovery and destruction (see Fig. 30-1). The former method leads to reuse of the chemical and hence has a cost benefit. The latter method includes converting the chemical to a harmless product or into a liquid or solid pollutant that can be treated with another well established technology. Chemical, thermal, and biochemical approaches could be followed for destroying VOCs and pollutants from air, while absorption, adsorption, and cryogenic methods could be adopted for recovery and reuse of the VOCs. Chemical, catalytic, and thermal methods are very effective and well established but have several disadvantages

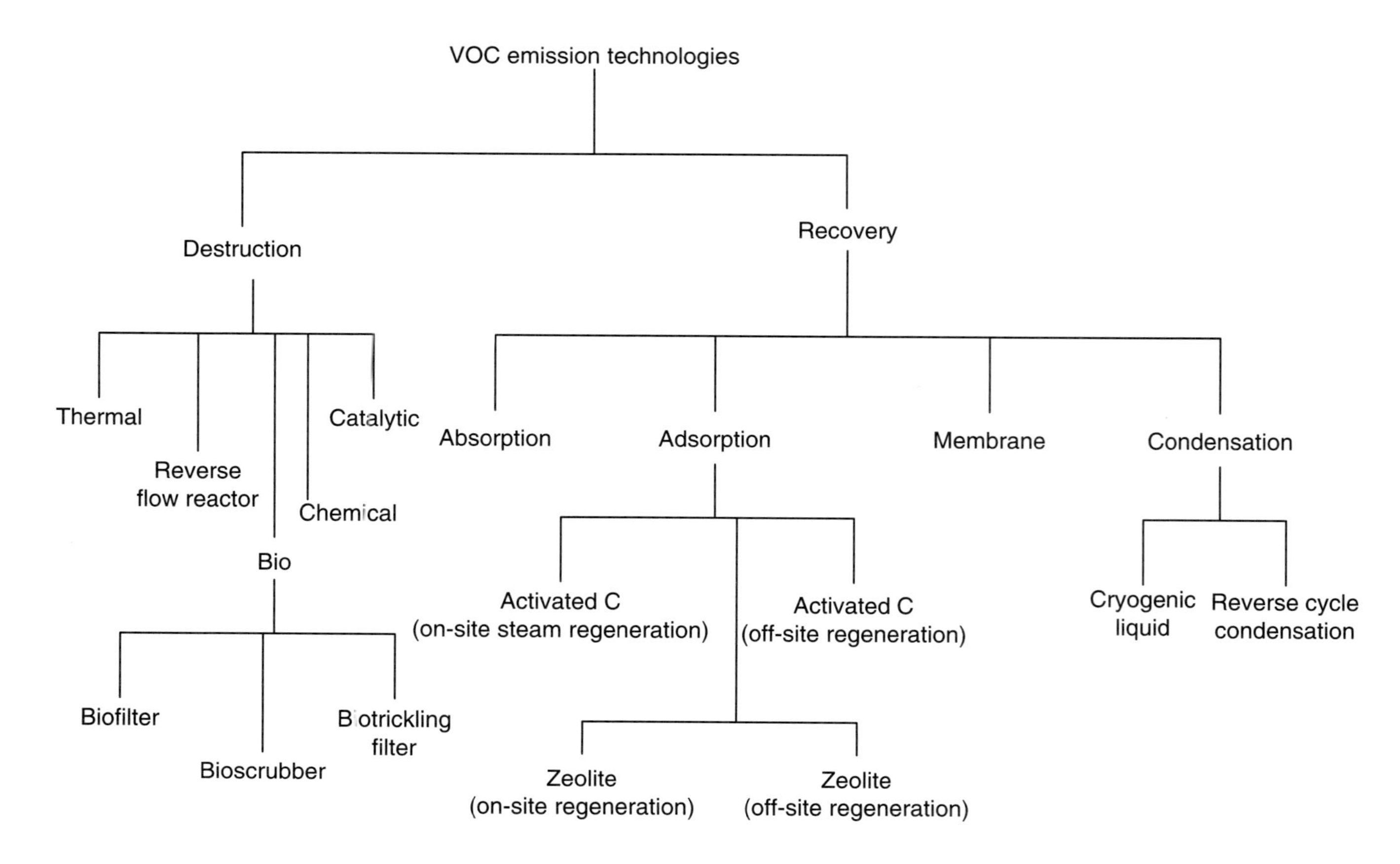

FIGURE 30-1. Various recovery and treatment procedures for VOCs.

such as high cost, conversion of one type of pollutant to another, and the possible generation of more toxic chemicals as the product. The physical methods are simple but need additional hardware for the regeneration of the absorbent or the adsorbent, and the recovery costs are high (Khan and Ghoshal, 2000). Recovery and reuse as solvent rather than burning as fuel is more economical in all cases (Spivey, 1988). Biofiltration is the cheapest and safest method, but can be slow and incomplete, and a colony of microorganisms is needed to treat a host of VOCs. The annual savings in operating cost when running a biofilter and a thermal oxidizer could be on the order of \$300,000 for a 85,000 m^3/h air stream containing 500 ppmv of VOC (Boswell, 2002). A typical biofilter, thermal oxidizer, and catalytic converter could cost \$280,000, \$300,000 to \$400,000, and \$325,000 to \$425,000, respectively. All these technologies are assumed to achieve a destruction and recovery efficiency of 90%. The absorption, adsorption, and biofiltration methods are operated at ambient conditions, but the first two methods need a high temperature operation to recover the adsorbent so it can be reused. The presence of moisture leads to a decrease in the efficiency of chemical methods.

Biotreatment Processes

Three basic aerobic methods are biofilter, biotrickling filter, and bioscrubber. The type of microorganism depends on the VOC that is being destroyed. All need 100% relative humidity, long contact times, and a community of microorganisms. The products of the process are CO_2, salts, water, and biomass. Moisture content, temperature, pH, nutrient amount, type of contaminants, presence of fine particles, and oxygen mass transfer rates play an important role in the biodegradation process. A warmer reactor can oxidize the contaminants faster, thereby increasing the destruction and the removal efficiency, but can also deactivate the sensitive microorganisms. The reactors can work over a wide range of pH conditions (from 2 to 9). Careful thought must be given if the contaminants are sulfur- or chlorine-containing compounds since acid is produced on destruction of such compounds, making the biomass highly acidic. If the biomass is too dry, growth of the microorganism stops and if too much water is present, washout of the biomass can occur. Also, shearing of the biofilm that has grown for a considerable amount of time is an issue.

Biofilter

A biofilter is a tube that is packed with material containing the microorganism, nutrients for its growth, and support material to hold the growing colony (Fig. 30-2). A nonbioactive humidification system maintains the moisture level. The support material prevents clogging of the reactor and also keeps the pressure drop low. A pre–particulate removal system is located before the biofilter. The conditioned gas stream is introduced from the bottom of

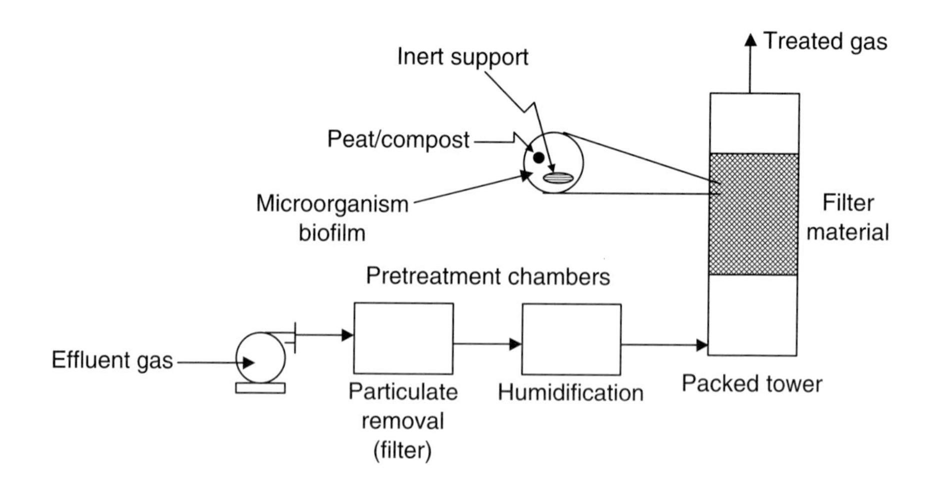

FIGURE 30-2. A typical biofilter.

a filter bed consisting of soil, peat, compost material, ceramic, calcium alginate, activated carbon, bark chips, wood chips, yard waste, or plastic. During the process, the contaminant may be adsorbed directly on the biofilm or dissolved in the aqueous film. Biofilters can also serve as odor preventers and can be installed at the exhaust side of waste treatment plants, sewer vents, etc. (Adler, 2001). Compared to the thermal oxidation process, which produces NOx and acid rain, biofilters are safer and cheaper to run. Bed drying, short circuiting, collapse of the packing, and blocking of the packed vessel are some issues that need to be addressed. Many VOCs, including ethanol, aldehydes, hydrogen sulfide, styrene, hydrocarbon solvents, and methyl methacrylate (MMA), have been successfully treated with biofilters, achieving 85 to 95% removal efficiencies. Efficiencies greater than 99.9% were achieved when the H_2S inlet concentrations were in the range 5 to 2,650 ppm. In laboratory experiments, removal of ethanol vapors was achieved using compost, granular activated carbon inoculated with different amounts of active biomass, and a mixture of compost and diatomaceous earth. Complete removal of H_2S from a stream containing 40 ppm in less than 30 s (empty bed residence time) has been reported in a biofilter packed to a height of 4 ft with synthetic inorganic media (hydrophilic mineral core) coated with hydrophobic material at 20-s residence time. Some of the microorganisms that have been found to sucessfully degrade VOCs and pollutants in air are listed in Table 30-1. Most of the microorganism found in the biofilters were bacteria that were predominantly coryneforms and endospore formers, and occasionally pseudomonads. Yeast and fungi are less abundant. In order to achieve a high degradation efficiency, an adaptation time for the microflora is necessary; during this time the organic loading is gradually increased.

TABLE 30-1
Microorganisms That Can Degrade Pollutants and VOCs in Air

Organism	*VOC/gas component*
Pseudomonas AM1	Methanol, formaldehyde
Pseudomonas aminovorans	Dimethyl amines
Pseudomonas putida,	Phenol
Trichosporon cutaneum	
Pseudomonas sp.	Benzene
Pseudomonas putida	Toluene
Sphingobacterium	
Turicella oritidis	
Bacteria + yeast	Xylene
Exophiala yeanselmei	Styrene
Rhodococcus sp.	Methyl ethyl ketone
Thiobacillus sp.	H_2S, methyl mercaptan, CS_2
Thiobacillus thioparus	
Pseudomonas putida	NH_3, H_2S
Arthrobacter oxydans	
Nocardia sp.	Aniline
Hypomicrobium sp.	Dimethyl sulfide
Pseudomonas fluorescence	*p*-Cresol
Pseudomonas sp.	*m*-Cresol
Chromobacterium violaceum	Indole
Bacteria	Methyl *tert* butyl ether

Exhaust air from paint booths is typically of high volume and contains low concentrations of VOCs. The energy costs for biofilters to carry out destruction of such pollutants are typically one-fourth to one-tenth the energy costs of thermal oxidation technologies, and the capital costs are about two-thirds to three-fourths that of competing technologies. Chlorinated solvents may produce acid, which would affect the growth of the microorganism. Nevertheless, biofilters are being used to treat such gases at VOC loadings of 500 to 1,500 ppm, achieving more than 85% reduction (Garner and Barton, 2002).

A mixture of benzene, toluene, ethyl benzene, and xylenes (BTEX) in air was effectively degraded in a biofilter packed with a mixture of compost (a mixture of yard waste and sewage sludge) and activated carbon (Abumaizar et al., 1998). The microorganisms preferentially utilized benzene, followed by toluene, ethylbenzene, and finally *o*-xylene. Removal efficiencies of greater than 90% were achieved for inlet concentrations of 200 ppm of each of the BTEX compounds and a gas loading rate of 17.6 $m^3/m^2 \cdot h$. The activated carbon helped in reducing the pressure drop in the bed and also acted as a buffer during shock pollutant loads. Because of the high adsorption capacity of activated carbon, the organic pollutants and oxygen got concentrated there, which in turn increased the growth of microorganisms in the vicinity.

Biotrickling Filter

A biotrickling filter also has packing material on which the microorganisms grow, but the water is made to trickle down the packing (about 1 to 20 m^3/m^2 day), while the gas is fed from the bottom (Fig. 30-3). The water is collected at the bottom and recycled back to the top of the column. Because of the flowing water, contaminants can be dissolved in it. The water phase is mobile here, whereas it is immobilized in the biofilter. The surface area for mass transfer is low in the latter, whereas it is high in the former (Soccol et al., 2003). Organic bedding material has several advantages over inorganic material, which include high absorbtivity, presence of nutrients, and better porosity. The microorganisms are immobilized on the filter packing by five different methods, including carrier binding, cross-linking, entrapment, microencapsulation, and membrane binding. Attached growth immobilization is found to be effective for treating fluids with several different contaminants, while the entrapment methods are suitable for few

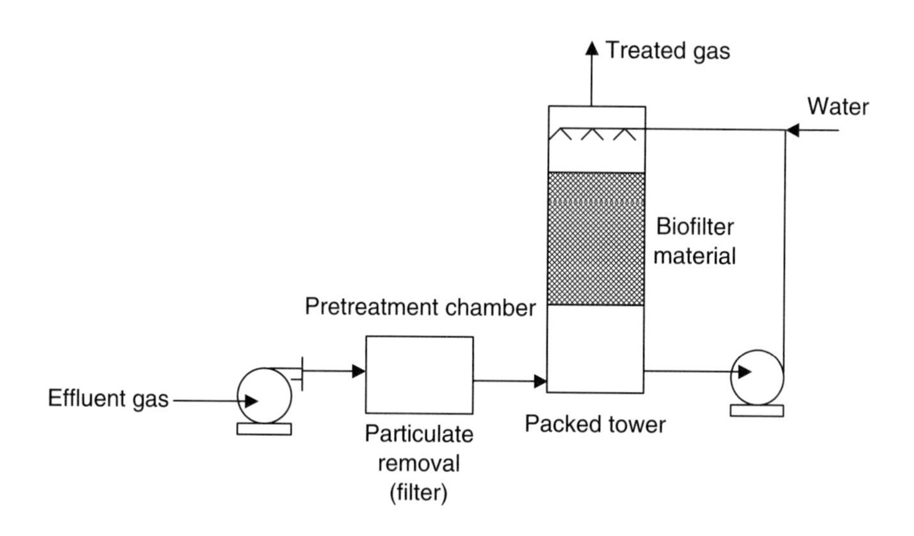

FIGURE 30-3. A typical biotrickling filter.

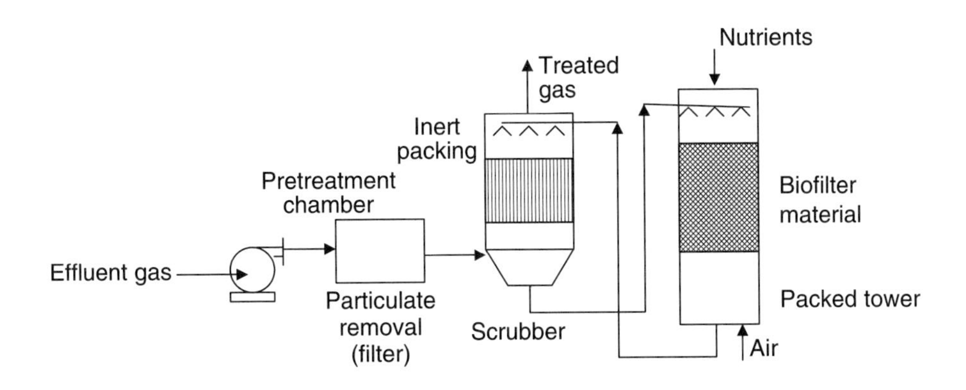

FIGURE 30-4. A typical bioscrubber.

contaminants (Cohen, 2001). The potential release of microorganisms to the atmosphere has been a concern for these technologies. Typical design parameters for these two filters are: a retention time in the range of 25 to 60 s, a typical reactor height around 0.5 to 1.5 m, an input concentration of about 0 to 1,000 ppmv VOC, waste air flow of 50 to 300,000 m^3/h, and an inlet oxygen concentration of 11 to 21%.

Bioscrubber

The third biooxidation process for treating VOCs is bioscrubbing, which consists of a twin reactor system that has water scrubbing and biooxidation vessels (Fig. 30-4). The microorganism is either suspended or attached onto a support, just like the previous cases in the biooxidation vessel. In the scrubber the VOCs and other gases get absorbed in the water medium, and are then destroyed in the second vessel by the microorganism, which is supplied with air and nutrient solutions. The water is recycled back to the scrubber.

Membrane Bioreactor

Membrane bioreactors combine membrane technology with biotechnology, where the membrane acts as a partition that separates the liquid and the gaseous medium (Fig. 30-5) (Reij et al., 1998). The microorganism grows on the liquid side of the membrane, which contains water and other nutrients required for its growth. The pollutants, gases, and oxygen approach the biofilm from the gas side after diffusing through the membrane and serve as the carbon source for the growth of the microbes. The membrane does not permit the microorganisms to pass through and contaminate the gas stream. The membrane is either hydrophobic, microporous, or dense. A hydrophobic

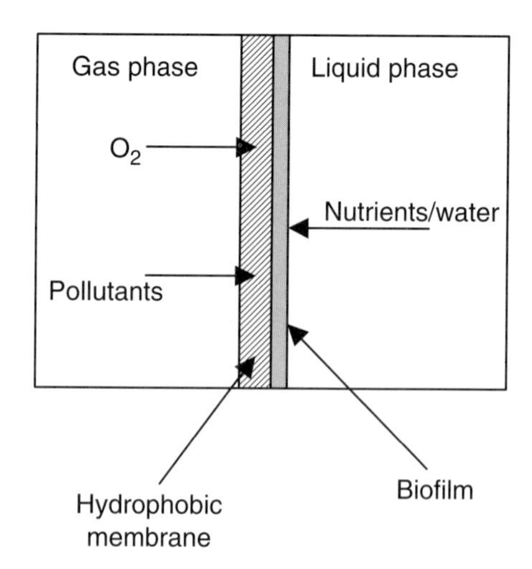

FIGURE 30-5. Membrane bioreactor.

membrane is a polymer matrix such as polypropylene or Teflon with pores of 0.01 to 1.0 μm diameter. Since the membrane material is hydrophobic, water does not enter the pores; VOCs or pollutants enter from the gaseous phase. The design generally consists of hollow fiber, spiral wound, and plate-and-frame modules. In the dense membrane there are no pores, but the solute dissolves and gets transported to the liquid side through diffusion. Silicone rubber (polydimethylsiloxane) has high oxygen permeability, and it is used as a dense membrane material for the aeration of wastewater.

Silicone tube dense membranes have been successfully used to remove several organics, including xylenes, *n*-butanol, dichloromethane, *n*-hexane, toluene, and dichloroethane, with activated sludge as the innoculum. A porous membrane has been used to destroy organics like toluene, dichloromethane, propene, and also NO (Reij et al., 1998). Generally the effluent concentration is on the order of 30 to 150 ppm.

The selectivity control exhibited by membranes is demonstrated in two examples; one is related to the treatment of vehicle exhaust gases containing NO and heavy metals, and the other is the degradation of chlorinated organics in the presence of acid vapors. In the first case, the membrane could successfully prevent heavy metals from entering and poisoning the biofilm. In the second case, silicone membranes, because of their selectivity for hydrophobic components, retain acid vapors (SO_2) that could hamper biodegradation of 1,2-dichloroethane (Freitas dos Santos et al., 1997).

Having both the aerobic and the anaerobic regions in the same biomembrane reactor was considered by Parvatiyar et al. (1996) for the treatment of trichloroethylene. On the gas side, the film near the membrane and where the oxygen concentration was high acted as the aerobic region, and the region farthest from it near the liquid side, where the oxygen concentration was zero, acted as the anaerobic zone. Both regions were able to degrade trichloroethylene completely.

One disadvantage of biofilters is that sulfates, chlorides, and nitrates accumulate when acidic gases are treated, which may inhibit the growth of microorganisms. This is circumvented to some extent by adding a neutralizing agent like lime, but it may not be possible to neutralize large amounts of acid gases. The presence of a water phase in the membrane bioreactor helps to wash out these inorganic acidic salts and prevent their accumulation.

Suspended Growth Reactors

A suspended growth reactor (SGR) is nothing but an agitated or unagitated gassed vessel containing the microorganism in a suspended state (in a nutrient medium). In an aerobic suspended growth reactor, VOCs and air are passed through an aqueous suspension of active microorganisms. Mass transfer of organic chemicals and oxygen from the gas to the liquid phase, where suspended active organisms biodegrade the contaminant of interest, is the crucial step. In a biofilter, the microorganism is attached to a support, whereas in this design the organism is kept suspended under agitation. Absence of plugging and better biomass and nutrient control are the advantages in this design.

SGR performance was comparable to that of a biofilter in treating gas containing toluene (Neal and Loehr, 2000). The percentage removals in a biofilter and an SGR were almost similar (almost 97%) for mass loadings in the range of 4 to 30 mg/L h. The support medium used in the biofilter was a 70:30 mixture of compost and perlite. The microorganisms for the SGRs were cultivated from toluene degraders and compost.

Treatment of Inorganic Gases

Sulfur compounds such as hydrogen sulfide, dimethyl sulfide, dimethyl disulfide, methane thiol, carbon disulfide, and carbonyl sulfide are produced by industries like aerobic wastewater treatment plants, composting plants, and rendering plants. Generally biofilters are used for odor reduction, as they are able to clean complex waste gases, but the absence of a water phase makes it unsafe for use in areas that may produce acidic byproducts. Different approaches for the degradation of dimethyl sulfide reported in the literature are shown in Table 30-2 (Bo et al., 2002). A membrane bioreactor containing a flat-plate composite assembly made up of polydimethylsiloxane and polysulfone membranes impregnated with ZrO fillers inoculated with *Hyphomicrobium* VS, a methylotrophic microorganism, was able to remove dimethyl sulfide from a gas stream (Bo et al., 2002). The removal efficiency

TABLE 30-2
Different Technologies and Microorganisms Used for Treating H_2S

Biofilter	Peat/night soil sludge
	Peat/*Thiobacillus thioparus* DW44
	Peat/*Hyphomicrobium* I55
	Bark/*Hyphomicrobium* MS3
	Compost/*Hyphomicrobium* MS3
	Compost/dolomite/*Hyphomicrobium* MS3
Biotrickling filter	Polypropylene/*Thiobacillus thioparus* TK-m
	Polyurethane/*Hyphomicrobium* VS
	Ceramic/activated sludge from wastewater treatment plant
Membrane bioreactor	*Hyphomicrobium* VS

of air contaminated with 38 mg/m^3 was found to be 99% for 24-s residence time.

Gases produced during the (1) hydrotreatment of oil fractions and natural gas and, (2) synthesis gases produced by coal gasification and fuel oil partial oxidation contain highly concentrated H_2S. Scrubbing this gas using ethanolamines at high pressures leads to recovery of H_2S, which is oxidized to produce elemental sulfur (Busca and Pistarino, 2003). The oxidation is performed chemically using ferric ions, which are reoxidized by air to complete the cycle. This reoxidation can be speeded up by using a microorganism *Thiobacillus ferrooxidans* in a biological reactor, such as in the EniTecnologie process (Gianna et al., 2002). Oxidation of a low concentration H_2S stream can also be performed biologically using *Thiobacillus thiooxidans* (Sublette and Sylvester, 1987), as in the Shell-THIOPAQ process (Kijlstra et al., 1999).

Biofilter technology has been successful for the treatment of waste gas containing H_2S at low concentrations of the contaminant and at high gas flow rates. The effect of inorganic packing material on the destruction efficiency has been studied, and the conclusion was that porous ceramic performed well (Hirai et al., 2001). Peat as a filtering material was able to degrade H_2S without the need to inoculate the filter with oxidizing microbes. Since peat was acidic (pH 4), it performed better when neutralized, removing 95% of the H_2S in 1 day of operation (Hartikainen et al., 2002). Sulfur is mineralized in biofilters, generating mainly sulfate ions, which remain in the biofilter. Acidification of the biofilter takes place only if the sulfur concentration is relatively high. When pellets made of pig manure and sawdust were used as the packing bed material, more than 90% removal efficiency was achieved (Busca and Pistarino, 2003). Sulfur dioxide was reduced to H_2S biochemically by contact with sulfate-reducing microorganisms in which *Desulfovibrio desulfuricans* was dominant. Subsequently the H_2S could be oxidized to sulfur by ferric sulfate, where ferrous ions were regenerated.

SO_2 from flue gases could be microbially oxidized to sulfate by *Thiobacillus ferrooxidans* (Gasiorek, 1994). The sulfate-reducing bacterium *Desulfovibrio desulfuricans* used $SO_2(g)$ as a terminal electron acceptor and converted SO_2 to H_2S (Dasu et al., 1993). The use of glucose as an electron donor in microbial SO_2-reducing cultures makes this process expensive. Heat and alkali pretreated sewage sludge was used as a carbon and energy source; it was found to reduce SO_2 completely in a continuous, anaerobic mixed culture. *Desulfotomaculum orientis* grown in batch cultures on a feed of SO_2, H_2, and CO_2 was also able to reduce SO_2 to H_2S completely at gas-liquid contact times of 1 to 2 s.

Treatment of NO gas poses several problems. Since the solubility of NO in water is very poor, bioprocesses that involve transfer of pollutants at low concentrations to the aqueous phase are not very efficient. One approach was to preconcentrate the gas using activated carbon and treat the desorbed, more concentrated gas using biological methods (Chagnot et al., 1998). Nitrate and nitrite ions, which are formed in this process, are destroyed by a denitrificating biomass involving *Thiobacillus denitrificans* in an anoxic medium grown on a sulfur-Maerl support. The presence of oxygen during adsorption leads to the formation of NO_2, which remains on the adsorbent, whereas NO does not. This technique is also well suited for gases like NH_3 and H_2S. Two heterotrophic bacteria, *Paracoccus denitrificans* and *Pseudomonas denitrificans*, have also been found in a batch culture with succinate, heat, alkali pretreated sewage sludge as carbon and energy sources, and NO as a terminal electron acceptor (Dasu et al., 1993).

References

Abumaizar, R. J., W. Kocher, and E. H. Smith. 1998. Biofiltration of BTEX contaminated air streams using compost-activated carbon filter media. *J. Hazardous Materials* 60: 111–126.

Adler, S. F. 2001. Biofiltration a primer. *Chem. Eng. Progr.* April, 33–42.

Bo, I. D., H. Van Langenhove, and J. Heyman. 2002. Removal of dimethyl sulfide from waste air in a membrane bioreactor. *Desalination* 148:281–287.

Boswell, A. 2002. Understand the capabilities of bio-oxidation. *Chem. Eng. Progr.* Dec., 48–54.

Busca, G., and C. Pistarino. 2003. Technologies for the abatement of sulphide compounds from gaseous streams: a comparative overview. *J. Loss Prevention Process Ind.* 16:363–371.

Chagnot, E., S. Taha, G. Martin, and J. F. Vicard. 1998. Treatment of nitrogen oxides on a percolating biofilter after preconcentration on activated carbon. *Process Biochem.* 33(6): 617–624.

Cohen Y. 2001. Biofiltration—The treatment of fluids by microorganism immobilized into the biofilter bedding material: A review. *Bioresource Tech.* 77:252–274.

Dasu, B. N., V. Deshmane, R. Shanmugasundram, C.-M. Lee, and K. L. Sublette. 1993. Microbial reduction of sulfur dioxide and nitric oxide. *Fuel* 72(12):1705–1714.

Freitas dos Santos, L. M., P. Pavasant, L. F. Strachan, E. N. Pistikopoulos, and A. G. Livingston. 1997. Membrane attached biofilms for waste treatment—fundamentals and applications. *Pure & Appl. Chem.* 69(11):2459–2469.

Garner, L. G., and T. A. Barton. 2002. Biofiltration for abatement of VOC and HAP emissions. *Metal Finishing* 100(11–12):12–18.

Gasiorek, J. 1994. Microbial removal of sulfur dioxide from a gas stream. *Fuel Processing Tech.* 40(2–3):129–138.

Gianna, R., M. Galileo, and A. Robertiello. 2002. Abbattimento chimico biologico di H_2S in correnti gassose. *Chim. Ind.* 84(11):39.

Hartikainen, T., P. J. Martikainen, M. Olkonnen, and J. Ruuskanen. 2002. Peat biofilters in long-term experiment for removing odorous sulphur compounds. *Water, Air and Soil Pollution* 133:335–348.

Hirai, M., M. Kamamoto, M. Yani, and M. Shoda. 2001. Comparison of the biological H_2S removal characteristics among four inorganic packing materials. *J. Biosci. Bioeng.* 91: 396–402.

Khan, F. I., and A. K. Ghoshal. 2000. Removal of volatile organic compounds from polluted air. *J. Loss Prevention Process Ind.* 13:527–545.

Kijlstra, W. S., P. F. A. van Grinsven, A. J. H. Janssen, and C. J. Buisman. (1999) New commercial process for H_2S removal from high pressure natural gas: the Shell-Thiopaq gas desulfurization process. www.shellglobalsolutions.com/gasevents/gas_treat/.

Neal, A. B., and R. C. Loehr. 2000. Use of biofilters and suspended-growth reactors to treat VOCs. *Waste Management* 20, 59–68.

Parvatiyar, M. G., R. Govind, and D. F. Bishop. 1996. Treatment of trichloroethylene (TCE) in a membrane biofilter. *Biotechnol. Bioeng.* 50:57–64.

Pinjing, H., S. Liming, Y. Zhiwen, and L. Guojian. 2001. Removal of hydrogen sulfide and methyl mercaptan by a packed tower with immobilized micro-organism beads. *Water Sci. Tech.* 44:327–333.

Reij, M. W., J T. F. Keurentjes, and S. Hartmans. 1998. Membrane bioreactors for waste gas treatment. *J. Biotech.* 59:155–167.

Soccol, C. R., A. L. Woiciechowski, L. P. S. Vandenberghe, M. Soares, G. K. Neto, V. T. Soccol. 2003. Biofiltration: an emerging technology. *Indian J Biotech.* 2:396–410.

Spivey, J. J. 1988. Recovery of volatile organics from small industrial sources. *Environ. Progr.* 7(1):31.

Sublette, K. L., and N. D. Sylvester. 1987. Oxidation of hydrogen sulfide by continuous cultures of Thiobacillus denitrificans. *Biotechnol. Bioeng.* 29:753–758.

Bibliography

Agarwal, S. K. 1998. *Environmental biotechnology*. New Delhi, India: APH Publishing Corp.

Engleman, V. S. 1994. Updates on choices of appropriate technology for control of VOC emissions. *Metal Finishing* 92(5A):238–251.

Malhotra, S., A. S. Tankhiwale, A. S. Rajvaidya, and R. A. Pandey. 2002. Optimal conditions for bio-oxidation of ferrous ions to ferric ions using *Thiobacillus ferrooxidans*. *Bioresource Tech.* 85:225–234.

Shareefdeen, Z, B. Hernes, D. Webb, and S. Wilson. 2003. H_2S removal in synthetic media biofilters. *Environ. Progr.* 22(3):207–213.

Index

F

G

H

I